Nonlinear Equations in Physics and Mathematics

NATO ADVANCED STUDY INSTITUTES SERIES

Proceedings of the Advanced Study Institute Programme, which aims at the dissemination of advanced knowledge and the formation of contacts among scientists from different countries

The series is published by an international board of publishers in conjunction with NATO Scientific Affairs Division

A Life Sciences B Physics	Plenum Publishing Corporation London and New York
C Mathematical and Physical Sciences	D. Reidel Publishing Company Dordrecht, Boston, and London
D Behavioral and Social Sciences	Sijthoff International Publishing Company Leiden
E Applied Sciences	Noordhoff International Publishing Leiden

Series C – Mathematical and Physical Sciences

Volume 40 – Nonlinear Equations in Physics and Mathematics

Nonlinear Equations in Physics and Mathematics

Proceedings of the NATO Advanced Study Institute held in Istanbul, Turkey, August 1-13, 1977

edited by

A. O. BARUT

Department of Physics, University of Colorado, Boulder, Colo., U.S.A.

D. Reidel Publishing Company

Dordrecht : Holland / Boston : U.S.A. / London : England

Published in cooperation with NATO Scientific Affairs Division

Library of Congress Cataloging in Publication Data

CIP

NATO Advanced Study Institute, Istanbul 1977.
Nonlinear equations in physics and mathematics.

(NATO advanced study institutes series : Series C, Mathematical and physical sciences ; v. 40)
'Published in cooperation with NATO Scientific Affairs Division.'
Bibliography: p.
Includes index.
1. Nonlinear theories –Congresses. 2. Mathematical physics–Congresses. 3. Mathematics–Congresses. I. Barut, Asim Orhan, 1926– II. North Atlantic Treaty Organization. Division of Scientific Affairs. III. Title. IV. Series.
QC20.7.N6N28 1977 530.1'5 78-12797
ISBN 90-277-0936-X

Published by D. Reidel Publishing Company
P.O. Box 17, Dordrecht, Holland

Sold and distributed in the U.S.A., Canada, and Mexico
by D. Reidel Publishing Company, Inc.
Lincoln Building, 160 Old Derby Street, Hingham, Mass. 02043, U.S.A.

Printed in The Netherlands

TABLE OF CONTENTS

PART II - SOLITONS

PART III - DISCRETE SYSTEMS AND CONTINUUM MECHANICS

PREFACE

This is the third Volume in a series of books devoted to the interdisciplinary area between mathematics and physics, all emanating from the Advanced Study Institutes held in Istanbul in 1970, 1972 and 1977. We believe that physics and mathematics can develop best in harmony and in close communication and cooperation with each other and are sometimes inseparable. With this goal in mind we tried to bring mathematicians and physicists together to talk and lecture to each other—this time in the area of nonlinear equations.

The recent progress and surge of interest in nonlinear ordinary and partial differential equations has been impressive. At the same time, novel and interesting physical applications multiply. There is a unifying element brought about by the same characteristic nonlinear behavior occurring in very widely different physical situations, as in the case of "solitons," for example.

This Volume gives, we believe, a very good indication over all of this recent progress both in theory and applications, and over current research activity and problems.

The 1977 Advanced Study Institute was sponsored by the NATO Scientific Affairs Division, The University of the Bosphorus and the Turkish Scientific and Technical Research Council. We are deeply grateful to these Institutions for their support, and to lecturers and participants for their hard work and enthusiasm which created an atmosphere of lively scientific discussions.

Boulder, April 1978 A. O. Barut

PART I

DYNAMICAL SYSTEMS AND INVERSE SCATTERING PROBLEMS

INTEGRABLE MANY-BODY PROBLEMS†

F. Calogero

Istituto di Fisica, Università di Roma, 00185 Roma
Istituto Nazionale di Fisica Nucleare

ABSTRACT. A survey of recent results on integrable many-body problems, mainly in one dimension.

1. INTRODUCTION

This is a summary of the three lectures I have given at the Advanced Study Institute on "Nonlinear Equations in Physics and Mathematics" (Istanbul, August 1977). It provides a survey of certain recent developments in the theory of integrable dynamical systems, consisting essentially in the display and analysis of a number of "many-body problems," mainly in one-dimension, that are exactly integrable. The selection of topics is strongly biased by my own research interests. The presentation is terse and it is mainly meant to serve as a guide to the literature. No novel results are reported, although some of the remarks towards the end appear here for the first time in print.

2. MANY-BODY PROBLEMS SOLVABLE BY THE LAX TRICK

In this section we concentrate on many-body models characterized by the Hamiltonian

---

† Lectures given at NATO Advanced Study Institute on Nonlinear Equations in Physics and Mathematics, Istanbul, August 1977.

A. O. Barut (ed.), Nonlinear Equations in Physics and Mathematics, 3–53.

$$H = \frac{1}{2}\sum_{j=1}^{n} p_j^2 + \sum_{j>k=1}^{n} V(x_{jk}), \qquad V(x) = V(-x). \tag{2.1}$$

Here, and always in the following,

$$x_{jk} \equiv x_j - x_k \quad , \tag{2.2}$$

and we have set to unity the mass of the particles. Note that we are assuming that the interaction act between every pair of particles; we are therefore excluding models of the Toda type [1], although the first application of the Lax approach (originally introduced in the context of "solvable" nonlinear partial differential equations [2]) to dynamical systems with a (finite or infinite) denumerable number of degrees of freedom, has been done (independently) by H. Flaschka [3] and S. V. Manakov [4] just in the context of the Toda many-body problem.

The Hamiltonian (2.1) yields, of course, the equations of motion

$$\dot{x}_j = p_j \quad , \qquad \dot{p}_j = -\sum_{\substack{k=1\\k\neq j}}^{n} V'(x_{jk}) \tag{2.3}$$

implying

$$\ddot{x}_j = -\sum_{\substack{k=1\\k\neq j}}^{n} V'(x_{jk}) \quad . \tag{2.4}$$

Here, and always in the following, dots indicate time-differentiation, and primes differentiation with respect to the argument of the function they are appended to.

The Lax trick consists in the identification of two matrices L and M, of rank n, depending explicitly on the variables x_j and p_j and such that the equations of motion (2.3) are equivalent to the matrix equation

$$\dot{L} = [M, L] . \tag{2.5}$$

Here, and always in the following, the symbol [A,B] indicates the commutator: $[A,B] \equiv AB-BA$. We assume moreover that L is hermitian and M anti-hermitian:

$$L^{+} = L, \qquad M^{+} = -M. \tag{2.6}$$

The matrix equation (2.5) yields n^2 scalar (real) equations; while the equations of motion (2.3) are 2n (real) equations. Thus for $n>2$ the possibility to write an equation such as (2.5) implies a restriction on the many-body model characterized by the Hamiltonian (2.1); indeed a corollary of (2.5) is the complete integrability of this many-body model, since (2.5) implies that the n eigenvalues $\lambda^{(j)}$ of L are constants of the motion.

[To prove this result, let $v^{(j)}$ be the normalized eigenvector of L corresponding to the eigenvalue $\lambda^{(j)}$:

$$Lv^{(j)} = \lambda^{(j)} v^{(j)} , \tag{2.7}$$

$$(v^{(j)}, v^{(j)}) = 1 , \tag{2.8}$$

so that

$$\lambda^{(j)} = (v^{(j)}, Lv^{(j)}). \tag{2.9}$$

Then by differentiation

$$\dot{\lambda}^{(j)} = \left(v^{(j)}, \dot{L}v^{(j)}\right) + \left(\dot{v}^{(j)}, Lv^{(j)}\right) + \left(v^{(j)}, L\dot{v}^{(j)}\right) \tag{2.10}$$

$$= \left(v^{(j)}, [M,L]v^{(j)}\right) + \lambda^{(j)}\left[(\dot{v}^{(j)}, v^{(j)}) + (v^{(j)}, \dot{v}^{(j)})\right] \tag{2.11}$$

$$= 0. \tag{2.12}$$

The first step follows from (2.5), (2.7) and the hermicity of L; the second step from (2.7) and the hermicity of L, and from the observation that the term multiplying $\lambda^{(j)}$ in the r.h.s. of (2.11) vanishes because it is the time derivative of the l.h.s. of (2.8)].

To identify a class of Hamiltonians of type (2.1) for which a Lax formula of type (2.5) holds introduce the ansatz [5]

$$L_{jk} = \delta_{jk} p_j + (1 - \delta_{jk})\alpha(x_{jk}) , \tag{2.13}$$

$$M_{jk} = \delta_{jk} \sum_{\substack{\ell=1 \\ \ell \neq j}}^{n} \beta(x_{j\ell}) - (1-\delta_{jk})\alpha'(x_{jk}) , \tag{2.14}$$

where δ_{jk} is of course the Kronecker symbol, $\delta_{jk}=1$ if $j=k$, $\delta_{jk}=0$ otherwise. It is then easily seen that (2.5) is equivalent to (2.3) with

$$V(x) = \alpha(x)\alpha(-x) + \text{const.}, \tag{2.15}$$

provided the functions $\alpha(x)$ and $\beta(x)$ satisfy the functional equations

$$\alpha'(y)\alpha(z) - \alpha(y)\alpha'(z) = \alpha(y+z)[\beta(y) - \beta(z)], \tag{2.16}$$

$$\beta(-x) = \beta(x). \tag{2.17}$$

It can be shown [6] that the more general solution for the potential $V(x)$ consistent with (2.15)-(2.17) is

$$V(x) = B\,\wp(ax|\omega,\omega') + \text{const.} \tag{2.18}$$

where

$$\wp(z|\omega,\omega') = z^{-2} + \sum_{n,m}{}' \left[(z - 2m\omega - 2n\omega')^{-2} - (2m\omega + 2n\omega')^{-2}\right] \tag{2.19}$$

is the Weierstrass function. The sum in (2.19) extends over all (positive and negative) integers, excluding the single (and singular) term with $n=m=0$ (the prime appended to the sum is a reminder of this).

While the potential $V(x)$ is unique [up to the constants in Equation (2.18)], there exist several solutions of the functional equations (2.16)-(2.17); for instance [7]

$$\alpha_1(x) = b\,\mathrm{dn}(a'x)/\mathrm{sn}(a'x), \tag{2.20}$$

$$\alpha_2(x) = b\,\mathrm{cn}(a'x)/\mathrm{sn}(a'x), \tag{2.21}$$

$$B = -\eta^2 b^2, \quad a = \eta a', \quad \eta = (\ell_1 - \ell_3)^{-\frac{1}{2}} \tag{2.22}$$

where $\mathrm{sn}(y)$, $\mathrm{cn}(y)$ and $\mathrm{dn}(y)$ are Jacobian elliptic functions [7]. The corresponding expressions of $\beta(x)$ are given by the general formula [6]

$$\beta(x) = \alpha(x)\alpha(-x)/\lim_{x\to 0}[x\alpha(x)] + \text{const.}, \tag{2.23}$$

that follows from (2.16). Note moreover that (2.16) is invariant under the transformation

$$\alpha(x) \longrightarrow \alpha(x)\exp(cx), \tag{2.24}$$

that clearly changes neither $V(x)$, Equation (2.15), nor $\beta(x)$, Equation (2.23).

In order that $V(x)$, Equation (2.18), be real, one of the two constants ω and ω' must be real and the other pure imaginary (they cannot be both real or both pure imaginary, otherwise the sum (2.19) diverges). In the following we concentrate on certain limiting cases, when one or both of these quantities diverge. These cases are particularly interesting since the corresponding many-body models have a more sensible physical interpretation, due to the asymptotic vanishing of the pair force (the other models might, however, also be of interest in solid state modeling). In particular for $\omega = i\omega' = \infty$ [8],

$$V(x) = g^2/x^2, \quad \alpha(x) = ig/x, \tag{2.25}$$

and for $\omega = \infty$, $\omega' = i\pi/2$ [9],

$$V(x) = g^2 a^2/\sinh^2(ax), \quad \alpha(x) = iga/\sinh(ax). \tag{2.26}$$

In this last equation the choice of $\alpha(x)$ corresponding to (2.20) [rather than (2.21)] has been made, so that $\alpha(x)$ vanish asymptotically (this is useful below); and the constant has been set to zero in (2.18), so that also $V(x)$ vanish asymptotically.

We have seen above that the Lax formula (2.5) characterizes isospectral flows, namely time-evolutions that maintain constant the eigenvalues of L; thereby producing, once L has been identified, n constants of the motion, and implying therefore the integrability of the corresponding dynamical system. It is sometimes more convenient to focus attention on the symmetric invariants I_m of the matrix L, defined by the formula

$$\det[L + \lambda \mathbb{1}] = \lambda^n + \sum_{m=1}^{n} I_m \lambda^{n-m}, \tag{2.27}$$

or on the traces J_m of L, defined by the formula

$$J_m = m^{-1} \operatorname{tr}[L^m], \quad m = 1, 2, \ldots, n. \tag{2.28}$$

Indeed these quantities are more readily expressible in terms of the matrix elements of L [and therefore of the dynamical variables p_j and x_j; see (2.13)] than the eigenvalues $\lambda^{(m)}$ that are the n roots of the secular equation in λ obtained equating (2.27) to zero; while clearly the time-independence of the n eigenvalues $\lambda^{(m)}$ also implies the constancy of the quantities I_m and J_m. It can moreover be shown that each of the sets $\{\lambda^{(m)}\}$, $\{I_m\}$, and $\{J_m\}$ is composed of n independent quantities (none of which can be written as a combination of the others with coefficients independent of the dynamical variables x_j and p_j); and moreover that the n quantities in each set are in involution, namely [10]

$$\{\lambda^{(j)},\lambda^{(k)}\} = \{I_j,I_k\} = \{J_j,J_k\} = 0, \qquad j,k = 1,2,\dots,n, \qquad (2.29)$$

where of course we define as usual

$$\{A,B\} \equiv \sum_{m=1}^{n}\left[(\partial A/\partial x_m)(\partial B/\partial p_m) - (\partial A/\partial p_m)(\partial B/\partial x_m)\right] \qquad (2.30)$$

to be the Poisson bracket of A and B.

2.1 Implications of the existence of n conserved quantities

In this subsection we show that the time-independence of the n eigenvalues $\lambda^{(j)}$ of the Lax matrix L may be exploited to get quite readily some conclusions about the corresponding many-body problem. We do this focussing attention on the (classical) one-dimensional many-body model characterized by the Hamiltonian (2.1) with

$$V(x) = g^2a^2/\sinh^2(ax). \qquad (2.1.1)$$

Since this is a repulsive potential, singular at zero separation, the typical physical phenomenon occurring in this system is a scattering process, with the coordinates of the particles characterized by the boundary conditions

$$x_j(t) \xrightarrow[t\to-\infty]{} \bar{p}_j t + \bar{a}_j, \quad \bar{p}_j \equiv p_j(-\infty), \qquad j = 1, 2, \dots, n, \qquad (2.1.2a)$$

$$x_j(t) \xrightarrow[t\to+\infty]{} \tilde{p}_j t + \tilde{a}_j, \quad \tilde{p}_j \equiv p_j(+\infty), \qquad j = 1, 2, \dots, n, \qquad (2.1.2b)$$

where

$$\bar{p}_j < \bar{p}_{j+1}, \qquad j = 1, 2, \dots, n-1, \qquad (2.1.3a)$$

and

$$\tilde{p}_j > \tilde{p}_{j+1}, \qquad j = 1, 2, \dots, n-1. \qquad (2.1.3b)$$

We are assuming here that the particles are labeled so that

$$x_j > x_{j+1} \quad , \quad j = 1, 2, \ldots, n-1; \tag{2.1.4}$$

note that the inequalities (2.1.4) are necessarily maintained throughout the motion, since the singular nature of the forces forbids each particle from overtaking any other particle. The inequalities (2.1.3) are consistent with (2.1.4) and with the asymptotic formulae (2.1.2), implying that the particles separate asymptotically both in the past and in the future.

The Lax matrix L corresponding to this problem reads [see (2.13) and (2.26)]

$$L_{jk}(t) = \delta_{jk} p_j + (1-\delta_{jk}) iga/\sinh[a(x_j - x_k)]. \tag{2.1.5}$$

Thus, since the particles separate asymptotically,

$$L_{jk}(\pm\infty) = \delta_{jk} p_j(\pm\infty). \tag{2.1.6}$$

The time-independence of the eigenvalues $\lambda^{(j)}$ of L implies therefore that the set of asymptotic momenta coincide:

$$\{p_j(+\infty),\ j = 1, 2, \ldots, n\} = \{p_j(-\infty),\ j = 1, 2, \ldots, n\}, \tag{2.1.7a}$$

or equivalently [see (2.1.2)]

$$\{\tilde{p}_j,\ j = 1, 2, \ldots, n\} = \{\bar{p}_j,\ j = 1, 2, \ldots, n\}; \tag{2.1.7b}$$

and this equation, together with the inequalities (2.1.3), implies the simple rule [11]

$$\tilde{p}_j = \bar{p}_{n+1-j} \quad , \quad j = 1, 2, \ldots, n. \tag{2.1.8}$$

As for the asymptotic positions, it appears [12] that

$$\tilde{a}_j = \bar{a}_{n+1-j} - \sum_{k=1}^{j-1} \Delta(\bar{p}_j - \bar{p}_k) + \sum_{k=j+1}^{n} \Delta(\bar{p}_j - \bar{p}_k) \tag{2.1.9}$$

where $\Delta(\bar{p}_j - \bar{p}_k)$ is the shift in the <u>two-body</u> scattering case with initial momenta $\bar{p}_j$ and $\bar{p}_k$, namely

$$\Delta(\bar{p}_1 - \bar{p}_2) = \tilde{a}_1 - \bar{a}_2 \tag{2.1.10}$$

for $n = 2$, and we are using the usual convention that sets to zero a sum if the upper limit is smaller than the lower limit. Note that Equations (2.1.8) and (2.1.9) imply that the asymptotic outcome of the scattering process is the same that would be produced by a sequence of two-body interactions. The appropriateness of this interpretation is underscored by the argument [12] yielding (2.1.9). It should perhaps be emphasized that these results are peculiar to the particular model we are considering, being a consequence of its integrability through the Lax approach; they would not hold if the potential (2.1.1) were replaced by another function, say $\sinh^{-4}(ax)$ in place of $\sinh^{-2}(ax)$, or if the coupling constants g^2 were different for different particle pairs.

The two-body shift $\Delta(p)$ can, of course, be easily computed:

$$\Delta(p) = (2a)^{-1}\log(1+g^2a^2/p^2). \tag{2.1.11}$$

Note that, in the special case $a=0$, $\Delta(p)$ vanishes, so that (2.1.9) becomes in this case the simple rule [13]

$$\tilde{a}_j = \bar{a}_{n+1-j}. \tag{2.1.12}$$

2.2 The quantal case

The transition from classical to quantal can be effected by the usual substitution

$$p_j \longrightarrow -i\partial/\partial x_j, \tag{2.2.1}$$

having of course set $\hbar=1$. The Hamiltonian (2.1) becomes then the linear operator

$$H = \frac{1}{2}\Delta + \sum_{j>k=1}^{n} V(x_{jk}) \tag{2.2.2}$$

with Δ the n-dimensional Laplace operator,

$$\Delta = \sum_{j=1}^{n} \partial^2/\partial x_j^2. \tag{2.2.3}$$

It is remarkable that some of the problems described above were first solved in the quantal context; certain properties (for instance, Equation (2.1.8) for the case $a=0$) were then conjectured to hold also in the classical case; but an explicit validation of the conjecture through the actual solution of the classical problem came only after several years (the historical development has been outlined above [8, 9]).

Here we indicate how the Lax approach can be applied also in the quantal context, focussing again for definiteness on the case discussed in the preceding subsection, with the potential (2.1.1) [9].

With the substitution (2.2.1) the diagonal elements of the matrix L of Equation (2.13) become differential operators, and each of them does not commute with the matrix elements of the same line and column [see (2.13)], although it does commute with the matrix elements of other lines and columns. Thus, while there are no anambiguous quantal analogs of the eigenvalues $\lambda^{(j)}$ (see, however, below), the symmetric invariants I_m become well-defined quantal operators under (2.2.1), since their definition through (2.27) implies that they never contain a product of noncommuting operators; so that there is no ambiguity in their case in the transition from the classical to the quantal context. Thus the quantities I_m obtained from (2.13), (2.27) and (2.2.1) are well-defined quantal observables. It can, moreover, be shown that these observables are constants of the motion—namely, that they commute with the Hamiltonian:

$$[H, I_m] = 0 \quad , \quad m = 1, 2, \ldots, n, \tag{2.2.4}$$

this result being a consequence of the quantal analog of the Lax formula (2.5) that reads now

$$i[H, L_{jk}] = \frac{1}{2} \sum_{\ell=1}^{n} \left\{ A_{j\ell} L_{\ell k} + L_{\ell k} A_{j\ell} - L_{j\ell} A_{\ell k} - A_{\ell k} L_{j\ell} \right\} ,$$

$$j, k = 1, 2, \ldots, n, \tag{2.2.5}$$

and that they provide a (complete) system of commuting observables, since they are clearly as independent in the quantal case as they were in the classical case, and they commute with each other,

$$[I_\ell, I_m] = 0 \quad , \quad \ell, m = 1, 2, \ldots, n, \tag{2.2.6}$$

this formula being clearly the quantal analog of the corresponding equation in the classical case [see (2.29)]. Of course, in the last three formulae the commutator in the l.h.s. refers to the operator character of the quantities H, I_m and L_{jk}, in the quantal context. Clearly these results imply a posteriori that the classical expressions of the eigenvalues $\lambda^{(j)}$ in terms of the symmetric invariants I_m provide an appropriate, nonambiguous, starting point for defining the (operators) $\lambda^{(j)}$ in the quantal context.

These properties allow to infer, also in the quantal case, the validity of some of the results already described in the classical case—for instance, the property of Equation (2.1.8). An equivalent reformulation of this remarkable property can be given in terms of wave propagation in an n-dimensional space, rather than in terms of a quantal many-body problem, the correspondence being of course through the (stationary) Schroedinger wave equation

$$H\Psi(x_1, x_2, \ldots, x_n) = E\Psi(x_1, x_2, \ldots, x_n) \qquad (2.2.7)$$

with H defined by (2.2.2)-(2.2.3) with (2.1.1), or rather through the time-dependent Schroedinger equation that obtains from (2.2.7) with the replacement of $\Psi(x)$ by $\Psi(x,t)$ and of E by $i\partial/\partial t$. It states that, in n-dimensional space, the plane wave

$$\exp\left(i \sum_{j=1}^{n} p_j x_i\right)$$

incoming from a distance on the (non-spherically-symmetrical) potential

$$V(x) = \sum_{j>k=1}^{n} g^2 a^2/\sinh^2[a(x_j - x_k)]$$

gets essentially reflected into the outgoing plane wave

$$\exp\left(i \sum_{j=1}^{n} p_{n+1-j} x_j\right),$$

<u>without diffraction</u>.

2.3 Particles of different types

We return hereafter to consider the classical case (although several of the results given below are also applicable in the quantal context). In this subsection we describe a trick that extends the many-body models considered above to models involving particles of different types [5]. We again focus for definiteness on the case with the potential (2.1.1).

We divide the total number n of particles in two groups, shifting $n_2 = n - n_1$ coordinates according to the rule

$$x_j \longrightarrow x_j + \frac{i}{2}\pi/a \quad \text{for} \quad j = n_1+1,\, n_1+2,\, \ldots,\, n. \qquad (2.3.1)$$

The resulting many-body model describes then n_1 particles of one type and n_2 particles of another type, having the same (unit) mass and interacting pairwise among themselves, with the potential

$$V_\ell(x) = g^2 a^2/\sinh^2(ax) \qquad (2.3.2a)$$

acting between equal particles, and the potential

$$V_d(x) = -g^2 a^2/\cosh^2(ax) \qquad (2.3.2b)$$

acting between different particles. Note that, while the potential acting between equal particles is repulsive and singular at zero separation, the potential acting between different particles is attractive and non-singular; thus different particles can go through each other, and can form bound states (classically, and also in the quantal context), implying a considerably richer phenomenology, when both types of particles are present, than in the case with only one type of particles.

An analysis of the argument that led to (2.1.8) implies, for instance, that in this case, if asymptotically there are no bound states (neither in the past nor in the future), (2.1.7) still holds, although it does no more imply (2.1.8) (the ordering of equal particles cannot change throughout the motion, but that of different particles can).

Classical two-body bound states can exist, with binding energy up to g^2a^2; this extremal value corresponds to tight binding of two different particles, with coinciding positions and speeds. And it is easy to show that if such a tightly bound state, composed say of particle 1 of type "plus" and particle 2 of type "minus," scatters on a third particle, say of type "plus," the final outcome is the return of particle 1 alone back to where it came from (with $\tilde{p}_1 = \bar{p}_3 = -2p_0$, in the notation of subsection 2.1), while 2 and 3 emerge instead tightly bound in the opposite direction (i.e., $\tilde{p}_2 = \tilde{p}_3 = \bar{p}_1 = \bar{p}_2 = p_0$; note that we are assuming to look at this system in its center-of-mass frame).

A problem that is still open is the existence of three-body bound states; indeed it is easy to see that there is an infinity of three-body configurations (with two particles equal and one different) all having the minimal potential energy $-g^2a^2$, interpolating with continuity between the (single) symmetrical configuration with the different particle at the center and the other two a distance $a^{-1}\text{arcsinh}(2^{-\frac{1}{2}})$ from it, and the cases with one of the equal particles tightly bound to the different particle, and the other infinitely far away from the bound pair. Thus it is not easy to decide whether a stable three-body bound state, that never decays into the clusterized configuration with two particles close together and the third far away, does or does not exist (the positive statement in Reference 5 was based on an incorrect computation and should therefore be disregarded); and a similar problem exists in the quantal context. I am inclined to conjecture that no three-body bound state exists, nor any other multi-body bound state (for a terse discussion of the case of large n, with $n_1 = n_2 = n/2$, see [5]); but the matter still awaits a firm conclusion that could of course be fairly easily obtained by a numerical analysis [48].

2.4 Lax matrices of higher rank

Let us continue to focus on the case discussed in the preceding subsection, although the results described below can be extended also to the more general case of the potential (2.18), suitably extended to the case of particles of different types [5] (indeed such an extension is eventually associated to nontrivial duplication formulae for Jacobian elliptic functions; see below).

We consider the case of 2n particles, the first n of one type and the remaining n of the other. For obvious symmetry reasons, a configuration characterized initially by the conditions

$$x_{n+j} = x_j \quad , \quad p_{n+j} = p_j \quad , \quad j = 1, 2, \ldots, n, \tag{2.4.1}$$

is maintained throughout the motion. It corresponds to a new many-body problem: n novel particles (each composed by a two-body tightly bound state) interacting via the potential

$$V_\ell(x) + V_d(x) = g^2(2a)^2/\sinh^2(2ax). \tag{2.4.2}$$

Note that we have written here [from (2.3.2)] the potential acting on each one of the two components of every tightly bound state; indeed only one of these coordinates need be kept track of since the other simply follows.

But (2.4.2) is, except for the replacement of a by 2a, just the original potential for the problem with only one type of particles. Thus, instead of generating a novel many-body problem, we have reproduced the original problem. But this implies a novel approach to its solution, in particular the possibility to define new matrices L and M given explicitly in terms of the Hamiltonian coordinates x_j and p_j and satisfying the Lax condition (2.5), but having now rank 2n. For instance, for the n-body potential (2.1.1), the new matrix L, of rank 2n, has the explicit structure

$$L_{jk} = \delta_{jk}p_j + i(1-\delta_{jk})\tfrac{1}{2}\, ag/\sinh[a(x_j-x_k)/2] \qquad (2.4.3a)$$

for $1 \leqq j \leqq n$ and $1 \leqq k \leqq n$ or $n < j \leqq 2n$ and $n < k \leqq 2n$,

$$L_{jk} = \tfrac{1}{2}\, ag/\cosh[a(x_j-x_k)] \qquad (2.4.3b)$$

for $n < j \leqq 2n$ and $1 \leqq k \leqq n$ or $1 \leqq j \leqq n$ and $n < k \leqq 2n$ [of course, in (2.4.3a) and (2.4.3b) x_j and p_j, as well as x_k and p_k, must be replaced by x_{j-n}, x_{k-n}, p_{k-n} if $j > n$ or $k > n$]. Thus the 2n eigenvalues of this matrix, as well as its symmetric invariants and the traces of its powers, are also constant of the motion; of course, only 2n of these can be independent. The relation of dependency among the eigenvalues, as well as the simultaneous validity of the Lax equation (2.5) for the original L, M matrices of rank n and the novel ones of rank 2n obtained by the procedure that we have just described, is clearly related to the duplication formulae for trigonometric (or, equivalently, hyperbolic) functions.

Clearly the procedure we have outlined can be iterated, and in this manner L, M pairs can be generated of rank 4n, 8n, 16n, and so on; they typically contain trigonometric (or, equivalently, hyperbolic) functions of argument $ax_{jk}/4$, $ax_{jk}/8$, $ax_{jk}/16$, and so on.

2.5 The integration technique of Olshanetsky and Perelomov

In this subsection we describe, in an elementary way, an elegant technique due to Olshanetsky and Perelomov [15] that reduces the determination of the trajectories of the particles interacting according to certain of the many-body models described above to the determination of the eigenvalues of a matrix of rank n given in completely explicit form in terms of the initial data (position and velocities of the particles). We focus on the simplest problem, namely, the model with n equal particles interacting via the two-body potential

$$V(x) = g^2/x^2 \tag{2.5.1}$$

(and possibly also with an oscillator potential; see below). It should be mentioned that results similar to those of Olshanetsky and Perelomov have been obtained by Moser [8] and Adler [16], and subsequently also by other researchers in the framework of a group-theoretical approach based on, or related to, the conformal group (see subsection 2.7.2 below). Analogous results to those described here have been extended by Olshanetsky and Perelomov [17], and also by Adler [16], to the n-body problem with the interparticle potential (2.1.1); the general case with the potential (2.18) has not yet been treated.

Let the t-dependent matrix y(t), of rank n, have the eigenvalues $x_j(t)$, and write

$$y(t) = u(t)x(t)u^+(t) \tag{2.5.2}$$

where

$$x(t) = \text{diag}[x_j(t)] \tag{2.5.3}$$

and

$$u(t)u^+(t) = u^+(t)u(t) = \mathbb{1}. \tag{2.5.4}$$

Differentiation of (2.5.2) yields

$$\dot{y}(t) = u(t)L(t)u^+(t) \tag{2.5.5}$$

with the matrix L(t) defined by

$$L = \dot{x} - [M,x] \tag{2.5.6}$$

where [see (2.5.4)]

$$M = \dot{u}^+u = -u^+\dot{u}; \tag{2.5.7}$$

and a second differentiation yields

$$\ddot{y} = u\{\dot{L} - [M,L]\}u^+. \tag{2.5.8}$$

Assume now that the matrix M has the form

$$M_{jk} = -ig\left\{\delta_{jk} \sum_{\ell=1}^{n}{}' (x_j-x_\ell)^{-2} - (1-\delta_{jk})(x_j-x_k)^{-2}\right\}, \tag{2.5.9}$$

as suggested by (2.14), (2.25) and (2.23); this of course implies a specific choice for the unitary matrix u [see (2.5.7)]; but this need not concern us at this stage.

Insertion of (2.5.9) into (2.5.6) yields easily

$$L_{jk} = \delta_{jk}\dot{x}_j + i(1-\delta_{jk})g/(x_j-x_k); \qquad (2.5.10)$$

note the consistency of this formula with (2.13), (2.3) and (2.5).

Insertion of this formula and of (2.5.9) into (2.5.8) yields finally

$$\ddot{y} = uDu^+ \qquad (2.5.11)$$

with

$$D = \mathrm{diag}\left[\ddot{x}_j - 2g^2 \sum_{k=1}^{n}{}' (x_j-x_k)^{-3}\right]; \qquad (2.5.12)$$

note, incidentally, that only at this stage the choice (2.5.9) for the diagonal elements of M plays a role.

Equations (2.5.2), (2.5.11) and (2.5.12) imply a biunivocal correspondence between the matrix equation

$$\ddot{y} = f(y) \qquad (2.5.13)$$

and the equations

$$\ddot{x}_j = f(x_j) + 2g^2 \sum_{k=1}^{n}{}' (x_j-x_k)^{-3}, \quad j = 1, 2, \ldots, n. \qquad (2.5.14)$$

Thus the integration of these "equations of motion" is related to the integration of the matrix equation (2.5.13), with the boundary conditions

$$y(0) = x(0) \quad , \quad \dot{y}(0) = L(0) \qquad (2.5.15)$$

implied by (2.5.11) and (2.5.5) [with the choice

$$u(0) = u^+(0) = \mathbb{1}; \qquad (2.5.16)$$

this last equation can be assumed, any choice of u at time zero being consistent with (2.5.7); indeed, (2.5.16) is the boundary condition needed, in conjunction with the differential equation (2.5.7), to specify uniquely u once M is given].

The simplest case in which (2.5.13) can be integrated is when f vanishes identically. Then

$$y(t) = x(0) + L(0)t \tag{2.5.17}$$

is an explicit representation of the matrix y at time t, in terms of the initial positions $x_j(0)$ and velocities $\dot{x}_j(0)$ of the particles [see (2.5.3) and (2.5.10)]; and the n eigenvalues at time t of this (hermitian) matrix coincide with the coordinates $x_j(t)$ satisfying the equations

$$\ddot{x}_j = 2g^2 \sum_{k=1}^{n}{}' (x_j - x_k)^{-3}, \tag{2.5.18}$$

which are just the equations of motion yielded by the Hamiltonian (2.1) with pair potential (2.5.1).

A second case in which (2.5.13) can be integrated explicitly corresponds to the choice $f(y) = \omega^2 y$. Then

$$y(t) = x(0)\cos(\omega t) + [L(0)/\omega]\sin(\omega t). \tag{2.5.19}$$

Thus the eigenvalues of this matrix are solutions $x_j(t)$ of the equations of motion

$$\ddot{x}_j = -\omega^2 x_j + 2g^2 \sum_{k=1}^{n}{}' (x_j - x_k)^{-3}, \tag{2.5.20}$$

corresponding to the Hamiltonian

$$H = \frac{1}{2} \sum_{j=1}^{n} \left(p_j^2 + \omega^2 x_j^2 \right) + g^2 \sum_{j>k=1}^{n} (x_j - x_k)^{-2}. \tag{2.5.21}$$

A trivially related system is the translation-invariant model with the oscillator potential acting between every particle pair rather than towards a fixed center, due to the identity

$$\sum_{j=1}^{n} x_j^2 = n^{-1} \sum_{j>k=1}^{n} (x_j - x_k)^2 + nX^2, \tag{2.5.22}$$

where X is the center-of-mass coordinate,

$$X = n^{-1} \sum_{j=1}^{n} x_j. \tag{2.5.23}$$

The Hamiltonian system (2.5.21) is also integrable. It was first solved in the quantal case, exhibiting its complete energy spectrum [8]; the conjecture [8] of complete periodicity of the corresponding classical motion, suggested by these quantal results, was validated by Adler [16], who was the first to treat this problem classically for arbitrary n by a slightly modified version of the Lax approach (as introduced by Moser [8]). The complete periodicity is, of course, explicitly evident also from the results that we have just described [see (2.5.19)]. [49]

Adler has also treated an analogous extension of the many-body problem with the two-body potential (2.1.1), characterized by the additional presence of an external (one-body) exponential potential [16].

2.6 Two-dimensional models

The many-body models we have discussed heretofore are one-dimensional. By a formal trick ("complexification") it is possible to generate from these certain "equations of motion" that resemble those appropriate to a two-dimensional many-body problem.

Consider for instance the classical many-body problem of n particles in the plane interacting pairwise via the pair force [18]

$$\vec{F}(\vec{r}) = \vec{f}(\vec{r}) - \omega^2 \vec{r}, \qquad (2.6.1)$$

where $\vec{r}$ is the interparticle vector and

$$f_x(\vec{r}) = Gr^{-6} x(x^2-3y^2) = Gr^{-3} \cos 3\varphi, \qquad (2.6.2a)$$

$$f_y(\vec{r}) = Gr^{-6} y(y^2-3x^2) = Gr^{-3} \sin 3\varphi. \qquad (2.6.2b)$$

Here f_x and f_y are the x and y components of the force f, while x and y are the cartesian coordinates of the interparticle vector $\vec{r}$, and r, φ are the polar coordinates of this same vector:

$$x = r \cos\varphi, \qquad y = r \sin\varphi. \qquad (2.6.3)$$

Note that the modulus of the force $\vec{f}$ depends only on the interparticle distance; but this force is not central, depending on the orientation of the interparticle vector with respect to the frame of reference. Only if $\vec{r}$ is parallel to the x or y axes, or if it makes with them a 45-degree angle, is the force directed along the interparticle line: repulsive in the former case, attractive in the latter (for positive G; the reverse if the sign of G is changed).

The equations of motion of this many-body problem read

$$\ddot{\vec{r}}_j(t) = \sum_{k=1}^{n}{}' \left[\vec{f}(\vec{r}_{jk}) - \omega^2 \vec{r}_{jk} \right], \qquad (2.6.4)$$

where, of course,

$$\vec{r}_{jk} \equiv \vec{r}_j - \vec{r}_k \qquad (2.6.5)$$

and

$$\vec{r}_j \equiv (x_j, y_j) \qquad (2.6.6)$$

is the two-dimensional coordinate of the jth particle. They can be solved noting their equivalence, through the position

$$z_j = x_j + iy_j \qquad (2.6.7)$$

(also implying, of course,

$$z_{jk} \equiv z_j - z_k = x_{jk} + iy_{jk} \quad , \quad |z_{jk}|^2 = x_{jk}^3 + y_{jk}^2 = r_{jk}^2 \Big), \qquad (2.6.8)$$

to the equations of motion,

$$\ddot{z}_j = \sum_{k=1}^{n}{}' \left(Gz_{jk}^{-3} - {}^2 z_{jk} \right), \qquad (2.6.9)$$

corresponding formally to the problem discussed in the preceding subsection (with $G = 2g^2$), namely to the model with the Hamiltonian

$$H = \frac{1}{2} \sum_{j=1}^{n} p_j^2 + \frac{1}{2} \sum_{j>k=1}^{n} \left(Gz_{jk}^{-2} + \omega^2 z_{jk}^2 \right). \qquad (2.6.10)$$

Clearly the consideration of complex z_j's (and p_j's) does not affect the (algebraic) methods discussed above. Thus we may conclude that the solution of the equations of motion (2.6.4), with given initial conditions $\vec{r}_j(0)$, $\dot{\vec{r}}_j(0)$, is provided, through (2.6.6) and (2.6.7), by the (complex) eigenvalues $z_j(t)$ of the explicit (non-hermitian) matrix

$$Y_{jk}(t) = \delta_{jk} \left\{ z_j(0) \cos(\bar{\omega} t) + \dot{z}_j(0) \sin(\bar{\omega} t)/\bar{\omega} \right\} +$$

$$+ (1-\delta_{jk}) i (G/2)^{\frac{1}{2}} \sin(\bar{\omega}t)/[\bar{\omega} z_{jk}(0)]. \quad (2.6.11)$$

We are assuming that the system is being investigated in its center-of-mass frame; and we have set

$$\bar{\omega} = n^{\frac{1}{2}} \omega \quad (2.6.12)$$

[see Equation (2.5.22); clearly the same result, but with ω in place of $\bar{\omega}$, would be appropriate if the external harmonic oscillator force $-\omega^2 \vec{r}_j$ appeared in (2.6.4) in place of the interparticle force $-\omega^2 \vec{r}_{jk}$].

If $\omega \neq 0$ the many-body system is confined to a finite region of the plane, around its center-of-mass; and the motion is completely periodic [see (2.6.11)].

If $\omega = 0$, the above formulae remain valid [with $\cos(\bar{\omega}t)$ in (2.6.11) replaced by 1, and $\sin(\bar{\omega}t)/\bar{\omega}$ by t]; in this case the system is not confined, and the typical phenomenon is a scattering process, with all particles widely separated at the beginning and at the end. It is very easy to prove from (2.6.11) that also in this case (as in the examples treated in previous subsections) no new momenta are created in the scattering process: namely that if the asymptotic (two-dimensional) momenta $\vec{p}_j$ and $\vec{p}_j$ are defined by the formulae

$$\vec{r}_j(t) \xrightarrow[t \to -\infty]{} \vec{p}_j{}' t + \vec{a}_j{}' \quad , \quad j = 1, 2, \ldots, n, \quad (2.6.13a)$$

$$\vec{r}_j(t) \xrightarrow[t \to +\infty]{} \vec{p}_j{}'' t + \vec{a}_j{}'' \quad , \quad j = 1, 2, \ldots, n, \quad (2.6.13b)$$

there holds the rule

$$\left\{\vec{p}_j{}' \ , \ j = 1, 2, \ldots, n\right\} = \left\{\vec{p}_j{}'' \ , \ j = 1, 2, \ldots, n\right\}: \quad (2.6.14)$$

the set of final momenta coincides with the set of initial momenta. But in this case there is no simple rule [such as (2.1.8)] relating each $\vec{p}_j{}''$ to a corresponding $\vec{p}_k{}'$.

Because the interparticle force $\vec{f}$, Equation (2.6.2), is singular at zero separation and is attractive for certain orientations of the interparticle vector, collapse may occur, with two particles coming, in a finite time, to the same position, acquiring an infinite kinetic energy at the time of encounter; while there is a natural continuation of the motion beyond the collapse, given by the above equations, the identification of the two particles gets lost in the cataclysm. But such events are not generic, originating from a set of initial conditions of vanishing measure relative to the whole set of permissible initial conditions. To convince the reader of this, and to display the richness of behaviors of this system, we describe below the results in the (trivial) two-body case (with $\omega = 0$). Let then $\vec{r} = \vec{r}_1 - \vec{r}_2$ indicate the relative coordinate, and define the parameters characterizing the asymptotic (relative) motion through

$$\vec{r}(t) \xrightarrow[t \to -\infty]{} \vec{v}'t + \vec{a}' , \qquad (2.6.15a)$$

$$\vec{r}(t) \xrightarrow[t \to +\infty]{} \vec{v}''t + \vec{a}'' . \qquad (2.6.15b)$$

There are then, given $\vec{v}$ and $\vec{a}$, three possible outcomes:

i) collapse occurs

ii) $\vec{v}'' = \vec{v}'$, $\vec{a}'' = \vec{a}'$ ("quasifree")

iii) $\vec{v}'' = -\vec{v}'$, $\vec{a}'' = -\vec{a}'$ ("exchange")

Let us first dispose of the cases in which the motion is effectively one-dimensional: if $v'_x = a'_x = 0$ or $v'_y = a'_y = 0$, collapse occurs for $G < 0$ and "exchange" prevails if $G > 0$; if $v_x = \epsilon v_y$ and $a_x = \epsilon a_y$ (with $\epsilon = +$ or $\epsilon = -$), collapse occurs if $G > 0$ and "exchange" prevails if $G < 0$.

The behavior in the case of truly two-dimensional motion depends on the value of the parameters

$$G_- = -\tfrac{1}{2} L^2 v'^4 / (2v'_x v'_y)^2 , \qquad (2.6.16a)$$

$$G_+ = \tfrac{1}{2} L^2 v'^4 / (v'^2_x - v'^2_y)^2 , \qquad (2.6.16b)$$

where L is the initial angular momentum of the pair,

$$L = a'_x v'_y - a'_y v'_x ; \qquad (2.6.17)$$

if G_- or G_+ coincide with the "coupling constant" G, collapse occurs; if $G_- < G < G_+$, "quasifree" behavior prevails; otherwise (namely, if $G < G_-$ or $G > G_+$), "exchange" prevails.

2.7 Miscellanea

In this last subsection we indicate tersely certain developments closely related to the topics discussed above, referring for more complete treatments to the original literature.

2.7.1 Models corresponding to symmetrical configurations and their relation to the root systems of semisimple Lie algebras

Consider a special "symmetrical" configuration of the model described by the Hamiltonian (2.1), for instance the case with an even number of particles, $n=2m$, and with the coordinates and momenta satisfying the (non-translation-invariant) restriction

$$x_{m+j} = -x_j \ , \ p_{m+j} = -p_j \ , \quad j = 1, 2, \ldots, m. \qquad (2.7.1.1)$$

Clearly this restriction is compatible with the flow (2.1), i.e., if it holds at one (say, initial) time t_o, it holds for all time. But solving the n-body problem characterized by the Hamiltonian (2.1) with this restriction is equivalent to solving the m-body problem characterized by the Hamiltonian

$$H = \tfrac{1}{2} \sum_{j=1}^{m} \left[p_j^2 + V(2x_j)\right] + \sum_{j>k=1}^{n} \left[V(x_j - x_k) + V(x_j + x_k)\right] . \qquad (2.7.1.2)$$

In this manner from one solvable model another one is generated, that is also solvable, but whose "physical" interpretation is marred by its lack of translation invariance, and in particular by the appearance of the sum of the interparticle distance in the argument of the potential.

A similar approach is possible starting with an odd number of particles, $n=2m+1$, and considering the symmetrical configuration with one particle fixed at the center and the rest disposed symmetrically, m to its left and m to its right [19].

While the integrable dynamical systems obtained by such tricks are not very appealing as models of many-body problems, their mathematical interest is highlighted by their relationship to the root systems of semisimple Lie algebras, discovered and analyzed by Olshanetzky and Perelomov, to whose papers the interested reader is referred [20].

2.7.2 Group-theoretical origin of the integrability of these dynamical systems

The conjecture that the remarkable properties of the models described in this section have a group-theoretical origin is very natural and was formulated at an early stage [21]. Initial progress towards such an understanding was due to Perelomov [22]; major progress has occurred recently, not only through the work of Perelomov and Olshanetsky mentioned above, but also through the work of Barucchi and Regge [23], who pointed out the relevance of the conformal group (analogous results have been obtained independently by Wojciechowski [24]), and more recently through the highly mathematical treatment of Kazhdan, Kostant and Sternberg [25].

2.7.3 Equilibrium configurations of certain integrable many-body problems: special properties

The dynamical evolution of the systems considered in this section is quite peculiar, as evidence by their integrability and by the specific properties displayed above in some specific cases. It is therefore not surprising that, for those systems that have equilibrium configurations, these are also characterized by remarkable mathematical properties that have been recently uncovered. We mention here only two cases.

The first result refers to the model characterized by the Hamiltonian

$$H = \tfrac{1}{2} \sum_{j=1}^{n} \left(p_j^2 + x_j^2\right) + \sum_{j>k=1}^{n} (x_j - x_k)^{-2}. \qquad (2.7.3.1)$$

Clearly the equilibrium configuration $p_j = 0$, $x_j = \bar{x}_j$, is characterized by the equations

$$\bar{x}_j = 2 \sum_{k=1}^{n}{}' (\bar{x}_j - \bar{x}_k)^{-3}, \quad j = 1, 2, \ldots, n. \qquad (2.7.3.2)$$

Here and below a prime appended to a sum indicates that the singular term must be omitted.

But it has been shown [26] that the quantities $\bar{x}_j$ are also the solutions of the equations

$$\bar{x}_j = \sum_{k=1}^{n}{}' (\bar{x}_j - \bar{x}_k)^{-1}, \quad j = 1, 2, \ldots, n, \qquad (2.7.3.3)$$

corresponding therefore also to the equilibrium configuration of the system characterized by the Hamiltonian

$$H = \tfrac{1}{2} \sum_{j=1}^{n} \left(p_j^2 + x_j^2\right) - \sum_{j>k=1}^{n} \log |x_j - x_k|. \qquad (2.7.3.4)$$

Moreover, these quantities $\overline{x}_j$ have a remarkable mathematical significance, coinciding simply with the n zeros of the Hermite polynomial of degree n:

$$H_n(\overline{x}_j) = 0, \qquad j = 1, 2, \ldots, n. \qquad (2.7.3.5)$$

These results are proved in Section 4 below.

The second result refers to the model characterized by the Hamiltonian

$$H = \tfrac{1}{2} \sum_{j=1}^{n} p_j^2 + \sum_{j>k=1}^{n} \sin^{-2}(x_j - x_k). \qquad (2.7.3.6)$$

It is easily seen that the equilibrium configuration corresponding to this model is given by the formula

$$\overline{x}_j = x_o + \pi j/n, \qquad j = 1, 2, \ldots, n. \qquad (2.7.3.7)$$

Consider now the L and M Lax matrices appropriate to the system characterized by the Hamiltonian (2.7.3.6). Drawing on the results of the first part of this section, they can be defined as follows:

$$L_{jk} = p_j \delta_{jk} + i(1-\delta_{jk}) \operatorname{corg}(x_j - x_k), \qquad (2.7.3.8)$$

$$M_{jk} = \delta_{jk} \sum_{\ell=1}^{n}{}' \sin^{-2}(x_j - x_\ell) - (1-\delta_{jk}) \sin^{-2}(x_j - x_k). \qquad (2.7.3.9)$$

Indicate with $\overline{L}$ and $\overline{M}$ the values of these matrices corresponding to the equilibrium configuration $p_j = 0$, $x_j = \overline{x}_j$ [with $\overline{x}_j$ given by (2.7.3.7)]. Clearly the Lax equation (2.5) [or rather its equivalence to the equations of motion implied by (2.7.3.6)] implies that $\overline{L}$ and $\overline{M}$ commute:

$$[\overline{L}, \overline{M}] = 0. \qquad (2.7.3.10)$$

Less obvious, and highly remarkable, is the fact that the two matrices $\overline{L}$ and $\overline{M}$ have a common set of eigenvectors, and that their corresponding eigenvalues $\overline{\lambda}_s$ and $\overline{\mu}_s$ are integers, being given by the simple formulae

$$\overline{\lambda}_s = 2s-n, \quad j = 1, 2, \ldots, n-1; \quad \overline{\lambda}_n = 0, \qquad (2.7.3.11)$$

$$\bar{\mu}_s = 2s(n-s), \quad j = 1, 2, ..., n. \tag{2.7.3.12}$$

For a proof of these results, and for a description of some implications of mathematical interest, we refer the interested reader to the original literature [27] (see also Subsection 3.3 below).

Results analogous to those just reported, but for other systems of the class considered above, are now being obtained; they will be reported elsewhere [28].

2.7.4 Numerical experiments [29]

The integrability of the one-dimensional n-body problem with quadratic and inversely-quadratic two-body potentials has also been tested numerically; surprisingly it is claimed that it holds even for fixed-end boundary conditions. It appears instead that the integrability disappears if the interaction acts only between nearest neighbors.

2.7.5 Infinitely many particles: thermodynamical limit and mathematical implications

Some of the one-dimensional many-body models described above have been investigated in the limit of infinitely many particles. Early study of the statistical mechanics of such models is due to Sutherland [9]. An exact treatment of the thermodynamical properties of the one-dimensional quantal many-body system with inverse-square two-body forces was then given by Marchiro and Presutti [30]. Identification of the classical limit of the (easily computable) quantal partition function for such a system (confined by an external oscillator potential) with the classical partition function has yielded the following mathematical result [31]:

$$I_n(g,\omega) \equiv \int_{-\infty}^{+\infty} dx_1 \ldots dx_n \exp\left[-\tfrac{1}{2}\omega^2 \sum_{j=1}^{n} x_j^2 - g^2 \sum_{j>k=1}^{n} (x_j - x_k)^{-2}\right]$$

$$= I_n(0,g) \exp\left[-\tfrac{1}{2} n(n-1) g\omega\right], \tag{2.7.5.1}$$

where, of course, $I_n(0,g)$ can be trivially evaluated.

An interesting investigation of the (Vlasov) dynamics of a one-dimensional fluid made up of particles interating via inverse-square pair potentials has been made by the Choodnovski brothers [32].

2.7.6 Solvable many-body models with three-body forces

It was noted by Wolfes that the quantal one-dimensional three-body problem with inverse-square pair potentials and with, in addition, a second potential inversely proportional to the square of the distance of each particle from the center-of-mass of the system is also solvable (even in the presence of an harmonic oscillator potential acting on each particle or between every pair of particles) [33]. The corresponding scattering process (when the harmonic potential is absent) has also been completely studied, always in the quantal context [34]. The corresponding classical problems are also integrable and can be given a group-theoretical interpretation [35]. No extension to the case of more than three particles has been considered (indeed, it is probably unfeasible, i.e., the corresponding n-body problem is not integrable for $n > 3$).

3. MOTION OF POLES OF NONLINEAR PARTIAL DIFFERENTIAL EQUATIONS AND RELATED MANY-BODY PROBLEMS

Recently a remarkable connection has been discovered between certain nonlinear partial differential equations and some of the dynamical systems described in the previous section. The basic idea goes as follows. Let $\varphi(x,t)$ satisfy a nonlinear evolution equation, say

$$\varphi_t(x,t) = F\left[\varphi(x,t), \varphi_x(x,t), \varphi_{xx}(x,t), \ldots\right], \tag{3.1}$$

where F is a polynomial in φ and its x-derivatives; and assume that there exist solutions of this equation that are (for all values of time!) rational in x, say

$$\varphi(x,t) = \sum_{j=1}^{n} [x-x_j(t)]^{-1} r_j(t). \tag{3.2}$$

[This implies, of course, a restriction on (3.1); see below.] Then to the time evolution of φ, determined by (3.1), there corresponds through (3.2) a time evolution of the quantities $r_j(t)$ and $x_j(t)$ characterized by equations of motion having the structure

$$\dot{r}_j(t) = f_j[x_k(t), r_k(t)], \quad j = 1, 2, \ldots, n, \qquad (3.3a)$$

$$\dot{x}_j(t) = g_j[x_k(t), r_k(t)], \quad j = 1, 2, \ldots, n. \qquad (3.3b)$$

But these equations correspond to the time evolution of a dynamical system with 2n degrees of freedom; thus a connection has been established between the (nonlinear) partial differential equation (3.1) and the time evolution (3.3) of a dynamical system.

This relatively trivial observation has acquired great importance due to the discovery by Airault, McKean and Moser [36] (motivated by previous work of Kruskal [37] and Thikstun [38]) of a remarkable instance of the connection described above, in which the partial differential equation is the celebrated Korteweg-de Vries equation and the dynamical system is closely related to one of those described in the preceding section (indeed, to the prototypical one: the one-dimensional many-body problem with a pair potential proportional to the inverse square of the interparticle distance). Further important progress has been made by the Choodnovsky brothers who, besides rediscovering independently many of the results of Airault, McKean and Moser, extended the treatment to other partial differential equations, including in particular the so-called Burgers-Hopf equation [40], [50].

But it is not our intention here to report these results which our interested reader shall find in the original literature [36], [40]. We merely mention, to motivate the developments described below, that the results outlined above, interesting as they are for the mathematical connections they uncover between two previously unrelated fields of research [36], [40], are however not entirely satisfactory as a tool to produce "solvable many-body problems." Indeed, while the dynamical systems they originate coincide with, or are closely related to, some of the one-dimensional many-body models considered in the preceding section, they are in addition characterized by certain restrictions on the relevant manifold of solutions (corresponding to restrictions on the initial conditions), that all but destroy the possibility of an associated "physical" interpretation in terms of a (one-dimensional) many-body problem.

The developments reported in this, and in the following section, are instead motivated by the search for many-body models whose solvability, without any restriction on the initial conditions (or with only weak restrictions, see below) can be demonstrated in the framework of the idea outlined at the beginning of this section. Indeed, a rich harvest of such models exists and the corresponding mathematical developments are remarkably elementary. A detailed treatment of these results has already been published [41]; our treatment below will be terse and selective—its main purpose being to introduce and outline the basic ideas.

3.1 A simple example

Consider the nonlinear partial differential equation

$$\varphi_t(x,t) + \varphi_x(x,t) + \alpha\varphi(x,t) + \varphi^2(x,t) = 0. \qquad (3.1.1)$$

With the simple change of dependent variable

$$\varphi(x,t) = 1/\psi(x,t), \qquad (3.1.2)$$

it becomes the linear equation

$$\psi_t(x,t) + \psi_x(x,t) - \alpha\psi(x,t) = 1. \qquad (3.1.3)$$

It is therefore immediately solvable:

$$\varphi(x,t) = \varphi(x-t,0)\exp(-\alpha t)\Big/\left\{1+\varphi(x-t,0)[1-\exp(-\alpha t)]/\alpha\right\}. \qquad (3.1.4)$$

Consider now a special solution $\varphi(x,t)$ having the form

$$\varphi(x,t) = \sum_{j=1}^{n} [x-x_j(t)]^{-1} r_j(t). \qquad (3.1.5)$$

The consistency of this ansatz with (3.1.1) is implied by the explicit form of the solution (3.1.4) (see below). On the other hand elementary algebra yields, from (3.1.5), the formulae

$$\varphi_x(x,t) = -\sum_{j=1}^{n} [x-x_j(t)]^{-2} r_j(t), \qquad (3.1.6)$$

$$\varphi_t(x,t) = \sum_{j=1}^{n} \left\{[x-x_j(t)]^{-2} r_j(t)\dot{x}_j(t) + [x-x_j(t)]^{-1}\dot{r}_j(t)\right\}, \qquad (3.1.7)$$

$$\varphi^2(x,t) = \sum_{j=1}^{n} \left\{[x-x_j(t)]^{-2} r_j^2(t) + \right.$$

$$+ [x-x_j(t)]^{-1} 2r_j(t) \sum_{k=1}^{n}{}' r_k(t) \Big/ [x_j(t) - x_k(t)] \Big\} . \tag{3.1.8}$$

Let us recall that (here, and always below) dots indicate time-differentiation, and a prime appended to a sum indicates that the singular term must be omitted. To obtain (3.1.8) we have squared (3.1.5), separated the diagonal from the nondiagonal terms in the resulting double sum, and used in the latter the identity

$$\Big\{[x-x_j(t)][x-x_k(t)]\Big\}^{-1} = [x_j(t)-x_k(t)]^{-1}\Big\{[x-x_j(t)]^{-1} - [x-x_k(t)]^{-1}\Big\}. \tag{3.1.9}$$

Insertion of (3.1.5)-(3.1.8) in (3.1.1) implies the equations

$$r_j(t) = 1 - \dot{x}_j(t), \quad j = 1, 2, \ldots, n, \tag{3.1.10}$$

$$\dot{r}_j(t) = -\alpha r_j(t) - 2r_j(t) \sum_{k=1}^{n}{}' r_k(t) \Big/ [x_j(t) - x_k(t)], \quad j = 1, 2, \ldots, n; \tag{3.1.11}$$

and insertion of (3.1.10) in (3.1.11) yields

$$\ddot{x}_j(t) = \alpha[1-\dot{x}_j(t)] + 2[1-\dot{x}_j(t)] \sum_{k=1}^{n}{}' [1-\dot{x}_k(t)] \Big/ [x_j(t) - x_k(t)], \quad j = 1, 2, \ldots, n. \tag{3.1.12}$$

These last equations clearly resemble the equations of motion of an n-body problem; but before introducing such an interpretation we prefer to define the new dependent variables

$$y_j(t) = x_j(t) - t, \quad j = 1, 2, \ldots, n, \tag{3.1.13}$$

so that in place of (3.1.12) we have

$$\ddot{y}_j(t) = -\alpha\dot{y}_j(t) + 2\dot{y}_j(t) \sum_{k=1}^{n}{}' \dot{y}_k(t) \Big/ [y_j(t)-y_k(t)], \quad j=1, 2, \ldots, n. \tag{3.1.14}$$

The change of variables (3.1.13) corresponds, of course, to a (Galilei) transformation to a frame moving with unit speed.

Viewed as the equations of motion of an n-body problem, (3.1.14) describes a system with velocity-dependent forces. The first term in the r.h.s. corresponds (for positive α) to a "friction-type" interaction (a "breaking" force proportional to the speed of each particle); the second term, a velocity-dependent two-body interaction, represents a force inversely proportional to the inter-particle distance and moreover proportional to the speed of each particle of the pair. The velocity-dependence of these forces implies, of course, that the equations of motion are not Galilei-invariant; they are, on the other hand, manifestly translation-invariant. Indeed the center-of-mass coordinate

$$Y(t) = n^{-1} \sum_{j=1}^{n} y_j(t) \tag{3.1.15}$$

satisfies the simple equation

$$\ddot{Y}(t) = -\alpha\dot{Y}(t); \tag{3.1.16}$$

in the particular case $\alpha = 0$ (see below), the center-of-mass moves with constant speed.

The proportionality of all forces to the speed of the particle they act upon implies that they cannot make a particle reverse its motion (since the forces vanish when the speed of the particle vanishes). Thus the sign of the speed of every particle remains unchanged throughout the motion. We discuss below only the case when all these speeds have the same sign (say, they are all positive, $\dot{y}_j > 0$, $j = 1, 2, \ldots, n$), this being a sufficient condition to exclude the occurrence of collapse (the singular two-body interaction being repulsive when acting between particles that move in the same direction).

The solution of the equations of motion (3.1.14) can be easily achieved through the solution (3.1.4) of the nonlinear partial differential equation (3.1.1), noting that the coordinates $x_j(t)$ [related to the $y_j(t)$'s by (3.1.13)] are the positions of the poles of $\varphi(x,t)$ as implied by the ansatz (3.1.5). There follows that the values of the coordinates $y_j(t)$ are the n solutions of the algebraic equation in y,

$$\sum_{j=1}^{n} \dot{y}_j(0)/[y-y_j(0)] = \alpha/[1-\exp(-\alpha t)], \tag{3.1.17}$$

where of course $y_j(0)$ and $\dot{y}_j(0)$, $j = 1, 2, \ldots, n$, are the initial positions, and speeds, of the particles.

The behavior of the "many-body system" (3.1.14) can thus be displayed analyzing the solutions of (3.1.17). A convenient technique to do this is to plot the l.h.s. of (3.1.17) as a function of y; note that this function vanishes as $y \to \pm\infty$ and it has n poles (with positive residues). The intersections of this graph with a horizontal straight line representing the r.h.s. of (3.1.17) (and moving vertically with time) provides then a convenient display of the motion of the particles. This analysis implies, for instance, the following asymptotic formulae (without loss of generality, we assume $y_{j+1} > y_j$):

$$\lim_{t \to +\infty} [y_j(t)] = b_j(\alpha), \quad j = 1, 2, \ldots, n, \tag{3.1.18}$$

$$\lim_{t \to -\infty} [y_j(t)] = a_{j-1}, \quad j = 2, 3, \ldots, n, \tag{3.1.19a}$$

$$y_1(t) = -(v/\alpha)\exp(-\alpha t) \quad 1+0[\exp(\alpha t)] \ , \quad t \to -\infty, \tag{3.1.19b}$$

where the quantities $b_j(\alpha)$ are the n solutions, ordered so that $b_{j+1}(\alpha) \quad b_j(\alpha)$, of the algebraic equation in b

$$\sum_{j=1}^{n} \dot{y}_j(0)/[b-y_j(0)] = \alpha, \tag{3.1.20}$$

while the quantities a_j are the n-1 (finite) solutions of the algebraic equation in a

$$\sum_{j=1}^{n} \dot{y}_j(0)/[a-y_j(0)] = 0, \tag{3.1.21}$$

and

$$v = \sum_{j=1}^{n} \dot{y}_j(0). \tag{3.1.22}$$

These results hold for $\alpha > 0$; if instead $\alpha < 0$, the same asymptotic results obtain, except for the exchange of t with -t.

For $\alpha = 0$ one gets instead, for $t \to -\infty$,

$$y_1(t) = vt + a_o + 0(|t|^{-1}), \tag{3.1.23a}$$

$$y_j(t) = a_{j-1} + 0(|t|^{-1}), \quad j = 2, 3, \ldots, n, \tag{3.1.23b}$$

and for $t \to +\infty$,

$$y_j(t) = a_j + O(t^{-1}), \quad j = 1, 2, \ldots, n-1, \tag{3.1.24a}$$

$$y_n(t) = vt + a_o + O(|t|^{-1}), \tag{3.1.24b}$$

where

$$a_o = \sum_{j=1}^{n} \dot{y}_j(0) y_j(0)/v \tag{3.1.25}$$

and v is defined by (3.1.22). This latter example is particularly amusing: it corresponds to a many-body problem with only inter-particle forces [see (3.1.14)], whose center-of-mass Y(t) moves freely with speed v/n:

$$Y(t) = Y(0) + (v/n)t; \tag{3.1.26}$$

in the remote past, it has n-1 particles almost at rest at the positions a_j and one particle coming in (say, from the far left) with velocity v; at any intermediate time, it has all the particles moving towards the right; in the remote future, it has again n-1 particles almost at rest exactly in the same positions as in the remote past except for the fact that each particle has moved one place to the right, the first particle settling down in the first location a_1, while the last is escaping to the right according to the same trajectory that the particle coming initially from the left would have followed had it been free to move through the others (as it would have been the case if the other n-1 particles had been exactly at rest initially).

3.2 Extension: nontranslation invariant models

In addition to Equations (3.1.6)-(3.1.8), the ansatz (3.1.5) implies the following formulae, all of which can be easily proved by trivial algebra:

$$x\varphi = \sum_{j=1}^{n} \left\{ (x-x_j)^{-1} x_j r_j + r_j \right\}, \tag{3.2.1}$$

$$x\varphi_x = -\sum_{j=1}^{n} \left\{ (x-x_j)^{-2} x_j r_j + (x-x_j)^{-1} r_j \right\}, \tag{3.2.2}$$

$$x^2\varphi_x = -\sum_{j=1}^{n} \left\{ (x-x_j)^{-2} x_j^2 r_j + (x-x_j)^{-1} 2x_j r_j + r_j \right\}, \tag{3.2.3}$$

$$x\varphi_t = \sum_{j=1}^{n} \left\{ (x-x_j)^{-2} \dot{x}_j x_j r_j + (x-x_j)^{-1} (\dot{x}_j r_j + x_j \dot{r}_j) + \dot{r}_j \right\}, \quad (3.2.4)$$

$$x\varphi^2 = \sum_{j=1}^{n} \left\{ (x-x_j)^{-2} x_j r_j^2 + (x-x_j)^{-1} \left[r_j^2 + 2x_j r_j \sum_{k=1}^{n}{}' r_k / (x_j - x_k) \right] \right\}, \quad (3.2.5)$$

$$x^2\varphi^2 = \sum_{j=1}^{n} \left\{ (x-x_j)^{-2} x_j^2 r_j^2 + (x-x_j)^{-1} \left[2x_j r_j^2 + 2x_j^2 r_j \sum_{k=1}^{n}{}' r_k / (x_j - x_k) \right] \right\} + \left(\sum_{j=1}^{n} r_j \right)^2 . \quad (3.2.6)$$

Thus to the nonlinear partial differential equation

$$\left(A_0 + A_1 x\right)\varphi_t + \left(B_0 + B_1 x\right)\varphi + \left(C_0 + C_1 x + C_2 x^2\right)\varphi_x + \left(D_0 + D_1 x + D_2 x^2\right)\varphi^2 = 0 \quad (3.2.7)$$

there corresponds for the poles $x_j(t)$ and residues $r_j(t)$ the $2n+1$ equations

$$r_j = \left[C_0 + C_1 x_j + C_2 x_j^2 - \dot{x}_j (A_0 + A_1 x_j) \right] \Big/ \left(D_0 + D_1 x_j + D_2 x_j^2 \right), \quad j = 1, 2, \ldots, n, \quad (3.2.8)$$

$$(A_0 + A_1 x_j)\dot{r}_j + A_1 \dot{x}_j r_j + \left[B_0 - C_1 + (B_1 - 2C_2) x_j \right] r_j + (D_1 + 2D_2 x_j) r_j^2 + 2\left(D_0 + D_1 x_j + D_2 x_j^2 \right) r_j \sum_{k=1}^{n}{}' r_k \Big/ (x_j - x_k) = 0, \quad j = 1, 2, \ldots, n, \quad (3.2.9)$$

$$A_1 \dot{R} + \left(B_1 - C_2 + D_2 R \right) R = 0, \quad (3.2.10)$$

where

$$R \equiv \sum_{j=1}^{n} r_j . \quad (3.2.11)$$

Of course in these equations $\varphi = \varphi(x,t)$ and the quantities r_j and x_j are time-dependent; and there is no a priori need to exclude that also the coefficients A_j, B_j, C_j and D_j depend on time.

Substituting (3.2.8) in (3.2.9) one gets n "equations of motion" for the n quantities x_j, that are, however, generally translation-noninvariant; while care can be taken of the constraint (3.2.10) by an appropriate (if need be, time-dependent) choice of A_1, B_1, C_2 and D_2, the simpler possibility being of course the choice $A_1 = D_2 = 0$, $B_1 = C_2$.

On the other hand the nonlinear partial differential equation (3.2.7) can again be linearized by the simple change of dependent variable (3.1.2), yielding now

$$\left(A_0 + A_1 x\right)\psi_t - \left(B_0 + B_1 x\right)\psi + \left(C_0 + C_1 x + C_1 x^2\right)\psi_x + D_0 + D_1 x + D_2 x^2 . \qquad (3.2.12)$$

The corresponding initial condition is clearly

$$\psi(x,0) = 1 \Big/ \sum_{j=1}^{n} [x - x_j(0)]^{-1} r_j(0) . \qquad (3.2.13)$$

Thus the function ψ is completely determined, at the initial time $t=0$, by the initial positions $x_j(0)$ and velocities $\dot{x}_j(0)$ [through (3.2.13) and (3.2.8)]; its subsequent time evolution is provided by the linear first-order partial differential equation (3.2.12); and the positions of its zeros $x_j(t)$ coincide, as implied by (3.1.5) and (3.1.2), with the solutions $x_j(t)$ of the "many-body problem" characterized by the equations of motion (3.2.8)-(3.2.9).

3.3 Another extension: circular and hyperbolic functions

Another extension of the approach of Sect. 3.1 is obtained by replacing the <u>ansatz</u> (3.1.5) with the position

$$\varphi(x,t) = \beta\rho(t) + \beta \sum_{j=1}^{n} r_j(t) \operatorname{cotg}\{\beta[x - x_j(t)]\} , \qquad (3.3.1)$$

where $\rho(t)$ is a function of t to be chosen appropriately, and β is a constant. Clearly $\beta = 0$ reproduces (3.1.5).

Referring to the original paper [41] for a more complete treatment, we outline here only one special application of this approach, namely an analysis of the many-body model characterized by the equations of motion

$$\ddot{y}_j(t) = 2\beta\dot{y}_j(t) \sum_{k=1}^{n}{}' \dot{y}_k(t)\cot g\{\beta[y_j(t)-y_k(t)]\},$$

$$j = 1, 2, \ldots, n. \tag{3.3.2}$$

Without loss of generality, one can clearly restrict attention to initial conditions such that

$$0 < y_1(0) < y_2(0) < \ldots < y_n(0) < \pi/\beta. \tag{3.3.3}$$

We moreover assume that all the initial speeds $\dot{y}_j(0)$ have the same (say, positive) sign; for the motivation and implications of this assumption we refer to the analogous discussion in Subsection 3.1 above.

It can be shown that the solutions $y_j(t)$ of the equations of motion (3.3.2) coincide with the n solutions of the equation in y

$$\sum_{j=1}^{n} \dot{y}_j(0)\cot g\{\beta[y-y_j(0)]\} = v \cot g(\beta v t) \tag{3.3.4}$$

where

$$v = \sum_{j=1}^{n} \dot{y}_j(0). \tag{3.3.5}$$

Thus the behavior of these solutions can conveniently be based on a (graphical) analysis of (3.3.4). It is easily seen that this implies that, as time proceeds, all particles always move towards the right (positive direction); at the time $t_1 = \delta$, where

$$\delta = \pi/\beta v, \tag{3.3.6}$$

$$y_n(t_1) = y_1(0) + \pi/\beta, \tag{3.3.7a}$$

$$y_j(t_1) = y_{j+1}(0), \quad j = 1, 2, \ldots, n-1; \tag{3.3.7b}$$

at the time $t_2 = 2\delta$,

$$y_n(t_2) = y_2(0) + \pi/\beta, \tag{3.3.8a}$$

$$y_{n-1}(t_2) = y_1(0) + \pi/\beta, \tag{3.3.8b}$$

$$y_j(t_2) = y_{j+2}(0), \quad j = 1, 2, \ldots, n-2; \tag{3.3.8c}$$

and so on. At the time $t_n = n\delta$

$$y_j(t_n) = y_j(0) + \pi/\beta, \quad j = 1, 2, \ldots, n, \tag{3.3.9}$$

namely, the system has recovered exactly the initial structure, having moved collectively to the right a distance π/β. Thereafter the process is repeated. Thus the system has an internal structure that oscillates periodically, with period

$$T = n\delta = n\pi/\beta v, \tag{3.3.10}$$

while it travels collectively with the constant speed

$$\dot{Y}(0) = v/n \tag{3.3.11}$$

of its center-of-mass.

Clearly to discuss this system it is convenient to go over to the variables

$$z_j(t) = y_j(t) - Vt, \tag{3.3.12}$$

where we have set

$$V = \dot{Y}(0) = v/n. \tag{3.3.13}$$

Then the equations of motion read

$$\ddot{z}_j(t) = 2\beta[V+z_j(t)] \sum_{k=1}^{n}{}' [V + z_k(t)]\text{cotg}\left\{\beta[z_j(t) - z_k(t)]\right\},$$

$$j = 1, 2, \ldots, n, \tag{3.3.14}$$

and, for these variables, the center-of-mass

$$Z(t) = n^{-1} \sum_{j=1}^{n} z_j(t) \tag{3.3.15}$$

must be chosen at rest:

$$\dot{Z}(0) = 0, \quad Z(t) = Z(0). \tag{3.3.16}$$

The solutions $z_j(t)$ are the roots of the equation in z

$$\sum_{j=1}^{n} [V + \dot{z}_j(0)] \cot g \left\{ \beta[z + Vt - z_j(0)] \right\} = nV \cot g(\beta nVt); \qquad (3.3.17)$$

as implied by the previous analysis, they are always all real provided

$$\dot{z}_j(0) > -V, \quad j = 1, 2, \ldots, n, \qquad (3.3.18)$$

and they all oscillate with the period [see (3.3.10) and (3.3.13)]

$$T = \pi/\beta V \qquad (3.3.19)$$

around the equilibrium positions

$$\bar{z}_j = z_o + j\pi/\beta n, \quad j = 1, 2, \ldots, n, \qquad (3.3.20)$$

where z_o is arbitrary.

Note that the many-body system (3.3.14) approximates, provided

$$|\dot{z}_j(t)| \ll V, \qquad (3.3.21)$$

the Hamiltonian system of n unit-mass particles interacting via the two-body periodic potential

$$W(z) = -2V^2 \log|\sin(\beta z)|, \qquad (3.3.22)$$

namely, the n-body system characterized by the Hamiltonian

$$H(p,q) = \frac{1}{2}\sum_{j=1}^{n} p_j^2 - 2V^2 \sum_{k>j=1}^{n} \log|\sin\beta(q_j - q_k)|. \qquad (3.3.23)$$

Condition (3.3.18) can, of course, be satisfied, provided the initial configuration of the system is sufficiently close to the equilibrium configuration.

These results are the basis of the drivation of the results described in the second part of Subsection 2.7 above.

Clearly the method of solution described above for the system (3.3.2) applies also if the parameter β is pure imaginary, $\beta = i\gamma$, in which case the equations of motion remain real bu the circular functions get replaced by hyperbolic functions. The behavior of the solutions changes of course drastically; details may be found in the original paper [41].

3.4 More extensions: particles of two different types, symmetrical configurations, two-dimensional models by complexification

Let us merely mention, referring for more details to the original paper [41], three additional avenues of extension of the results described above, all three of which have been already discussed in the preceding section (see, in particular, Subsections 2.3, 2.7.1 and 2.6).

The first extension is the possibility to introduce models involving particles of two different kinds, by modifying the models of the previous subsection. The trick consists, of course, in shifting appropriately (say, by $\pi/2\beta$) the coordinates of n_1 of the n particles.

The second extension corresponds to the consideration of symmetrical configurations of the many-body problem, that preserve their symmetry throughout the motion. It turns out that, rather than generating novel solvable systems, this trick serves in this case to display some remarkable properties of the models discussed above.

The third extension to be mentioned is the possibility to generate two-dimensional models by complexification, i.e., by the trick already described in Subsection 2.6 above.

3.5 More general partial differential equations

We have discussed in Subsection 3.1 the solvable many-body problems that correspond to the motion of the poles of rational solutions, of type (3.1.5), of the very simple nonlinear partial differential equation (3.1.1), or of its variant (3.2.7); and in the subsequent Subsections 3.2-3.4 we have outlined various extensions that give rise to other solvable models. A natural question suggested by these results, as well as by the original findings of Airault, McKean and Moser [36] and of the Choodnovsky brothers [40], is whether the same approach can be applied to other nonlinear partial differential equations in order to relate them to "many-body problems." If only models free of any constraint on the initial data are sought, then it appears that the answer is negative; there are indeed many other partial differential equations that can be related to a many-body problem by an ansatz such as (5.1.5) (or some extension of it), but they generally imply some additional constraint on the poles besides that implied by the equations of motion. For a justification of this assertion, we refer the interested reader to Subsection 2.5 of the original paper [41].

4. MOTION OF ZEROS OF LINEAR EVOLUTION EQUATIONS AND RELATED INTEGRABLE MANY-BODY PROBLEMS

The developments described in the preceding section (and the original results of the Choodnovsky brothers [40]) suggest to investigate the motion of the zeros of (special) solutions of linear partial differential equations. We describe below tersely the basic idea of this approach, and we discuss some examples that illustrate its power. For a more detailed treatment, including several other examples and extensions analogous to those discussed above, the interested reader is referred to the original paper [41].

4.1 Basic ansatz and formulae

Let us consider a polynomial of degree n in x, with n zeros at the positions $x_j(t)$:

$$\psi(x,t) = \prod_{j=1}^{n} [x-x_j(t)], \qquad (4.1.1)$$

This representation immediately implies the following formulae:

$$\psi_x = \psi \sum_{j=1}^{n} (x-x_j)^{-1}, \qquad (4.1.2)$$

$$\psi_t = -\psi \sum_{j=1}^{n} (x-x_j)^{-1}\dot{x}_j, \qquad (4.1.3)$$

$$\psi_{xx} = 2\psi \sum_{j=1}^{n} (x-x_j)^{-1} \sum_{k=1}^{n}{}' (x_j-x_k)^{-1}, \qquad (4.1.4)$$

$$\psi_{xt} = -\psi \sum_{j=1}^{n} (x-x_j)^{-1} \sum_{k=1}^{n}{}' (\dot{x}_j+\dot{x}_k)/(x_j-x_k), \qquad (4.1.5)$$

$$\psi_{tt} = \psi \sum_{j=1}^{n} (x-x_j)^{-1}\left[-\ddot{x}_j + 2\dot{x}_j \sum_{k=1}^{n}{}' \dot{x}_k/(x_j-x_k)\right], \qquad (4.1.6)$$

$$x\psi_x - n\psi = \psi \sum_{j=1}^{n} (x-x_j)^{-1} x_j, \qquad (4.1.7)$$

$$x\psi_{xx} = 2\psi \sum_{j=1}^{n} (x-x_j)^{-1} x_j \sum_{k=1}^{n}{}' (x_j-x_k)^{-1}, \tag{4.1.8}$$

$$x\psi_{xt} = -\psi \sum_{j=1}^{n} (x-x_j)^{-1} x_j \sum_{k=1}^{n}{}' (\dot{x}_j+\dot{x}_k)/(x_j-x_k), \tag{4.1.9}$$

$$x^2\psi_{xx} - n(n-1)\psi = 2\psi \sum_{j=1}^{n} (x-x_j)^{-1} x_j^2 \sum_{k=1}^{n}{}' (x_j-x_k)^{-1}, \tag{4.1.10}$$

$$x\left[x^2\psi_{xx} - 2(n-1)x\psi_x + n(n-1)\psi\right] =$$

$$= 2\psi \sum_{j=1}^{n} (x-x_j)^{-1} x_j^2 \sum_{k=1}^{n}{}' x_k/(x_j-x_k), \tag{4.1.11}$$

$$x\left[x\psi_{xt} - (n-1)\psi_t\right] =$$

$$= -\psi \sum_{j=1}^{n} (x-x_j)^{-1} x_j \sum_{k=1}^{n}{}' (\dot{x}_j x_k + x_j \dot{x}_k)/(x_j-x_k). \tag{4.1.12}$$

In all these equations, of course, $\psi \equiv \psi(x,t)$ and $x_j \equiv x_j(t)$.

These formulae imply that, to the linear partial differential equation for ψ,

$$\left[A_0 + A_1 x + A_3 x^2 + A_3 x^3\right]\psi_{xx} + \left[B_0 + B_1 x - 2(n-1)A_3 x^2\right]\psi_x$$
$$+ C\psi_{tt} + \left[E - (n-1)D_2 x\right]\psi_t + \left[D_0 + D_1 x + D_2 x^2\right]\psi_{xt}$$
$$- \left[n(n-1)(A_2 - A_3 x) + nB_1\right]\psi = 0, \tag{4.1.13}$$

there corresponds for the n quantities x_j the system of n ordinary differential equations

$$C\ddot{x}_j + E\dot{x}_j = B_0 + B_1 x_j + \sum_{k=1}^{n}{}' (x_j-x_k)^{-1}\left[2\left(A_0 + A_1 x_j + A_2 x_j^2 + A_3 x_j x_k\right) +\right.$$

$$+ 2C\dot{x}_j\dot{x}_k - (\dot{x}_j+\dot{x}_k)(D_0+D_1x_j) - D_2x_j(\dot{x}_jx_k+x_j\dot{x}_k)\Big] ,$$

$$j = 1, 2, \ldots, n. \qquad (4.1.14)$$

These equations may be interpreted as the equations of motion of an n-body problem; the motion of the corresponding particles coincides then with the motion of the zeros of the polynomial (4.1.1), that may be identified as the solution of the linear partial differential equation (4.1.13) characterized by the initial data

$$\psi(x,0) = \prod_{j=1}^{n} [x-x_j(0)], \qquad (4.1.15a)$$

$$\psi_t(x,0) = -\psi(x,0) \sum_{j=1}^{n} [x-x_j(0)]^{-1} \dot{x}_j(0), \qquad (4.1.15b)$$

where, of course, $x_j(0)$ and $\dot{x}_j(0)$ are the initial positions and velocities of the particles.

The equations of motion (4.1.14) are too general to allow a transparent physical interpretation; below we shall consider some simpler instances that obtain when several of the quantities A_s, B_s, C_s, D_s and E in Equations (4.1.13)-(4.1.14) are set to zero (note incidentally that the developments described above do not require these quantities to be time-independent); but let us mention that even more general models can be written [41].

4.2 Equations of motion of first order: a simple example. Properties of the zeros of the classical polynomials

Consider the special case when all the quantities A_s, B_s, C and D_s vanish, except for $A_0=\frac{1}{2}$, $B_1=-1$ and $E=i$, so that (4.1.14) becomes

$$i\dot{x}_j(t) = -x_j + \sum_{k=1}^{n}{}' (x_j-x_k)^{-1}, \quad j = 1, 2, \ldots, n, \qquad (4.2.1)$$

and (4.1.13) becomes

$$i\psi_t + \tfrac{1}{2}\psi_{xx} - x\psi_x + n\psi = 0. \qquad (4.2.2)$$

The first-order equations (4.2.1) are not directly interpretable in terms of a many-body problem; but by differentiating them and using them again it is easily seen [40] that they imply

$$\ddot{x}_j = -x_j + 2 \sum_{k=1}^{n}{}' (x_j - x_k)^{-3}, \quad j = 1, 2, \ldots, n, \tag{4.2.3}$$

namely, exactly the equations of motion corresponding to the (ubiquitous) n-body problem characterized by the Hamiltonian

$$H(p,x) = \frac{1}{2} \sum_{j=1}^{n} (p_j^2 + x_j^2) + \sum_{j>k=1}^{n} (x_j - x_k)^{-2}. \tag{4.2.4}$$

Thus the motion of the zeros of a polynomial solution of (4.2.2) corresponds to the motion of the particles of the one-dimensional n-body problem characterized by the Hamiltonian (4.2.4); but only for the (unphysical) initial conditions corresponding to the validity of (4.2.1) at $t=0$ [the validity of (4.2.1) for $t>0$ being then clearly implied by (4.2.3)]. Indeed this is just one example of the many-body problems <u>with constraints</u> that we mentioned in the opening part of the preceding section.

Some interesting mathematical results can, however, be easily derived from the equations written above (although these results are not new, their derivation here is somewhat simpler than that given previously [26], [41]). Note first of all that (4.2.2) suggests another convenient representation for ψ, namely,

$$\psi(x,t) = 2^{-n} \sum_{m=0}^{n} c_m(t) H_{n-m}(x), \tag{4.2.5}$$

where $H_m(x)$ is the Hermite polynomial of degree m [42]. Indeed, the insertion of this <u>ansatz</u> in (4.2.2) yields for the coefficients $c_m(t)$ the simple equations

$$\dot{c}_m(t) = imc_m(t), \quad m = 0, 1, 2, \ldots, n, \tag{4.2.6}$$

implying, of course,

$$c_m(t) = c_m(0)\exp(imt), \quad m = 0, 1, 2, \ldots, n. \tag{4.2.7}$$

The simultaneous validity of (4.1.1) and (4.2.5) induces a (nonlinear) relationship among the n quantities x_j, $j=1,2,\ldots,n$, and the n quantities c_m, $m=1,2,\ldots,n$ [while $c_0=1$; note the consistency of this formula with (4.2.5)-(4.2.7)]. The remarkable property of this nonlinear relationship is to transform the nonlinear equations of motion (4.2.1) into the linear, and trivially solvable, Equations (4.2.6). An analogous role played by this same transformation is displayed in the following subsection.

Consider now the special solution of (4.2.1) corresponding to equilibrium. Clearly the corresponding values $x_j = \bar{x}_j$ must then satisfy the two systems of nonlinear equation

$$\bar{x}_j = \sum_{k=1}^{n}{}' (\bar{x}_j - \bar{x}_k)^{-1}, \quad j = 1, 2, \ldots, n, \qquad (4.2.8)$$

$$\bar{x}_j = 2\sum_{k=1}^{n}{}' (\bar{x}_j - \bar{x}_k)^{-3}, \quad j = 1, 2, \ldots, n, \qquad (4.2.9)$$

as implied by (4.2.1) and (4.2.3); moreover, they must coincide with the n zeros of the Hermite polynomial of order n,

$$H_n(\bar{x}_j) = 0, \quad j = 1, 2, \ldots, n, \qquad (4.2.10)$$

as implied by (4.2.5)-(4.2.7) (to the equilibrium configuration of the x_j's there must correspond a time-independent ψ).

Summing up, we may therefore now assert:

i) that the zeros $\bar{x}_j$ of the Hermite polynomial $H_n(x)$ satisfy the system of nonlinear algebraic equations (4.2.8), and therefore coincide with the equilibrium positions of the one-dimensional n-body problem characterized by the Hamiltonian

$$H = \frac{1}{2}\sum_{j=1}^{n} (p_j^2 + x_j^2) - \sum_{j>k=1}^{n} \log|x_j - x_k|; \qquad (4.2.11)$$

ii) that these same zeros $\bar{x}_j$ also satisfy the system (4.2.9), and therefore coincide with the equilibrium positions of the one-dimensional n-body problem characterized by the Hamiltonian (4.2.4); and,

iii) that therefore the equilibrium configurations of the n-body problems characterized by the Hamiltonians (4.2.8) and (4.2.11) coincide. The first of these results is actually quite old, having been discovered by Stieljes almost a century ago [43]; the other results are more recent [26].

It is, of course, also possible to derive (4.2.9) directly from (4.2.8) [44]; and many other analogous relations can be obtained for the zeros of other classical polynomials (Laguerre, Jacobi) and of linear combinations of these polynomials [45], and also for the zeros of Bessel functions and of combinations of Bessel functions [46].

It is, moreover, also possible to derive additional results, of a quite novel type, concerning the zeros of the classical polynomials; results that also obtain from the possibility to solve the "equations of motion" (4.2.1) by transforming them into the linear uncoupled system (4.2.6). Let us for this purpose analyze the behavior of (4.2.1) in the neighborhood of the equilibrium configuration $x_j = \bar{x}_j$. To this end we set

$$x_j(t) = \bar{x}_j + \epsilon y_j(t), \quad j = 1, 2, \ldots, n, \tag{4.2.12}$$

in (4.2.1), and we expand in ϵ, retaining only linear terms.

We get in this manner

$$\dot{y}(t) = i(A - \mathbb{1}) y(t), \tag{4.2.13}$$

that we have conveniently written in compact notation, introducing the vector $y(t) \equiv \{y_j(t)\}$ and the (hermitian) matrix A with elements

$$A_{jk} = \delta_{jk} \sum_{\ell=1}^{n}{}' (\bar{x}_j - \bar{x}_\ell)^{-2} - (1 - \delta_{jk}) (\bar{x}_j - \bar{x}_k)^{-2}, \tag{4.2.14}$$

while $\mathbb{1}$ is clearly the unit matrix, of elements δ_{jk}.

A comparison of (4.2.13) with (4.2.6) implies that the eigenvalues of the matrix $A - \mathbb{1}$ must coincide with the integers $m = 1, 2, \ldots, n$, and therefore that [41], [47] <u>the eigenvalues a_m of the hermitian matrix A are the natural numbers from 0 to $n-1$</u>:

$$a_m = m - 1, \quad m = 1, 2, \ldots, n. \tag{4.2.15}$$

This is clearly a nontrivial statement concerning the zeros x_j of the Hermite polynomial $H_n(x)$, out of which the matrix A is constructed: note incidentally that the order n does not appear explicitly in the definition of this matrix, but only by setting its rank (while, of course, the values $\bar{x}_j$ of the zeros of H_n do depend on n, even though we have omitted for notational simplicity to indicate explicitly this dependence).

The conjecture that the requirement that a matrix of rank n, defined in terms of n (unknown) quantities $\bar{x}_j$ as detailed by (4.2.14), have the eigenvalues (4.2.15), imply that the $\bar{x}_j$'s coincide, up to a translation, with the zeros of the Hermite polynomial $H_n(x)$, has been formulated [46]; it is easy to verify it by direct explicit computation for $n = 2$ and $n = 3$.

Analogous results [41], [46] for the zeros of Laguerre and Jacobi polynomials (namely, for all the classical polynomials) can also be obtained by similar techniques.

4.3 Equations of motion of second order: a simple example

Consider the special case when all the quantities A_s, B_s, D_s and E of Subsection 4.1 vanish, except for $A_0 = \frac{1}{2}$, $B_1 = -1$, and $C = 1$, so that (4.1.14) becomes

$$\ddot{x}_j = -x_j + \sum_{k=1}^{n}{}' (1+2\dot{x}_j\dot{x}_k)/(x_j - x_k), \quad j = 1, 2, \ldots, n, \tag{4.3.1}$$

and (4.1.13) becomes

$$\psi_{tt} + \frac{1}{2}\psi_{xx} - x\psi_x + n\psi = 0. \tag{4.3.2}$$

Note that, if

$$|\dot{x}_j| << 1, \quad j = 1, 2, \ldots, n, \tag{4.3.3}$$

the equations of motion (4.3.1) approximate the equations of motion of the one-dimensional many-body system characterized by the Hamiltonian (4.2.11). Of course, the condition (4.3.3) i fulfilled (for all time) provided the initial conditions are sufficiently close to the equilibrium configuration $x_j = \bar{x}_j$, $\dot{x}_j = 0$ discussed in the preceding subsection.

It is not known whether the system characterized by the Hamiltonian (4.2.11) is integrable. The equations of motion (4.3.1) correspond instead to an integrable Hamiltonian system.

Indeed, proceeding in close analogy to the treatment of the preceding subsection, it is immediately seen that the position

$$\prod_{j=1}^{n} [x - x_j(t)] = 2^{-n} H_n(x) + \sum_{m=1}^{n} b_m(t) H_{n-m}(x), \tag{4.3.4}$$

implies for the quantities $b_m(t)$ the linear decoupled equations of motion

$$\ddot{b}_m(t) + m b_m(t) = 0, \quad m = 1, 2, \ldots, n. \tag{4.3.5}$$

These equations are derivable from the Hamiltonian

$$\mathcal{H}(p,b) = \frac{1}{2} \sum_{m=1}^{n} \left(p_m^2 + m b_m^2\right) \tag{4.3.5'}$$

that clearly yields

$$p_m = \dot{b}_m, \quad \dot{p}_m = -m b_m, \quad m = 1, 2, \ldots, n, \tag{4.3.6}$$

and are, of course, explicitly solved by the formula

$$b_m(t) = b_m(0)\cos(m^{\frac{1}{2}}t) + m^{-\frac{1}{2}}\dot{b}_m(0)\sin(m^{\frac{1}{2}}t). \qquad (4.3.7)$$

The requirement that (4.3.4) hold for all x implies, of course, n (nonlinear) algebraic equations relating the n b_m's to the n x_j's. Clearly these relations induce a biunivocal correspondence between these two sets of quantities (for an explicit representation, see below); indeed, the change of variables from the set $\{x_j;\ j=1,2,\ldots,n\}$ to the set $\{b_m;\ m=1,2,\ldots,n\}$ can be viewed as the transition from the <u>individual coordinates</u> x_j of the particles to appropriate <u>collective coordinates</u> b_m describing the <u>normal modes</u> of the many-body system.

The Hamiltonian (4.3.5) defines an integrable system, the n quantities

$$h_m = \frac{1}{2}\left(p_m^2 + mb_m^2\right), \quad m = 1, 2, \ldots, n \qquad (4.3.8)$$

providing clearly n constants of motion. Therefore, the many-body problem characterized by the equations of motion (4.3.1) is also Hamiltonian and integrable, being related by a contact transformation [the change of variable (4.3.4) from the x_j's to the b_m's] to the integrable Hamiltonian (4.3.5). Of course, the momenta p_j' conjugate to the variables x_j in the Hamiltonian yielding the equations of motion (4.3.1) must be appropriately defined following the standard procedure: define from the Hamiltonian $\mathcal{H}$ the Lagrangian $\mathcal{L}$,

$$\mathcal{L} = \frac{1}{2}\sum_{m=1}^{n}\left(\dot{b}_m^2 - mb_m^2\right); \qquad (4.3.9)$$

replace in it the coordinates b_m and their time derivatives $\dot{b}_m$ by their expressions in terms of the coordinates x_j and the velocities $\dot{x}_j$, thereby obtaining the novel Lagrangian $L(x,\dot{x})$; and then define the novel momenta p_j' (conjugate to the coordinates x_j) by the formula

$$p_j' = \partial L(x,\dot{x})/\partial\dot{x}_j. \qquad (4.3.10)$$

Note, incidentally, that explicit expressions of the coordinates b_m's in terms of the x_j's are formally given by the formula

$$b_m(t) = \left[\pi^{\frac{1}{2}}\, 2^{n-m}(n-m)!\right]^{-1} \times$$

$$\times \int_{-\infty}^{+\infty} dx\, H_{n-m}(x) \prod_{j=1}^{n} [x-x_j(t)], \qquad m = 1, 2, \ldots, n, \qquad (4.3.11)$$

that is clearly implied by (4.3.4); and using this formula, and the one that obtains from it by differentiation (with respect to time), one can obtain from (4.3.8) explicit expressions of the constants of motion and of course from (4.3.5) an explicit expression of the Hamiltonian $H(p',x)$ [first replacing in (4.3.5) p_m by its expression $\dot{b}_m$, see (4.3.6); then replacing the b_m's by their expression (4.3.11) in terms of the x_j's, and the $\dot{b}_m$'s by their expression in terms of the x_j's and p_j's, obtained from (4.3.11) and (4.3.10)].

A more direct route to identify n constants of motion associated to the equations (4.3.1) would emerge from the identification of a Lax pair of matrices, as described in Section 2.

Let us emphasize that, although there exist for the system (4.3.1) n global constants of motion, the fact that the angular frequencies $\omega_m = m^{\frac{1}{2}}$ of the "normal modes" associated to this system are not commensurate implies that the generic motion fills the whole configuration space (or rather, the whole region of configuration space available compatibly with the given total energy).

A transparent explicit display of the motion corresponding to the equations (4.3.1) can in some cases be provided. Suppose, for instance, the initial conditions are such that

$$b_m(0) = \dot{b}_m(0) = 0, \qquad m = 1, 2, \ldots, n-1. \qquad (4.3.12)$$

Then throughout the time evolution the coordinates x_j coincide with the n roots of the equation in x

$$H_n(x) = A \cos [n^{\frac{1}{2}} (t-\bar{t})], \qquad (4.3.13)$$

with A and $\bar{t}$ expressed by trivial formulas in terms of $b_n(0)$ and $\dot{b}_n(0)$ [see (4.3.4) and (4.3.7)]; and a graphical display of these roots is conveniently obtained drawing a graph of the Hermite polynomial $H_n(x)$ (i.e., of the l.h.s. of this equation), and representing the r.h.s. as a vertical straight line that oscillates with the circular frequency $n^{\frac{1}{2}}$, whose intersection with the graph of the nth Hermite polynomials represents the position of the particles as a function of time. From this analysis it is easy to infer certain qualitative aspects of the motion of the particles, and also to understand and analyze the possible occurrence of collapse.

It is also possible to rederive [41] from an analysis of the motion of the system (4.3.1) in the neighborhood of its equilibrium configuration the results concerning the zeros of Hermite polynomials obtained in the preceding subsection.

4.4 Concluding remarks

The last two sections describe essentially some of the results of Reference 41, although in the last two subsections the expert and attentive reader will notice some novelties. We repeat that, for a deeper and more extended treatment, the reader is referred to the original paper [41]; although the results given here should certainly have been sufficient to provide a good introduction to the power, and limitations, of the basic approach underlying these results.

We end these Lecture Notes mentioning that there appears, in the framework of the approach outlined above, ample scope for the derivation of additional results. The most promising avenue of current research in that direction is the extension to problems involving infinitely many degrees of freedom.

REFERENCES AND FOOTNOTES

1. M. Toda, Prog. Theor. Phys. Suppl. 45, 1974 (1970).
2. P. D. Lax, Commun. Pure Appl. Math. 21, 467 (1968).
3. H. Flaschka, Phys. Rev. B9, 1924 (1974); Prog. Theor. Phys. 51, 703 (1974).
4. S. V. Manakov, Soviet Phys. JETP 40, 269 (1975) [Russian original: Zurn. Eksp. Teor. Fiz. 67, 543 (1974)].
5. F. Calogero, Lett. Nuovo Cimento 13, 411 (1975). Note the misprints that mar Equations (3), (11) and (13) of this paper: in the first case, a minus sign should separate the terms appearing in the r.h.s.; in the second, a wrong typographical character has been used, as it is apparent from what follows; in the third, the u.c. K without argument should be a l.c. κ. Moreover, the letter a in Equations (10) should be replaced by a $=(e_1-e_3)^{\frac{1}{2}}a$ to achieve notational consistency with Equation (11). Analogous results were (subsequently but independently) obtained by S. Wojciechowski, Phys. Lett. 59A, 84 (1976) and ICTP preprint 76/103 (to appear in Acta Physica Polonica, May 1978).
6. F. Calogero, Lett. Nuovo Cimento 16, 77 (1976). Essentially the same result has been proved independently also

by H. Airault (private communication), by M. A. Olshanetsky and A. M. Perelomov, Inventiones math. 37, 93 (1976), and by S. Pikuiko, Funk. Anal. Appl.

7. We use for elliptic functions the notation of A. Erdélyi (editor), Higher Transcendental Functions, McGraw Hill, New York, 1953, vol. II.

8. This problem was first solved in the quantal case: for n = 3 by C. Marchioro, J. Math Phys. 11, 2193 (1970) [using the results given, for the case with in addition an oscillator potential, by F. Calogero, J. Math. Phys. 12, 2197 (1969)], and for arbitrary n by F. Calogero, J. Math. Phys. 12, 419 (1971). In the classical case it was then solved: for n = 3 by C. Marchioro (unpublished) and by D. C. Khandekar and S. V. Lawande, Amer. J. Phys. 40, 458 (1972); for arbitrary n, by J. Moser, Adv. in Math. 16, 197 (1975). In this important paper Moser introduced for the first time the Lax technique in the context of (classical) many-body problems with pair interactions. Actually the (classical!) problem of three bodies interacting on the line via inverse-square two-body potentials was already solved by C. Jacobi, Problema trium corporum mutis attractionibus cubus distantiarum inverse proportionalibus recta linea se moventium, Gesammelte Werke, Bd. IV, Berlin, 1866 (I wish to thank A. M. Perelomov for pointing out this reference).

9. This case was first solved in the classical and quantal cases by F. Calogero, C. Marchioro and O. Ragnisco, Lett. Nuovo Cimento 13, 383 (1975). In the classical case this treatment involved merely a trivial modification of the case $V(x) = g^2/\sin^2 x$ as first treated, in the classical case, by J. Moser, loc. cit. [the first treatment of the $V(x) = g^2/\sin^2 x$ potential is due, in the quantal context, to B. Sutherland, Phys. Rev. A5, 1375 (1972)]. In the quantal case (that had not been considered by Sutherland, his approach to the $V(x) = g^2/\sin^2 x$ case not being extendable to the $V(x) = g^2/\sinh^2 x$ case), the treatment provided the first instance of utilization of the Lax approach in a quantal context. This treatment has now been extended and improved by M. A. Olshanetsky and A. M. Perelomov, Lett. Math. Phys. 2, 7 (1977).

10. For the potential (2.25) this was originally proved by Moser, loc cit., and for the potential (2.26) by Calogero, Marchioro and Ragnisco, loc. cit.; indeed in these cases the proof is very simple. The proof in the general case of the potential (2.18) is more difficult and has been given

independently by S. Wojciechowski, Lett. Nuovo Cimento 18, 103 (1977) [who took advantage of an elegant technique introduced, in the context of the problem with the potential of Equation (2.25), by K. Sawada and T. Kotera, J. Phys. Soc. Japan 39, 1614 (1975)] and by A. M. Perelomov, Lett. Math. Phys. 1, 531 (1977).

11. This remarkable result was first proved in the quantal case for the potential $V(x) = g^2/x^2$ [corresponding to (2.1.1) with $a = 0$] by C. Marchioro (for $n = 3$) and by F. Calogero (for arbitrary n); the conjecture that it hold also in the classical case was validated, for arbitrary n, by Moser [for the potential $V(x) = g^2/x^2$] and by Calogero, Marchioro and Ragnisco [for the potential (2.1.1); in the classical and quantal cases]. See References 8 and 9 above.

12. P. Kulish, "Factorization of scattering characteristics and integrals of motion," in Nonlinear Evolution Equations Solvable by the Spectral Transform, F. Calogero (editor), Proceedings of the Symposium held at the Accademia dei Lincei in Rome, June 1977, a volume in the Series "Research Notes in Mathematics," Pitman Publishing, London, 1978.

13. First noted, for $n = 3$, by Marchiro and by Khande-kar and Lawande, and proved for arbitrary n by Moser; see Reference 8.

14. F. Calogero, Lett. Nuovo Cimento 16, 22 (1976).

15. M. A. Olshanetsky and A. M. Perelomov, Lett. Nuovo Cimento 16, 333 (1976).

16. M. Adler, "Some finite dimensional integrable systems and their scattering behavior," in Proceedings of the Conference on the Theory and Applications of Solitons, H. Flaschka and D. W. McLaughlin (editors), Rocky Mountain Mathematics Consortium, Arizona State University, Tempe, 1978; Comm. Math. Phys. 55, 195 (1977).

17. M. A. Olshanetsky and A. M. Perelomov, Lett. Nuovo Cimento 17, 97 (1976).

18. F. Calogero, Lett. Nuovo Cimento 16, 35 (1976). Equation (10) of this paper should be corrected as implied by Equations (2.6.11)-(2.6.12) below (note also the change of notation).

19. One can actually start with a model having $2m + \nu$ particles, and then let ν of them sit at the origin while the remaining 2m move symmetrically around the origin. Indeed ν need not even be integer; this requires, of course, some appropriate procedures, but the fact that the dynamics of the particles fixed at the origin is in fact ignorable makes it

plausible that such more general cases are in fact also treatable. For explicit results, see the paper by Olshanetsky and Perelomov (following reference).

20. M. A. Olshanetsky and A. M. Perelomov, Lett. Math. Phys. **1**, 187 (1976); A. M. Perelomov, ITEF preprint 27, 1976 (unpublished); and their papers quoted above under References 6, 9, 10, 15 and 17.
21. See second paper by F. Calogero in Reference 8.
22. A. M. Perelomov, Theor. Math. Phys. **6**, 263 (1971) [Russian original: Teor. Mat. Fiz. **6**, 364 (1971)]. See also the papers by P. Gambardella, J. Math. Phys. **16**, 1172 (1975) and by J. D. Louck, M. Moshinsky and K. B. Wolf, J. Math. Phys. **14**, 692 and 696 (1973); M. Moshinsky, J. Patera and P. Wintermitz, J. Math. Phys. **16**, 82 (1975); M. Moshinsky and J. Patera, J. Math. Phys. **16**, 1866 (1975). A recent contribution reporting progress along this line is due to S. Wojciechowski, "The SU(2) symmetry group generators for the classical Calogero system" (ICTP preprint, submitted to Phys. Lett.).
23. G. Barucchi and T. Regge, J. Math. Phys. **18**, 1149 (1977).
24. S. Wojciechowski, Phys. Lett. **64A**, 273 (1977).
25. D. Kazhdan, B. Kostant and S. Sternberg, "Hamiltonian group action and dynamical systems of Calogero type" (Harvard preprint, 1977, to be published).
26. F. Calogero, Lett. Nuovo Cimento **20**, 251 (1977).
27. F. Calogero and A. M. Perelomov, Commun. math. Phys. (in press); Linear Algebra Appl. (submitted).
28. S. Ahmed, M. Bruschi, F. Calogero and A. M. Perelomov (to be published).
29. G. Casati and J. Ford, J. Math. Phys. **17**, 494 (1976).
30. C. Marchioro and E. Presutti, Lett. Nuovo Cimento **4**, 488 (1970).
31. G. Gallavotti and C. Marchioro, J. Math. Anal. Appl. **44**, 661 (1975).
32. D. V. Choodnovsky and G. V. Choodnovsky, Lett. Nuovo Cimento **19**, 300 (1977).
33. J. Wolfes, J. Math. Phys. **15**, 1420 (1974).
34. F. Calogero and C. Marchioro, J. Math. Phys. **15**, 1425 (1974).
35. M. A. Olshanetsky and A. M. Perelomov, private communication.
36. H. Airault, H. P. McKean and J. Moser, Commun. Pure Appl. Math. **30**, 95 (1977).

37. M. D. Kruskal, "The Korteweg-de Vries equation and related evolution equations," in Nonlinear Wave Motion, A. Newell (editor), Lectures in Appl. Math. 15, Amer. Math. Soc., Providence, R.I., 1974, pp. 61-83.
38. W. R. Thikstun, J. Math. Anal. Appl. 55, 335 (1976).
39. The recent literature on this topic includes too many entries (over 10^3); the interested reader may trace enough relevant papers from those quoted above.
40. D. V. Choodnovsky and G. D. Choodnovsky, Nuovo Cimento 40B, 339 (1977).
41. F. Calogero, Nuovo Cimento 43B, 117 (1978).
42. The definition of Hermite polynomials that we employ is the standard one, see f.i. Higher Transcendental Functions, A. Erdélyi (editor), McGraw Hill, New York, 1953, vol. 2, hereafter referred to as HTF.
43. G. Szegö, Orthogonal Polynomials, Amer. Math. Soc. Colloquium Publications 23, New York, N.Y., 1939, Subsection 6.7.
44. F. Calogero, Lett. Nuovo Cimento 20, 489 (1977).
45. S. Ahmed, M. Bruschi and F. Calogero, Lett. Nuovo Cimento (in press); S. Ahmed and M. Bruschi, Lett. Nuovo Cimento (in press).
46. F. Calogero, Lett. Nuovo Cimento 20, 254 and 476 (1977); S. Ahmed and F. Calogero, Lett. Nuovo Cimento (in press); P. Sabatier, Lett. Nuovo Cimento (in press).
47. F. Calogero, Lett. Nuovo Cimento 19, 505 (1977).

FOOTNOTES AND REFERENCES ADDED IN PROOFS

48. This conjecture has now been validated by an analysis based on the method of solution of Olshanetsky and Perelomov (see Subsection 2.5): M. A. Olshanetsky and V.-B. K. Rogov, "Bound states in completely integrable systems with two types of particles," to be published in Ann. Inst. H. Poincaré, 1978.
49. A. M. Perelomov (private communication; to be published) has recently discovered the following simple and remarkable relation between the solutions $x_j(t)$ of (2.5.18) and $\tilde{x}_j(t)$ of (2.5.20):

$$\tilde{x}_j(t) = \cos(\omega t)\, x_j(\mathrm{tg}(\omega t)/\omega).$$

50. Important progress in the direction outlined above has been recently made by I. M. Krichever, "Rational solutions of the Kadomtzev-Petiavshvili equation and many-body problems," Funct. Anal. Appl. (in press).

INVERSE SCATTERING PROBLEMS FOR NONLINEAR APPLICATIONS†

P. C. Sabatier
Département de Physique Mathématique‡
Université des Sciences et Techniques du Languedoc
34060 Montpellier-Cedex (France)

ABSTRACT. This lecture contains a unified review of the methods of solution of inverse scattering problems, in view of applications to nonlinear evolution equations.

1. INTRODUCTION

Inverse scattering problems (I.S.P.) in quantum mechanics have been the most deeply studied. Besides many other I.S.P. can be reduced to similar schemes. Even the generalized Zacharov-Shabat problem, whose physical meaning has almost nothing to do with quantum mechanics, can be related to these schemes. Hence we would like better to give an approach to these schemes rather than a study of particular inverse problems. Thus we shall give a parallel study of the methods of solution for some important I.S.P. Besides, among the possible methods, our choice will be directed altogether by our desire of showing parallel approaches and by eventual applications to nonlinear problems.

---

† Lecture given at the NATO Advanced Study Institute on Nonlinear Equations in Physics and Mathematics, Istanbul, August 1977.

‡ Physique Mathématique et Théorique, Equipe de Recherches associée au C.N.R.S. This work has been done as part of the program "Recherche Cooperative sur Programme n° 264 : Etude interdisciplinaire des problèmes inverses."

A. O. Barut (ed.), Nonlinear Equations in Physics and Mathematics, 55–80.

Since the main object of this lecture is to present I.S.P. in a Summer School, there is no point in giving original results. The Section 2 is a presentation of well-known scattering problems, the Sections 3 and 4 a presentation of well-known inverse scattering methods. The method of obtaining the transformation operators is due to several authors (in particular A. Y. Povzner, B. M. Levitan, I. M. Gelfand, V. A. Marchenko). The method of proving the completeness of the set of wave functions goes back to E. C. Titchmarsh. As for the method we present here to obtain the integral equations from the completeness relations, it is due to R. G. Newton. For references on classical I.S.P. (and additional details), let us be allowed to refer the reader to a book recently published in collaboration with K. Chadan [1]. For references and additional details on the generalized Zacharov-Shabat problem, we refer to a well-known paper by M. J. Ablowitz, D. J. Kaup, A. C. Newell and H. Segur [2].

In the Section 5, we use a simple generalization of the well-known Lax's method [3] to present the inverse scattering transform (I.S.T.) and nonlinear evolution equations. This choice is only due to a pedagogical purpose and should not conceal the greater efficiency of other methods, in particular the one of Ablowitz et al. [2], and the generalized Wronskian method by Calogero and Degasperis [4]. The Section 6 contains as a conclusion a diagram of the standard approach to I.S.P. in view of I.S.T.

Needless to say, the existence of a standard approach to I.S.P. is known. That they can be treated in such a similar way is, however, not so much well known—in particular in what concerns the energy dependent I.S.P. and the generalized Zacharov-Shabat problem. In deep, it is due to the fact that they are equivalent, as it is shown by Jaulent in a joint lecture to the Summer School [5].

2. SCATTERING PROBLEMS BEFORE NONLINEAR APPLICATIONS

2.1 Preparation

The fundamental origin of inverse scattering problems is obvious. Assume for instance that a collision problem is described by the Schrödinger equation

$$-\Delta\Psi + V(r)\Psi = E\Psi \tag{2.1}$$

for the wave function Ψ, which should belong to L_2 on any bounded set in $\mathbb{R}_3$, with real $V(\underset{\sim}{r})$, and such that $\Psi(\underset{\sim}{k}, \underset{\sim}{r})$ has the following asymptotic form:

$$\Psi(\underset{\sim}{k}, \underset{\sim}{r}) = \exp[i\,\underset{\sim}{k}\cdot\underset{\sim}{r}] + \frac{e^{ikr}}{r} A(\hat{r}, \underset{\sim}{k}) + o(r^{-1}) \tag{2.2}$$

where we use the notation v for the length of a vector $\underset{\sim}{v}$, $\hat{v}$ for $\underset{\sim}{v}/v$. From the cross section $\sigma(\hat{r}, \underset{\sim}{k})$, which is experimentally observed, and is equal to the modulus square of $A(\hat{r}, \underset{\sim}{k})$, it is possible in principle to obtain A by taking into account the unitarity. But this constraint is not sufficient to guarantee that the inverse problem $A \to V$ is well-posed. In fact, it is overdetermined and its solution, which is certainly complicated, is not completely fixed as yet. On the other hand, there are three one-dimensional problems which can be considered as reductions of the one above and are completely solved in the following sense: One can answer all these mathematical questions in a very clarge class $\mathcal{V}$ of functions V:

Existence of a solution
Uniqueness of solutions (in a given class of functions)
Construction of solutions (in a given class of functions)
Approximation theory (in a set of equivalent solutions)
Stability

<u>The one-dimensional problem on the line</u> is merely the reduction of the big problem to one dimension so that (2.1) becomes

$$-\frac{d^2\varphi}{dx^2} + V(x)\,\varphi(x) = k^2\varphi(x) \qquad -\infty < x < +\infty \tag{2.3}$$

and the scattering solution whose incoming part from $x = -\infty$ is "normalized" has the following asymptotic behavior:

$$\varphi_+(k,x) \sim \begin{cases} e^{ikx} + r_\ell(k)e^{-ikx} & x \to -\infty \\ t(k)e^{ikx} & x \to +\infty \end{cases} \tag{2.4}$$

whereas the one incoming from $x = +\infty$, $\varphi_-(k,x)$, is equal to $e^{-ikx} + r(k)e^{ikx}$ for $x \to +\infty$, and $t(k)e^{-kx}$ for $x \to -\infty$. The "reflection coefficient" to the left $r_\ell(k)$ and the "transmission coefficient" $t(k)$ are the input information in the inverse problem.

During a long time, this problem was essentially studied for its applications to electromagnetic problems. Nonlinear applications made it more popular.

An extension to the case where V linearly depends on k has also been completely solved:

$$-\frac{d^2\varphi^{\pm}(x)}{dx^2} + V^{\pm}(k,x)\varphi^{\pm}(x) = k^2\varphi^{\pm}(x) \tag{2.5}$$

with

$$V^{\pm}(k,x) = U(x) \pm kQ(x). \tag{2.6}$$

Again the scattering solutions, corresponding respectively to the ± potentials, are defined by

$$\varphi^{\pm}(k,x) \sim \begin{cases} e^{ikx} + r_{\ell}^{\pm}(k)e^{-ikx} & x \to -\infty \\ t^{\pm}(k)e^{ikx} & x \to +\infty \end{cases} \tag{2.7}$$

and similarly for the solutions incoming from $x=+\infty$. The two other one-dimensional inverse problems come from the expansion of Ψ [in Equation (2.1)] into spherical harmonics when V depends only on r. This yields partial wave equations

$$-\frac{d^2\varphi_{\ell}(k,r)}{dr^2} + V(r)\varphi_{\ell}(k,r) = \left(k^2 - \frac{\ell(\ell+1)}{r^2}\right)\varphi_{\ell}(k,r) \tag{2.8}$$

where $\varphi_{\ell}(k,r)$ should satisfy the boundary condition $\varphi_{\ell}(k,0)=0$, and has the asymptotic behavior:

$$\varphi_{\ell}(k,r) = e^{i\delta_{\ell}(k)} \sin[kr - \ell(\pi/2) + \delta_{\ell}(k)]. \tag{2.9}$$

If the input information is $\ell \to \delta_{\ell}(k)$ for a fixed k, we have the inverse problem at fixed energy. If it is $k \to \delta_{\ell}(k)$ for a fixed ℓ, we have the inverse problem at fixed ℓ. Both these problems are underdetermined. However, an additional knowledge of all bound states, plus parameters that are associated with them, make the inverse problem at fixed ℓ a well-posed problem.

Several extensions of radial problems have been completely solved—in particular, in the fixed ℓ case, nonlocal problems, relativistic problems, matrix problems, coupled channel problems (but with an unphysical input), and the problems whose potentials linearly depend on k.

Of a different physical nature is the Zacharov-Shabat generalized problem, which is the scattering problem associated with the system of equations:

$$\left.\begin{aligned} \frac{dy_1}{dx} + iky_1 &= q^+(x)y_2 \\ \frac{dy_2}{dx} - iky_2 &= q^-(x)y_1 \end{aligned}\right\} \qquad (2.10)$$

Following Calogero and de Gasperis, we find it convenient to introduce Pauli matrices

$$\sigma_1 = \begin{pmatrix} 0 & 1 \\ 1 & 0 \end{pmatrix}, \quad \sigma_2 = \begin{pmatrix} 0 & -i \\ i & 0 \end{pmatrix}, \quad \sigma_3 = \begin{pmatrix} 1 & 0 \\ 0 & -1 \end{pmatrix},$$

and set

$$\left.\begin{aligned} q_1 &= \tfrac{1}{2}(q^+ + q^-); \quad -iq_2 = \tfrac{1}{2}(q^+ - q^-) \\ D^{\pm}_{2s} &= i\frac{d}{dx} - k\sigma_3 - iq_1\sigma_1 \mp iq_2\sigma_2 \\ H^{\pm}_{2s} &= i\sigma_3\frac{d}{dx} + q_1\sigma_2 \mp q_2\sigma_1 \end{aligned}\right\}. \qquad (2.11)$$

Thus (2.10) simply becomes

$$D^{+}_{2s}Y = 0 \quad \text{or} \quad H^{+}_{2s}Y = kY. \qquad (2.12)$$

It is sometimes convenient to treat simultaneously the problems with ±. In any case, the scattering solution corresponding to (2.4) is defined by

$$\psi^{\pm}_{+}(k,x) = \begin{cases} r^{\pm}_{\ell}(k)\begin{pmatrix} 1 \\ 0 \end{pmatrix} e^{-ikx} + \begin{pmatrix} 0 \\ 1 \end{pmatrix} e^{ikx} & x \to -\infty \\ t^{\pm}(k)\begin{pmatrix} 0 \\ 1 \end{pmatrix} e^{ikx} & x \to +\infty \end{cases} \qquad (2.13)$$

2.2 Nature of the direct scattering problem

Let us first notice that in all these scattering problems there is an essential parameter: $(\ell + \frac{1}{2})$ or $(\ell + \frac{1}{2})^2$ in the fixed k problem, k, or E, in the fixed ℓ problem. All these problems are linear, and the unknown parameter, the potential, appears as a linear operator, which is simply a multiplication in all but non-local problems. Hence we can write as a general equation

$$\left.\begin{aligned} \underset{\sim}{\mathcal{L}}\psi &= 0 \\ \underset{\sim}{\mathcal{L}} &= \underset{\sim}{\mathcal{L}}_o + \underset{\sim}{v} \end{aligned}\right\} \qquad (2.14)$$

where $\underset{\sim}{v}$ is the "potential operator" to be determined. Let us discard for a moment the problems with $\underset{\sim}{v}$ depending on k. In all the others we can write

$$\underset{\sim}{\mathcal{L}}_o = \underset{\sim}{L}_o - E \qquad (2.15)$$

where E is an essential parameter, and $\underset{\sim}{L}$ is a linear formal operator that does not depend on E. Of course, $\underset{\sim}{L}$ will be defined as $\underset{\sim}{\mathcal{L}}+E$. In many cases of interest, e.g., the Schrödinger operator with a real potential, $\underset{\sim}{L}$ is a formally self-adjoint operator

$$\underset{\sim}{L} = -\Delta + \underset{\sim}{v} \qquad (2.16)$$

but it is not in others, e.g., the Zacharov-Shabat generalized equation.

In all cases, the scattering problem is a problem with boundary-conditions. The conditions are on all directions in the third problem, at the origin in radial problem, at the two ends in one-dimensional problem.

So as to achieve the definition of the problem, one has to define completely the operator $\underset{\sim}{L}$ by defining properly the class of functions on which it should be applied. This class $\mathcal{C}$ is a subset of the set $\mathcal{B}$ of all the physical functions. ($\mathcal{B}$ is usually the set of functions that are locally square integrable and bounded at ∞). Giving both $\underset{\sim}{L}$ and $\mathcal{C}$ defines an operator L. For instance, proper boundary conditions enable us to define from a formal self-adjoint operator $\underset{\sim}{L}$ a self-adjoint operator L (it suffices to impose homogeneous boundary conditions on the frontier of the domain and double differentiability inside). When this is done, it is usually possible to solve

$$(L_o - E)\mathcal{R}_o = I \qquad (2.17)$$

where I is the identity operator. $\mathcal{R}_o$ is the resolvent of the unperturbed operator, and its kernel is the so-called free Green's function. Conversely, this Green's function completely determines the boundary conditions. The general problem

$$(L - E)\psi = 0 = (L_o - E + V)\psi \tag{2.18}$$

with the same boundary conditions is easily transformed by using $\mathcal{R}_o$ into an integral equation

$$\psi = \psi_o + \mathcal{R}_o V\psi \tag{2.19}$$

which enables one to construct the scattering function ψ from the unperturbed one ψ_o. A similar equation gives $\mathcal{R} = (L - E)^{-1}$ from $\mathcal{R}_o$:

$$\mathcal{R} = \mathcal{R}_o + \mathcal{R}_o V\mathcal{R}. \tag{2.20}$$

Like L, $\mathcal{R}$ is here an operator which acts in $\mathcal{B}$. It exists (i.e., L - E is inversible) except for a set of values of E called its spectrum. Now, in all scattering problems, there exists a special class $\mathcal{C}_o$ of functions ($\mathcal{C}_o \subset \mathcal{B}$), whose L_2-norm on the domain of definition of L is finite and which is really the physical functions of the problem. $\mathcal{C}_o$ is a Hilbert space. The points E_i at which

$$(L - E_i)\psi_{E_i} = 0 \tag{2.21}$$

has a solution in $\mathcal{C}_o$ belong to the spectrum of L - E, and are isolated points in usual cases. They are called the eigenvalues of L and form the point spectrum of the problem. When the Lippmann-Schwinger equation can be reduced to a Fredholm equation, its Fredholm determinant vanishes at these points.

2.3 General method to study the direct scattering problem

So as to study the scattering problem, it is very convenient to introduce solutions of (2.18) that are not defined by boundary conditions at several points but one only ("initial conditions"). Such is the regular solution in the radial problems (initial condition at 0), or the Jost solutions (initial condition at ∞, i.e., scattering solutions going to a purely incoming or purely outgoing wave normalized to 1). The interest of these solutions is that the corresponding resolvent is triangular. Thus Equation (2.19) is Volterra's. This makes it easy to study analytic properties of the solution. Besides, by comparing the asymptotic behavior of two sets of independent solutions, one obtains coefficients (the Jost functions) which are narrowly related to the determinant of Equation (2.20). Indeed, two sets of independent solutions and their Jost functions enable one to construct the Green's function. And since our study gives the analytic properties of all these functions, it gives those of the Green's function—in particular its singularities—which are readily related to the spectrum of L. The study is achieved by showing that the scattering amplitude is readily related to the Jost functions.

Such a method of study applies to all the scattering problems we have quoted, including the Zacharov-Shabat problem and also the three-dimensional problem, provided some generalizations are done (e.g., Jost functions become operators).

Example. The simplest and very well-known example to illustrate the method is the S-wave scattering problem [Equation (2.8) with $\ell = 0$]. The scattering physical solution is defined by (2.9). The solutions defined by initial conditions are the regular solution [$\varphi(k,0) = 0$, $\varphi'(k,r) = 1$ for $r = 0$], which is the solution of the Volterra's equation:

$$\varphi(k,r) - k^{-1}\sin(kr) = k^{-1}\int_0^r \sin[k(r-\rho)]V(\rho)\varphi(k,\rho)d\rho \qquad (2.22)$$

and the Jost solutions $f(\pm k,r)$, which are asymptotic to $\exp[\pm ikr]$, and solutions of the Volterra integral equations:

$$f(k,r) - e^{ikr} = k^{-1}\int_r^\infty \sin[k(\rho-r)]V(\rho)f(k,\rho)d\rho\,. \qquad (2.23)$$

The integral equations are readily obtained from (2.8). For

$$\int_0^\infty \rho\,|V(\rho)|\,d\rho < \infty,$$

their iterative solutions uniformly converge and define $\varphi(k,r)$ as an even entire function of k, $f(k,r)$ as a holomorphic function in $\mathrm{Im}\,k > 0$. The Jost function $F(k)$ is defined by the equality:

$$k\varphi(k,r) = \tfrac{1}{2}i\left\{F(k)\,f(-k,r) - F(-k)\,f(k,r)\right\} \qquad (2.24)$$

or

$$F(k) = W\left\{f(k,r)\,,\,\varphi(k,r)\right\} \qquad (2.25)$$

and thus $F(k)$ is holomorphic in $\mathrm{Im}\,k > 0$. It follows from (2.24) that $\varphi(k,r)$ belongs to $L_2(0,\infty)$ (bound state) if and only if $F(k)$ is zero in $\mathrm{Im}\,k > 0$, and this implies also $\mathrm{Re}\,k = 0$. Thus the point spectrum is made of these zeros. They are obviously the only singularities (poles) of the Green's function for the scattering solution, which is given by

$$R_E(r,r') = \frac{\varphi(k,r_<)f(k,r'_>)}{F(k)} . \tag{2.26}$$

Results in other examples are sometimes more complicated but not really more difficult to obtain.

3. STRUCTURE OF THE I.S.P. SOLUTION: THE TRANSFORMATION OPERATOR

There are many approaches to I.S.P. Some of them do not yield a complete solution. We discard them. Others yield a complete solution but are founded on properties of partial differential equations in which the spectral properties, which will be essential for nonlinear applications, are concealed. For our purpose (nonlinear applications), we limit our study to I.S.P. in which E, or k, is the essential parameter, and to methods founded on spectral properties. Actually, these properties will appear through analytic properties of certain functions, as we have already seen in the analysis of Section 2.3.

With these restrictions in mind, we can say that in all the inverse scattering problems of interest there are two essential keys: the transformation operator, which enables one to construct all functions of interest, and the integral equation, which enables one to construct the transformation operator from the input information.

The transformation operator. This operator transforms zero potential solutions ψ^o into the corresponding general solutions

$$\psi = \underset{\sim}{T}\psi^o . \tag{3.1}$$

Typically, $\underset{\sim}{T}$ is of the form

$$\underset{\sim}{T} = \underset{\sim}{1} + \underset{\sim}{K} \tag{3.2}$$

where $\underset{\sim}{K}$ is a triangular integral operator that does not depend on k.

So as to construct T, one first chooses a solution ψ (resp. ψ^o in the case without interaction) defined by the wave equation and convenient initial conditions. Thus ψ is obtained from ψ^o through a Volterra's integral equation which enables us to derive also the analytic properties of ψ in the k-plane. One is done if these properties are such that a tauberian theorem insures that the k-Fourier-transform of ψ is triangular. It is interesting to notice that ψ is not always one of the solutions used in direct scattering theory (see, e.g., the three-dimensional inverse problem).

Examples. (a) It can be seen from (2.22) that $\varphi(k,r) - k^{-1}\sin kr$ is an even entire function of exponential type r, and is L^2 on the real k axis. Thus the Paley-Wiener theorem readily yields the (Povzner-Levitan) transformation operator $\underset{\sim}{T}$:

$$(\underset{\sim}{T} f)_r = f(r) + \int_0^r K(r,\rho)\, f(\rho) d\rho . \qquad (3.3)$$

(b) For the same problem, let us set

$$h(k,r) = f(k,r) - \exp[ikr]. \qquad (3.4)$$

It follows from (2.23) that $h(k,r)$ is holomorphic in $\operatorname{Im} k > 0$ and, for $r > 0$, an L^2-function of k on any line parallel to the real axis, and such that

$$\int_{-\infty}^{+\infty} |h(\sigma + i\tau)|^2 d\sigma = 0(\exp[-2\tau r]). \qquad (3.5)$$

Hence, according to Levin's theorem, the Fourier transform of h is triangular, and

$$f(k,r) = e^{ikr} + \int_r^{\infty} A(r,t) e^{ikt} dt \qquad (3.6)$$

is the desired (Marchenko's) representation.

(c) In the same way, the purely outgoing solution of (2.3) is defined from (2.4) as:

$$f_+(k,x) = [t(k)]^{-1} \varphi_+(k,x). \qquad (3.7)$$

Its Volterra integral equation is

$$f_+(k,x) = e^{ikx} - \int_x^{\infty} \frac{\sin k(x-t)}{k} V(t)\, f_+(k,t) dt. \qquad (3.8)$$

Assume that

$$\int_{-\infty}^{\infty} (1+|x|)\,|V(x)|\,dx$$

is finite. Then the iteration series for the solution of (3.8) converges for any k such that $\operatorname{Im} k \geqq 0$, and yields the bound:

$$|f_+(k,x) - e^{ikx}| \leqq C\,\frac{\exp[-\operatorname{Im} k\, x]}{1+|k|}. \qquad (3.9)$$

Again, Levin's theorem applies and gives a representation of $f_+(x,t)$ quite similar with (3.6) or, after setting $A_+(x,t) = \frac{1}{2}B_+(x,((t-x)/2)))$:

$$f_+(k,x) = e^{ikx}\left[1 + \int_0^{\infty} B_+(x,y)e^{2iky}dy\right]. \qquad (3.10)$$

Clearly when one can guess, by analogy, the transformation formula that is convenient for a certain type of solution, it is often possible to write it down and prove it <u>afterwards</u> by using the differential equations and Fourier transforms but avoiding "strong" theorems. This was widely done in our last two examples:

(d) The outgoing Jost solution $f_+^{\pm}(k,x)$ of (2.5), which is equal to $[t^{\pm}(k)]^{-1}\varphi_+^{\pm}(k,x)$, has the representation formula

$$f_+^{\pm}(k,x) = F^{\pm}(x)e^{ikx} + \int_x^{\infty} A^{\pm}(x,t)e^{ikt}dt \qquad (3.11)$$

where

$$F_+^{\pm}(x) = \exp\left[\pm\frac{i}{2}\int_x^{\infty} Q(t)\,dt\right]. \qquad (3.12)$$

(e) In the general Zacharov-Shabat problem, the outgoing Jost solution $F_+^{\pm}(k,x)$, which is equal to $t^{\pm}(k)\psi_+^{\pm}(k,x)$ [according to (2.13)], can be managed in the same way to obtain the following representation:

$$F_+^{\pm}(k,x) = \begin{pmatrix}0\\1\end{pmatrix}e^{ikx} + \int_x^{\infty} K^{\pm}(x,y)e^{iky}dy. \qquad (3.13)$$

Needless to say, there exists similar representations for the Jost functions f_- or $f_-^{\pm}$, which are related to φ_- like f_+ is to φ_+.

A common point to all these transformation operators is that their trace readily yields the potential. This is easy to show by inserting the transformation formula into the Volterra equation. For instance, inserting (3.10) into (3.8) readily yields

$$B_+(x,0) = \int_x^\infty V(t)\,dt \tag{3.14}$$

whereas the Povzner-Levitan transformation kernel yields

$$K(r,r) = \tfrac{1}{2}\int_0^r V(t)\,dt. \tag{3.15}$$

Very similar results, although they are more complicated, hold in the three-dimensional case. As for our last examples, the results are no more complicated. For instance, in the Zacharov-Shabat problem, if $K_1^\pm$ and $K_2^\pm$ are the component of $K^\pm$, it is easy to show that

$$K_1^\pm(x,x) = -\tfrac{1}{2}q^\pm(x) \tag{3.16}$$

$$K_2^\pm(x,x) = \tfrac{1}{2}\int_x^\infty q^+(y)q^-(y)\,dy$$

$$= \tfrac{1}{2}\int_x^\infty \left(q_1^2(y) + q_2^2(y)\right)dy. \tag{3.17}$$

4. STRUCTURE OF THE I.S.P. SOLUTION: THE INTEGRAL EQUATION

The integral equation connects the transformation kernel and the kernel of a symmetric operator, called the "symmetric kernel," and which should eventually be related to the physical information. Since we choose to study only I.S.P. in which the spectrum of the operator is essential, we shall construct the symmetric operator from a set of parameters associated to the spectrum and called the spectral data.

Now, more or less obviously, the construction of this operator and of the integral equation is always founded on the "completeness" of a set of eigenvectors (eigenvectors of L, defined in $\mathcal{B}$). The meaning of this property is that this set is total in the set of solutions of the physical problem. The importance of the property is most clear in a method which has been fully developed by R. G. Newton and in which the starting point is precisely a "completeness relation" of the set of regular wave functions corresponding to the discrete spectrum ($E = E_1$, $E_i < 0$) and the continuous spectrum ($E > 0$) of the Schrödinger operator. Let us show this on the simplest example:

Example. In the S-wave I.S.P., the "completeness relation" can be written:

$$\int_{-\infty}^{+\infty} \varphi(E,r)\,\varphi(E,r')\,d\rho(E) = \delta(r-r') \tag{4.1}$$

where the "spectral density" $d\rho(E)$ is equal to

$$d\rho(E) = \begin{cases} \pi^{-1} E^{\frac{1}{2}} \, | F(E^{\frac{1}{2}}) |^{-2} & (E \geqq 0) \\ \sum_j \left[\int_0^\infty |\varphi(E_j,r)|^2 dr \right]^{-1} \delta(E-E_j) & (E < 0) \end{cases} . \tag{4.2}$$

(We assume, for the sake of simplicity, that there is no bound state at $E = 0$.)

Now, the transform (3.1) holds for all the functions φ_k^{o} [we use the notation φ_k for $\varphi(k,r)$]. Let T^* be the adjoint operator of $\underset{\sim}{T}$ ["adjoint" with respect to the scalar product in $L_2(\mathbb{R})$]. We apply $\underset{\sim}{T}$ on the left and $\underset{\sim}{T}^*$ on the right of the symmetric kernel, defined by:

$$g(r,r') = \int d(\rho-\rho_o)\varphi_k^{o}\bar{\varphi}_k^{o} . \tag{4.3}$$

Since, in the operator sense,

$$\bar{\varphi}_k^{o}\underset{\sim}{T}^* = \overline{\underset{\sim}{T}\varphi_k^{o}} = \bar{\varphi}_k , \tag{4.4}$$

we easily get

$$\underset{\sim}{T}\,g\,\underset{\sim}{T}^* = \int d\rho\,\varphi_k\bar{\varphi}_k - \underset{\sim}{T}\underset{\sim}{T}^* = \underset{\sim}{1} - \underset{\sim}{T}\underset{\sim}{T}^* . \tag{4.5}$$

Let us now recall that $\underset{\sim}{T} = \underset{\sim}{1} + \underset{\sim}{K}$, where $\underset{\sim}{K}$ is triangular. It follows that $\underset{\sim}{T}^* = \underset{\sim}{1} + \underset{\sim}{K}^*$ where $\underset{\sim}{K}^*$ is also triangular (on the complementary domain). Besides, $\underset{\sim}{T}$ is inversible and $\underset{\sim}{T}^{-1} = \underset{\sim}{1} + \underset{\sim}{K}'$, where $\underset{\sim}{K}'$ is triangular like $\underset{\sim}{K}$. We therefore obtain from (4.5) the equality

$$\underset{\sim}{T}\,\underset{\sim}{g} = \underset{\sim}{K}'^* - \underset{\sim}{K} = \underset{\sim}{g} + \underset{\sim}{K}\underset{\sim}{g} \tag{4.6}$$

which can be written, for $r \geqq r'$,

$$K(r,r') + g(r,r') + \int_0^r K(r,\rho)g(\rho,r')d\rho = 0 \tag{4.7}$$

and which is the famous Gelfand-Levitan equation. A similar derivation can be done for the Marchenko equation. We start from the completeness relation for Jost solutions:

$$(2\pi)^{-1}\int_{-\infty}^{+\infty} f(k,r)\Big[f(-k,r') - S(k)f(k,r')\Big]dk = \delta(r-r') \tag{4.8}$$

(which has been written here in the simple case of no bound state). Then we apply $T \dots T^*$ to the symmetric kernel

$$A_o(r+r') = \int_{-\infty}^{+\infty} f^o(k,r)\,(S(k) - 1)\, f^o(k,r')dk \tag{4.9}$$

[where $f^o(k,r) = e^{ikr}$]. Noticing that $\underset{\sim}{T}$ is a real operator and that, for real k, $\overline{f(k,r)} = f(-k,r)$, we easily obtain as above

$$\underset{\sim}{T}A_o\underset{\sim}{T}^* = \underset{\sim}{T}\underset{\sim}{T}^* - \underset{\sim}{1}, \tag{4.10}$$

from which the Marchenko's equation readily follows

$$A_o(r+t) + \int_r^{\infty} A(r,s)\,A_o(s+t)ds = A(r,t). \tag{4.11}$$

There is however another, simpler, derivation of (4.11) from (4.8) by playing with Fourier transforms. Let us now study more completely the one-dimensional problem on the line to understand how the completeness comes in even if it needs not to be explicitly used.

The wave equation is (2.3). Using the outgoing Jost solution (1.4) together with the incoming one, we can introduce the S-matrix whose elements are

$$\left.\begin{aligned} S_{11} &= S_{22} = t(k) \\ S_{12} &= r_\ell(k) \\ S_{21} &= r(k). \end{aligned}\right\} \tag{4.12}$$

The equality $S_{11}=S_{22}$ comes from the time reversal invariance. Because of the flux conservation, S is unitary and

$$|S_{11}(k)| = \left(1 - |S_{12}(k)|^2\right)^{\frac{1}{2}} . \tag{4.13}$$

Coefficients relating two fundamental sets of solutions, like the Jost functions, can be defined and easily related to the previously defined coefficients:

$$f_-(k,x) = c_{11}(k)\, f_+(k,x) + c_{12}(k)\, f_+(-k,x) \tag{4.14}$$

$$c_{11}(k) = \frac{r(k)}{t(k)} = \frac{S_{21}(k)}{S_{22}(k)} \tag{4.15}$$

$$c_{12}(k) = \frac{1}{t(k)} = \frac{1}{2ik} W[f_-(k,x), f_+(k,x)]. \tag{4.16}$$

From the last equality (4.16) and Volterra's equations satisfied by f_- and f_+, it is readily seen that $c_{12}(k)$ is holomorphic for $\mathrm{Im}\, k \geqq 0$ and goes to 1 for $|k| \to \infty$. Besides, it has zeros on the positive imaginary axis, and there only. At these zeros, f_- and f_+ are proportional and since obviously f_+ exponentially decreases for $x \to +\infty$, f_- for $x \to -\infty$, we see that we have a bound state. Let us now introduce the Green's function

$$G(k,x,y) = \begin{cases} -\dfrac{t(k)}{2ik} f_-(k,x)\, f_+(k,y) & (y > x) \\[2ex] -\dfrac{t(k)}{2ik} f_-(k,y)\, f_+(k,x) & (x < y). \end{cases} \tag{4.17}$$

Again, we see that G is holomorphic in $\mathrm{Im}\, k \geqq 0$, with poles precisely at the bound states (and a pole at $k=0$). From bounds like (3.9), we easily derive for G the bound

$$|G(k,x,y)| \leqq c\,|k|^{-1} \exp[-|x-y|\,\mathrm{Im}\, k] \qquad (|k| \to \infty). \tag{4.18}$$

We are now able to derive the completeness relation. We will do it here because it is a good example—usually not given in the literature. Let $\psi(x)$ be a function of finite support and belong to C_2. Riemann-Green's relation readily yields

$$k^{-1}\psi(x) = k^{-1}\int_{-\infty}^{+\infty} G(k,x,y)\left\{-\psi''(y) + [V(y) - k^2]\psi(y)\right\}dy. \tag{4.19}$$

Integrating over a half-circle in the upper k-plane, and using

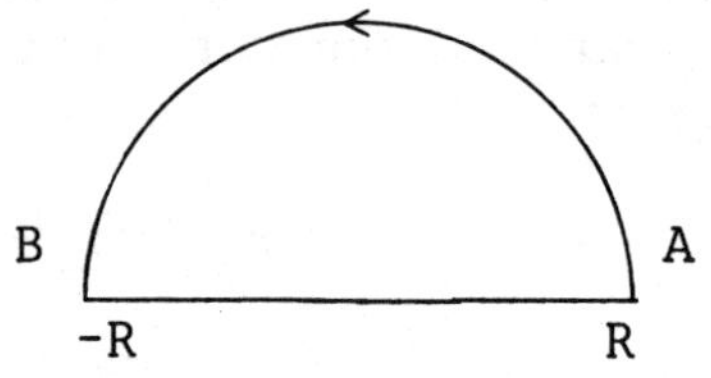

Cauchy's theorem, we obtain

$$\psi(x) = \frac{1}{i\pi}\int_{-\infty}^{+\infty} k\,dk \int_{-\infty}^{+\infty} G(k,x,y)\,\psi(y)\,dy$$

$$+ \sum_n \int_{-\infty}^{+\infty} G'(ik_n,x,y)\,\psi(y)\,dy \tag{4.20}$$

where the ik_n's are the bound states. The first term in (4.20) can as well be written, using (4.17),

$$\frac{1}{2\pi}\int_{-\infty}^{+\infty} dk\,t(k) \int_{-\infty}^{x} f_-(k,x)\,f_+(k,y)\,\psi(y)\,dy$$

$$+ \frac{1}{2\pi}\int_{-\infty}^{+\infty} dk\,t(k) \int_{x}^{\infty} f_-(k,y)f_+(k,x)\,\psi(y)\,dy \tag{4.21}$$

and, using (4.14), it is not difficult to show that the second term in (4.21) is equal to the same term with permutating x and y in f_- and f^+ so that the first term in (4.20) reduces to

$$\frac{1}{2\pi}\int_{-\infty}^{+\infty} dk\,t(k)\,f_-(k,x) \int_{-\infty}^{+\infty} f_+(k,y)\,\psi(y)\,dy.$$

Using the proportionality of f_- and f_+ at the bound states, it is also possible to reduce the second term, obtaining the completeness relation:

$$\frac{1}{2\pi}\int_{-\infty}^{+\infty} t(k)\, f_-(k,x)\, f_+(k,y)dk + \sum_n C_n f_-(ik_n,x)\, f_+(ik_n,y) = \delta(x-y).$$

(We use C_n and C_n', C_n'' below for constants whose exact value is not useful enough to be written here.) The completeness relation can be put in a form which is similar to (4.8) by using (4.14):

$$(2\pi)^{-1}\int_{-\infty}^{+\infty} f_+(k,x)\left[f_+(-k,y) - S_{21}(k)\, f_+(k,y)\right]$$
$$+ \sum_n C_n' f_+(ik_n,x)\, f_+(ik_n,y) = \delta(x-y) \qquad (4.22)$$

from which we easily obtain the Marchenko's equation in the one-dimensional case [Equation (4.27) below], just as we did in the radial case.

Now there is an alternative derivation of the integral equation. Starting from (4.14), using (4.15) and multiplying by exp[ikx], we obtain the equivalent form:

$$e^{ikx} S_{22}(k) f_-(k,x) = e^{2ikx} S_{21}(k) e^{-ikx} f_+(k,x) + e^{ikx} f_+(-k,x).$$

Setting then

$$s_{21}(k) = \int_{-\infty}^{+\infty} A_+(t) e^{-2ikt} dt \qquad (4.23)$$

$$e^{-ikx} f_+(k,x) - 1 = \int_{-\infty}^{+\infty} B_+(x,y) e^{2iky} dy$$
$$(4.24)$$
$$e^{ikx} S_{22}(k) f_-(k,x) - 1 = \int_{-\infty}^{+\infty} \tilde{B}_+(x,y) e^{2iky} dy = g_+(k,x),$$

we obtain

$$\tilde{B}_+(x,-y) = A_+(x+y) + B_+(x,y) + \int_{-\infty}^{+\infty} A_+(x+t+y) B_+(x,t) dt. \qquad (4.25)$$

From (3.8) it is not difficult to show that $B_+(x,y)$ and $\tilde{B}_+(x,y)$ are

L_2 in y for each fixed x. Besides the right-hand-side of (4.24) corresponding to B^+ is analytic for $\mathrm{Im}\,k>0$, the one corresponding to $\tilde{B}_+$ is analytic except at the poles of $S_{22}(k)$ [i.e., of t(k)]. Hence, $B_+(x,y)$ is zero for $y<0$, $\tilde{B}_+(x,y)$ is equal for $y<0$ to:

$$\begin{aligned}\tilde{B}_+(x,y) &= -i\sum_n \mathrm{Res}\; g_+(k,x)\big|_{ik_n} e^{-2k_n y} = \\ &= -\sum_n C_n'' e^{-2k_n(x+y)}\left(1+\int_0^\infty B_+(x,t)e^{-2k_n t}dt\right)\end{aligned} \tag{4.26}$$

so that we obtain from (4.25), for $y>0$, the Marchenko's equation of the one-dimensional case:

$$B_+(x,y)+\int_0^\infty B_+(x,t)\,\Omega_+(x+t+y)dt+\Omega_+(x+y) = 0 \qquad (y>0) \tag{4.27}$$

with

$$\Omega_+(t) = \sum_j C_j'' e^{-2k_j t} + \pi^{-1}\int_{-\infty}^{+\infty} s_{21}(k)e^{-2ikt}dk.$$

Clearly, it is only apparently that this method is independent of the completeness relations. They are concealed in the possibility of inverting the Fourier transform and the poles of t(k) in the complex plane.

4.1 The generalized Zacharov-Shabat problem

Remember that in this problem we introduced the Jost solutions $F_+^{\pm}(k,x)$ and $F_-^{\pm}(k,x)$:

$$\left.\begin{aligned} F_+^{\pm}(k,x) &\sim \begin{pmatrix}0\\1\end{pmatrix}e^{ikx} \qquad (x\to\infty) \\ F_-^{\pm}(k,x) &\sim \begin{pmatrix}1\\0\end{pmatrix}e^{-ikx} \qquad (x\to-\infty)\end{aligned}\right\} . \tag{4.28}$$

It is easy to see that $\sigma_1\bar{F}_+^{\mp}(-k,x)$ and $\sigma_1\bar{F}_-^{\mp}(-k,x)$ are also solutions of (2.12), with the asymptotic behaviors:

$$\left.\begin{aligned} \sigma_1 F_+^{\mp}(-k,x) &\sim \begin{pmatrix}1\\0\end{pmatrix} e^{-ikx} & (x \to \infty) \\ \sigma_1 F_-^{\mp}(-k,x) &\sim \begin{pmatrix}0\\1\end{pmatrix} e^{ikx} & (x \to +\infty) \end{aligned}\right\} . \tag{4.29}$$

$F_\pm^\pm(k,x)$ and $\sigma_1 F_\pm^\mp(-k,x)$ form fundamental systems of equations for Equation (2.12). They can be studied through their Volterra equation, giving analyticity in the domain $\operatorname{Im} k > 0$. One can define from them the scattering matrix

$$S^\pm(k) = \begin{pmatrix} T^\pm(k) & R^\pm(k) \\ R_\ell^\pm(k) & T^\pm(k) \end{pmatrix} \tag{4.30}$$

where $R^\pm(k)$, resp. $R_\ell^\pm(k)$, the reflection coefficient to the right (resp. to the left), and $T^\pm(k)$, the transmission coefficient, are complex numbers defined by the equalities:

$$\left.\begin{aligned} F_-^\pm(k,x) &= \frac{R^\pm(k)}{T^\pm(k)} F_+^\pm(k,x) + \frac{1}{T^\pm(k)} \sigma_1 F_+^\mp(-k,x) \\ F_+^\pm(k,x) &= \frac{R_\ell^\pm(k)}{T^\pm(k)} F_-^\pm(k,x) + \frac{1}{T^\pm(k)} \sigma_1 F_-^\mp(k,x) \end{aligned}\right\} . \tag{4.31}$$

It is easy to show the conservation formula:

$$S^+(k)\,{}^t S^-(-k) = I. \tag{4.32}$$

Since $\operatorname{Tr}(-k\sigma_3 + iq_1\sigma_1 \pm iq_2\sigma_2) = 0$, the Wronskian of two solutions of (2.12) is independent of x and so the quantities $(R^\pm(k))/(T^\pm(k))$, $(R_\ell^\pm(k))/(T^\pm(k))$ and $1/(T^\pm(k))$ can be expressed as Wronskians of solutions, and their analyticity in $\operatorname{Im} k > 0$ can be proved. The function $1/(T^\pm(k))$ is analytic for $\operatorname{Im} k > 0$ and continuous for $\operatorname{Im} k \geqq 0$. With an additional assumption as below, one can prove that $1/(T^\pm(k))$ have a finite number of simple zeros, $N^\pm$, located at the points $k_n^\pm$, which correspond to the bound states. Constants $C_n^\pm$ are associated to the residues at their poles:

$$C_n^\pm = \frac{iR^\pm(k_n^\pm)}{T^\pm(k_n^\pm)} \; \frac{1}{[(d/dk)\cdot(1/(T^\pm(k)))]_{k=k_n^\pm}} . \tag{4.33}$$

So as to derive the equations connecting the transformation operator which appears in (3.13) to these scattering data, one inserts (3.13) into (4.31a) [whose two sides are multiplied by $T^{\pm}(k)$]. Making use then of Fourier integral formulas and contour integration in $\operatorname{Im} k > 0$, we obtain the fundamental coupled integral equations

$$\sigma_1 K^{\pm}(x,y) = \binom{0}{1} p^{\mp}(x+y) + \int_x^{\infty} \bar{p}^{\mp}(y+u) \bar{K}^{\mp}(x,u) du \qquad y > x \tag{4.34}$$

where

$$p^{\pm}(x) = \sum_{n=1}^{N^{\pm}} C_n^{\pm} e^{ik_n^{\pm} x} - \frac{1}{2\pi} \int_{-\infty}^{+\infty} R^{\pm}(k) e^{ikx} dk. \tag{4.35}$$

Note that (4.34) represents four scalar integral coupled equations with unknowns $K_1^+(x,y)$, $K_2^+(x,y)$, $K_1^-(x,y)$, $K_2^-(x,y)$. Either of the restrictions

$$q^- = -\overline{q^+} \quad (\Rightarrow q_1 \ \& \ q_2 \ \text{imaginary})$$

or

$$q^- = \overline{q^+} \quad (\Rightarrow q_1 \ \& \ q_2 \ \text{real})$$

and (4.36)

$$\int_{-\infty}^{+\infty} q^+(x)\, dx < 0{,}523$$

is sufficient to guarantee that (K^+, K^-) is uniquely defined as a solution of (4.34).

4.2 The problem where V depends linearly on k

Again in this problem, the Jost solutions

$$f_+^{\pm}(k,x) \sim e^{ikx} \qquad (x \to +\infty)$$

$$f_-^{\pm}(k,x) \sim e^{-ikx} \qquad (x \to -\infty) \tag{4.37}$$

enable us to define the S-matrix:

$$S^{\pm}(k) = \begin{pmatrix} t^{\pm}(k) & r^{\pm}(k) \\ r_{\ell}^{\pm}(k) & t^{\pm}(k) \end{pmatrix} \tag{4.38}$$

where $r^{\pm}(k)$ [resp. $r_{\ell}^{\pm}(k)$], the reflection coefficient to the right (resp. to the left), and $t^{\pm}(k)$, the transmission coefficient, are defined by the equalities:

$$f_{-}^{\pm}(k,x) = \frac{r^{\pm}(k)}{t^{\pm}(k)} f_{+}^{\pm}(k,x) + \frac{1}{t^{\pm}(k)} f_{+}^{\mp}(-k,x) \tag{4.39}$$

$$f_{+}^{\pm}(k,x) = \frac{r_{\ell}^{\pm}(k)}{t^{\pm}(k)} f_{-}^{\pm}(k,x) + \frac{1}{t^{\pm}(k)} f_{-}^{\mp}(-k,x). \tag{4.40}$$

The function $1/(t^{\pm}(k))$ is analytic for $\operatorname{Im} k > 0$ and continuous for $\operatorname{Im} k \geqq 0$. We now make a further somewhat technical bound states assumption. All the zeros of $1/(t^{\pm}(k))$ $(\operatorname{Im} k > 0)$ are simple; they are not on the real axis (except maybe at 0), and $t^{\pm}(k)$ is bounded as $|k| \to 0$. Then one can prove that $1/(t^{\pm}(k))$ $(\operatorname{Im} k > 0)$ each has a finite number of simple zeros, $N^{\pm}$, located at the points $k_n^{\pm}$ $(n = 1, \ldots, N^{\pm})$. The corresponding functions $f_{+}^{\pm}(k_n^{\pm}, x)$ are the only $L^2(R)$ solutions of (2.5) for $\operatorname{Im} k \geqq 0$ and are the bound states. The constants C_n that appear in the fundamental equation are again given by (4.33). Now let us insert the transformation formula (3.11) into the formula (4.39) whose two sides have been multiplied by $t^{\pm}(k)$ and make use of Fourier integral formulas and contour integration in the complex plane. We obtain the coupled integral equation

$$A^{\pm}(x,y) = F^{\mp}(x) p^{\mp}(x+y) + \int_{x}^{\infty} p^{\mp}(y+u) A^{\mp}(x,u)\, du, \qquad y \geqq x, \quad x \in R \tag{4.41}$$

where

$$p^{\pm}(x) = \sum_{n=1}^{N^{\pm}} C_n^{\pm} e^{ik_n^{\pm} x} - \frac{1}{2\pi} \int_{-\infty}^{\infty} r^{\pm}(k) e^{ikx}\, dk \tag{4.42}$$

Note that S and (p^{+}, p^{-}) are bijectively related via Fourier transforms. Furthermore, from (3.11) and (3.12) we derive the following additional equations (in which the data do not occur)

$$\left.\begin{aligned} h^+(x)F^-(x) &= h^-(x)F^+(x) \\ F^+(x)F^-(x) &= 1 \end{aligned}\right\} \tag{4.43}$$

These Equations (4.41) and (4.43) form the fundamental system.

5. CONSTRUCTION OF NONLINEAR EQUATIONS

Let L be the linear operator of a scattering problem and $\mathcal{S}$ its spectrum, which can contain a continuous part and a point spectrum. Let B be the set of physical functions. [An example is the Schrödinger operator $d^2/dx^2 + V(x)$, $V(x)$ real, continuous, and

$$\int_{-\infty}^{+\infty} (1 + |x|)\ |V(x)|\, dx$$

finite, and where B is the set of twice differentiable functions, bounded at ∞.] Now, let M be an operator defined on B and with the following properties:

(1) There exists an operator M^{-1}, defined on B and such that, for any vector ψ of B,

$$M M^{-1}\psi = M^{-1}M\psi = \psi. \tag{5.1}$$

(2) For any vector ψ of B, $M\psi$ and $M^{-1}\psi$ belong to B.

(3) The operator $L' = M L M^{-1}$ has the same structure as L, i.e., it preserves the kind of scattering problem which is studied.

Clearly L is equal to $M^{-1}L'M$. Thus the transforms $L \to L'$ and $L' \to L$ are of the same nature. If E is an eigenvalue of L, it follows from (1) <u>and</u> (2) that it is an eigenvalue of L', corresponding to the acceptable eigenvector $M\psi$. Thus M is an "isospectral" transformation. As an example, for the Schrödinger operator, the isospectral transformations are the unitary transformations. They preserve (1), (2) and also the self-adjointness of the Schrödinger operator, i.e., a large part of (3).

Let us now study a class of operators M satisfying (1), (2), (3) and depending on one parameter t. From the two equalities

$$\left.\begin{aligned} M(t)M^{-1}(t) &= M^{-1}(t)M(t) = I \\ L(t) &= M(t)L(0)M^{-1}(t) \end{aligned}\right\} \tag{5.2}$$

we readily derive:

$$\frac{dL}{dt} = [B,L] \tag{5.3}$$

where

$$B = \frac{dM}{dt} M^{-1} . \tag{5.4}$$

If it is possible to identify a class of operators B such that the conditions (1) and (2) are satisfied by M, the formula (5.3) can be used to achieve the determination of B so that (3) is satisfied. As an example, if L is the Schrödinger operator, it suffices that B is antisymmetric and the formula (5.3) reduces to

$$\frac{\partial V}{\partial t} = \left[B, \frac{\partial^2}{\partial x^2} + V(x,t)\right] . \tag{5.5}$$

This means that, in this example, the commutator on the right-hand-side should reduce simply to a multiplication operator. The most important point is that (5.3) [or (5.5)] are usually nonlinear relations for the interaction term of the scattering theory. Thus, if one knows how to construct this interaction for any value of t, one knows how to solve a nonlinear evolution equation. But this is the case if one knows how to solve the inverse scattering problem from the scattering data and if the evolution of the scattering data is trivial. Since $\psi(x,t) = M(t)\psi(x,0)$, ψ is solution of the equation

$$\frac{\partial \psi(x,t)}{\partial t} = B\psi(x,t) \tag{5.6}$$

and the asymptotic form of (5.6) gives the evolution of scattering data. It is particularly simple when $B\psi(x \to \infty, t)$ is dominated by a term that does not depend on t. The best known example is the Schrödinger operator, with

$$V(x,t) = \frac{1}{6} u(x,t)$$

and

$$B = -4 \frac{\partial^3}{\partial x^3} - u \frac{\partial}{\partial x} - \frac{1}{2} \frac{\partial u}{\partial x} .$$

Then we readily obtain from (2.4):

$$\Psi(x,t,k) \sim \begin{cases} e^{ikx} + e^{-8ik^3t} r_0(k) e^{-ikx} & (x \to -\infty) \\ t_0(k) e^{ikx} & (x \to +\infty) \end{cases} \tag{5.7}$$

and similar evolutions for the parameters associated to bound states. From this very simple evolution of scattering data one can describe the evolution of $V(x,t)$. In this case, as is well known, (5.5) is nothing but the Korteweg-de Vries equation.

Thus we see how scattering and inverse scattering problems can be associated to nonlinear equations. The method we have described is consistent and self-contained for self-adjoint operators (Lax's method). For others, the class of transformations satisfying (1), (2), (3), has to be identified, sometimes by very different means. But we think that the present presentation has some good points (in addition to be short and simple):

(a) One understands why radial inverse problems are not very useful for nonlinear applications: simple integrodifferential operators B usually take $\psi(x,t)$ out of the space of functions that verify a boundary condition at a given finite point, so that the condition (2) is usually violated.

(b) On the other hand, one is led to study the "one-dimensional" or "three-dimensional" extension of "exotic" radial problems, which may have interesting applications. There is a large number of miscellaneous radial inverse problems which have not been used as yet for nonlinear applications and which could be after they are extended.

(c) This approach also suggests to study isospectral transformation operators as a subclass of the transformation operators T introduced in Section 3, and to introduce simple generalizations using transformation operators that modify a bit the spectrum.

These studies are the object of the current work of the author.

6. A CONCLUDING DIAGRAM

As a matter of conclusion, we give on the following page a diagram which summarizes what can be called a standard approach of I.S.P. in view of I.S.T.

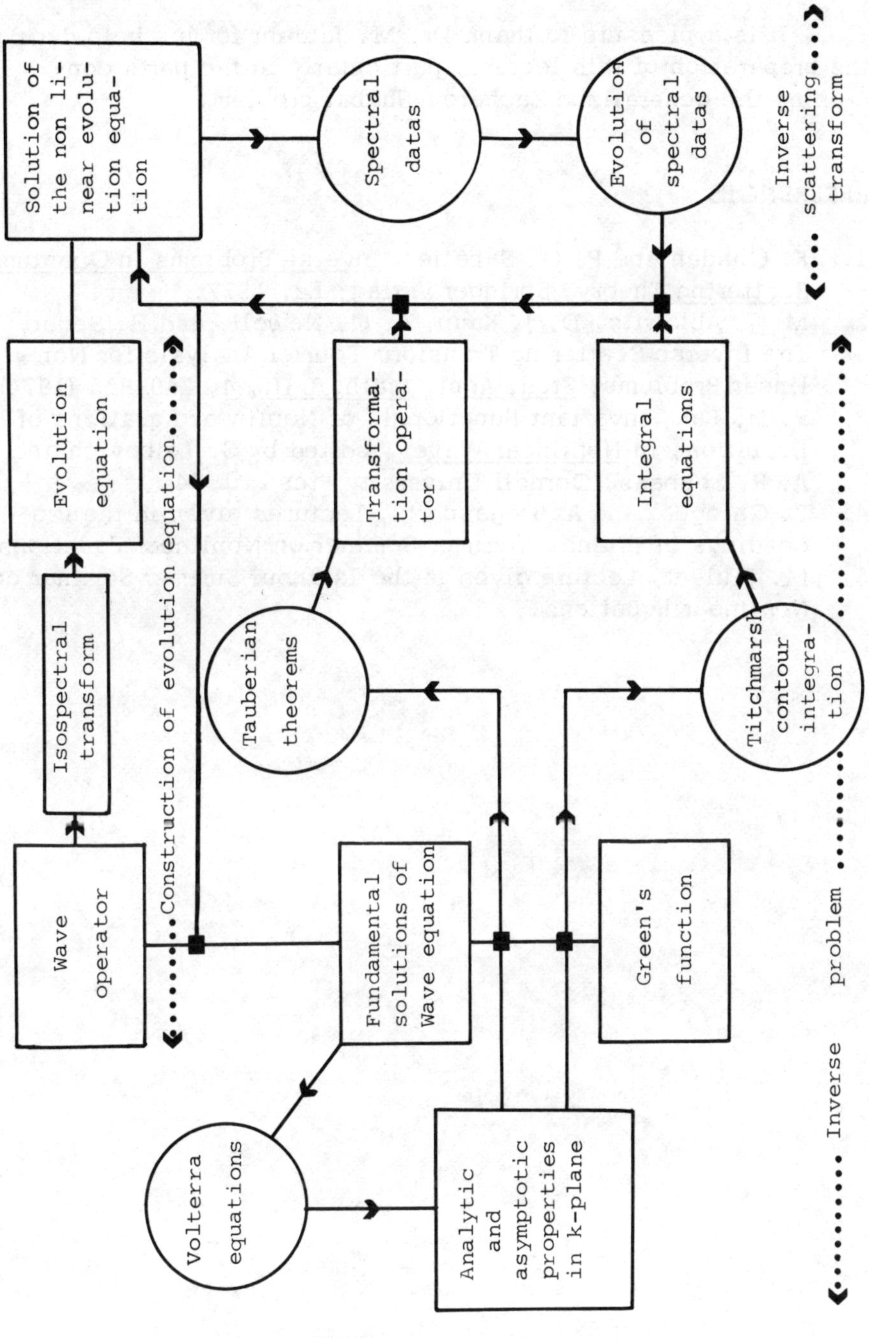
Wave operator
Isospectral transform
Evolution equation
Solution of the non linear evolution equation
Construction of evolution equation
Fundamental solutions of Wave equation
Tauberian theorems
Transformation operator
Spectral datas
Green's function
Titchmarsh contour integration
Integral equations
Evolution of spectral datas
Volterra equations
Analytic and asymptotic properties in k-plane
Inverse problem
Inverse scattering transform

ACKNOWLEDGEMENTS

It is a pleasure to thank Dr. M. Jaulent for his help during the preparation of this lecture, particularly in the parts concerning the generalized Zacharov-Shabat problem.

REFERENCES

1. K. Chadan and P. C. Sabatier, Inverse Problems in Quantum Scattering Theory, Springer Verlag, Ed. 1977.
2. M. J. Ablowitz, D. J. Kaup, A. C. Newell, and H. Segur, The Inverse Scattering Transform Fourier Analysis for Nonlinear Problems, Stud. Appl. Math. L III, 4, 249-315 (1974).
3. P. D. Lax, Invariant Functionals of Nonlinear Equations of Evolution, in Nonlinear Waves, edited by S. Leibovich and A. R. Seebass, Cornell University Press, 1974.
4. F. Calogero and A. Degasperis, Lectures given in the Proceedings of Istanbul Summer Seminar on Nonlinear Equations.
5. M. Jaulent, Lecture given in the Istanbul Summer Seminar on Nonlinear Equations.

SOLUTIONS OF NONLINEAR EQUATIONS SIMULATING PAIR PRODUCTION AND PAIR ANNIHILATION†

A. O. Barut

Department of Physics
The University of Colorado
Boulder, Colorado 80309

ABSTRACT. Exact solutions of nonlinear evolution equations are constructed which correspond to a swelling lump which then divides itself into two (or more) solitary waves moving in opposite directions or, conversely, two (or more) solitary waves coming together and annihilating each other with an exponential decay of the amplitude for large times. A particle motion can be associated to the phenomenon.

By now many curious solutions of nonlinear equations are known. We add to this list a class of solutions which correspond to phenomena of the type of pair annihilation, pair production, or cell division. The analysis shows that this class of equations or solutions are very rich.

Consider the nonlinear equation of the type

$$u_t - u_x = F(u), \tag{1}$$

or, by changing variable, $\xi = t - x$, $\eta = t + x$, simply

$$u_\xi = F(u). \tag{2}$$

This equation is integrated immediately to give

† Lecture given at NATO Advanced Study Institute on Nonlinear Equations in Physics and Mathematics, Istanbul, August 1977.

A. O. Barut (ed.), Nonlinear Equations in Physics and Mathematics, 81–84.

$$\int \frac{du}{F(u)} = \xi + v(\eta) , \tag{3}$$

where v is an arbitrary function.

We now choose F(u) such that, for example,

$$\int \frac{du}{F(u)} = \cosh^{-1}\left(\frac{1}{\sqrt{u}}\right)^{+C} = \xi + v(\eta) , \tag{4}$$

i.e.,

$$F(u) = -2u\sqrt{1-u} .$$

Hence the solution is

$$u = \frac{1}{\cosh(\xi+v(\eta))} . \tag{5}$$

This is a localized shape as a function of its arguments like the shape of the soliton of the KdV equation.

Consider now a particularly simple choice

$$v(\eta) = P(\eta) = \text{polynomial in } \eta .$$

Then the argument of cosh-function in (5) is

$$\xi+v(\eta) = (t-x) + P(t+x)) . \tag{6}$$

We can look now at the wave u(x,t) from a moving frame. By puting $x' = t+x$, then

$$\xi + v(\eta) = 2t - x' + P(x') . \tag{6'}$$

The zeros of $2t - x' + P(x')$ correspond to maxima of u which move with time t. For example, for $P(x') = ax'^2$, we have the following motions:

(1) $a > 0$: For $t < 1/8a$, two distinct peaks of u at $x'_{\pm} = 1/2a(1 \pm \sqrt{1-8at})$ move toward each other, overlap and become one peak at $x = 1/2a$ at time $t = 1/8a$. This single peak then begins slowly disappearing for $t > 1/8a$, and the maximum value at $x = 1/2a$ decays for large times exponentially as e^{-4t}.

(2) $a < 0$: The opposite behavior occurs: The solution shows a peak building up for times $t < -1/8|a|$, and the maximum at $x = -1/2|a|$ increases to the value 1 at $t = -1/8|a|$. The solitary waves then begin to split and then become two parts moving away from each other.

The solution (5) has the symmetry property under

$$t \to -t, \quad x \to -x. \tag{7}$$

Although the solution (5) has no poles or zeros, we can associate a soluble two-body motion with the maxima of the solitary waves given by

$$x_i(t) = \frac{1}{2a}\left(1 + (-1)^i\sqrt{1-8at}\right), \quad i = 1, 2;$$

as is done in the general study of moving poles and zeros [1]. The center of mass remains constant, but the relative coordinate $r = x_1 - x_2$ satisfy (for the choice (4))

$$\ddot{r} = -\frac{8}{a^2}\frac{1}{r^3},$$

i.e., an interparticle force increasing like $1/r^3$. It is interesting that after the collision the relative coordinate becomes purely imaginary.

REMARKS

(1) The first order Equation (1) can be, of course, imbedded into a second-order equation

$$\Box u = u_{tt} - u_{xx} = G(u)(u_t + u_x)$$

where $G = F'$.

(2) With $u = \sin^2\phi$, Equations (1) and (4) can also be written as

$$\phi_t - \phi_x = -\sin\phi, \quad \text{or} \quad \phi_{tt} - \phi_{xx} = -\cos\phi(\phi_t + \phi_x).$$

(3) By taking higher polynomials P in (6), it is easy to study the motion of several solitary waves, their production and annihilation.

(4) Because of the symmetry of the solution with respect to the annihilation point, we can define energy conservation if oppositely moving peaks are assigned opposite energies—analog to the negative energy solutions of the Dirac equation. In fact, the present approach can be generalized to the Dirac equation with nonlinear interactions.

ACKNOWLEDGMENT

The idea that one should look for solutions of the type of pair annihilation was first suggested to me by Gary Bornzin (private communication, 1974).

REFERENCES

1. M. D. Kruskal, Lectures in Appl. Math. **15**, Amer. Math. Soc., 61-83 (1974).
2. F. Calogero, Motion of Poles and Zeros of Special Solutions of Nonlinear and Linear Partial Differential Equations and Related "Solvable" Many-Body Problems, Nuovo Cim. **43B**, 177-241 (1978).
3. F. Calogero, these Proceedings.

THE TWO-TIME METHOD APPLIED TO SLOWLY EVOLVING OSCILLATING SYSTEMS†‡

J. Robert Buchler

Department of Physics and Astronomy
University of Florida, Gainesville, Florida 32611

ABSTRACT. The two-time asymptotic formalism is adapted to the study of an autonomously evolving n-dimensional nonlinear oscillating system. The method is designed to obtain the evolution when the smallness of the ratio of period to evolutionary time make a straightforward integration impossible. The approach is also ideally suited for exploring the phase-space of the dynamical system and in particular for studying the vibrational stability both to soft and to hard oscillations.

1. INTRODUCTION

We investigate the oscillatory quasiperiodic motion of a dynamical system governed by the equations

$$\frac{d^2y}{dt^2} + g(y,z) = 0 \tag{1a}$$

$$\frac{dz}{dt} = \epsilon h(y,z), \tag{1b}$$

---

† NATO Advanced Study Institute on Nonlinear Problems in Physics and Mathematics, Istanbul, 1-13 August 1977.

‡ Work supported in part by the National Science Foundation (Grant Number AST75-05012).

A. O. Barut (ed.), Nonlinear Equations in Physics and Mathematics, 85–95.

where $y = (y_1 y_2 \ldots y_n) \in D_y \subset R^n$, $z = (z_1 \ldots z_s) \in D_z \subset R^s$, $h \in D_h \subset R^s$, where the function $g(y,z)$ is assumed to be derivable (at fixed z) from a potential $V(y,z)$,

$$g(y,z) = \frac{\partial}{\partial y} V(y,z), \tag{2}$$

where by D we denote bounded domains and where ϵ is usually a small parameter of the system. The functions $g(y,z)$ and $h(y,z)$ are assumed to be sufficiently smooth ($\in C^2$). In the limit $\epsilon = 0$ the system is Hamiltonian and is assumed to admit at least one stable, strictly periodic solution of period P with two rest-points per period (i.e., points where the total kinetic energy vanishes). We also assume that this solution is nondegenerate and stable, i.e., that the associated variational problem in the sense of Lyapunov has only two null Poincaré indices corresponding to the period P and that all the other 2n-2 indices are negative. In addition, this solution is supposed to be known or calculable as a function of t and the fixed variable z and will henceforth be referred to as the adiabatic (or isentropic) oscillation (for further discussion see Buchler and Perdang, 1977).

For the purpose of illustration we shall put the problem into the astrophysical context in which it has actually arisen. Stars at one time or another in their evolution may run into a stage where they become vibrationally unstable; i.e., where a disturbance will cause the star to oscillate with increasing amplitude until nonlinear effects cause it to reach a stable oscillation (limit cycle). If an infinitesimal disturbance is sufficient to lead to a finite oscillation, we talk about soft oscillations, whereas hard oscillations require a finite kick for their excitation. Under the neglect of rotation and magnetic fields, spherical symmetry can be assumed and the behavior of the star is governed in a Lagrangian description by the equations:

$$\frac{d^2r}{dt^2} = -4\pi r^2 \frac{\partial p}{\partial m} - \frac{Gm}{r^2}, \tag{3a}$$

$$\frac{dS}{dt} = h, \tag{3b}$$

$$\frac{dX_\alpha}{dt} = R_\alpha . \tag{3c}$$

Here r(m) is the radius enclosing the mass m, p (ρ, S) is the pressure, G the gravitational constant, S the entropy and h its sinks and sources. The density ρ is related to the radius by the subsidiary relation:

$$\frac{\partial r}{\partial m} = \frac{1}{4\pi r^2 \rho} . \tag{4}$$

The set $X = \{X_\alpha\}$ represents the composition variables and $R = \{R_\alpha\}$ the corresponding nuclear production rates. When these equations are discretized properly (e.g., Cox and Giuli, 1968), they can be cast into the form of Equation (1). Physically, this means that the star is divided into n mass zones which behave like point masses coupled by nonlinear springs whose characteristics depend on additional variables S and X. With each of the three equations [(3a), (3b), and (3c)] is associated a time-scale: The dynamic time, t_{dyn} (i.e., the time for the system to react to a pressure disturbance) corresponds to Euler's Equation (3a) and we may take $t_{dyn} \cong P$, the period of oscillation. The thermal history, governed by Equation (3b), evolves on the Kelvin-Helmholtz time, t_{KH}; whereas the nuclear time scale, t_{nuc}, is associated with the overall nuclear evolution given by the set of Equations (3c). Some important phases of stellar evolution are characterized by the fact that $t_{dyn} \ll t_{evol}$ (i.e., t_{KH} and t_{nuc}); we note that often also $t_{nuc} \gg t_{KH}$, so that the nuclear network (3c) can be dropped and X can be considered constant for the study at hand. When we describe the physical variables in dimensionless form and require them to be of order one, then the parameter ϵ appears in Equation (1) as the small ratio $\epsilon = P/t_{evol}$.

In such cases where ϵ is small it is prohibitive to follow the evolution of the system through a straightforward integration of the system (1) because the average state of the system hardly changes over one period. The frequently used expedient procedure of artificially suppressing the oscillations through the omission of the acceleration (so-called quasistatic evolution) with the hope of reproducing the average evolution if fraught with danger and has been shown to be incorrect at least in some cases (Buchler, Perdang and Yueh, 1977). Similarly, the artificial scaling up of the parameter ϵ to make the numerical integration feasible may misrepresent the nature of the solution.

2. THE TWO-TIME METHOD

It is our purpose to describe a mathematically exact as well as practical scheme to deal with the solution of system (1) when the parameter ϵ is small. To that effect we make use of a two-time asymptotic technique which is a generalization to autonomous systems of type (1) of a procedure described in Cole (1968). For further details and discussion we refer the reader to Buchler et al. (1977), Buchler (1977), and Buchler and Perdang (1977).

We look for solutions of Equations (1) of the form

$$y(t) = y(t^*, \tau), \quad z(t) = z(t^*, \tau), \tag{5}$$

where we have introduced the <u>long</u> time $\tau = \epsilon t$ and the <u>short</u> time t^* through the relation

$$\frac{dt^*}{dt} = \phi(\tau). \tag{6}$$

The function $\phi(\tau)$ is to be chosen such that the motion at fixed τ becomes strictly periodic in t^* (as opposed to quasiperiodic in the real time variable t). Anticipating, we note that it is this strict periodicity which will allow us to clearly and simply define vibrational stability coefficients where other methods have led to a great deal of confusion.

We now try to construct an asymptotic expansion of the form

$$y(t^*, \tau) = y_o(t^*, \tau) + \epsilon y_1(t^*, \tau) + \cdots \tag{7a}$$

$$z(t^*, \tau) = z_o(t^*, \tau) + \epsilon z_1(t^*, \tau) + \cdots . \tag{7b}$$

Substitution of (7) and (6) into (1) leads to a hierarchy of systems of partial differential equations which up to order ϵ are given by

$$\phi^2 \frac{\partial^2}{\partial t^{*2}} y_o + g(y_o, z_o) = 0 \tag{8a}$$

$$\phi \frac{\partial}{\partial t^*} z_o = 0 \tag{8b}$$

and

$$\phi^2 \frac{\partial^2}{\partial t^{*2}} y_1 + \frac{\partial g}{\partial y_o} \cdot y_1 = - \frac{\partial g}{\partial z_o} \cdot z_1$$

$$- \left(2 \frac{d\phi}{d\tau} \frac{\partial y_o}{\partial t^*} + \phi \frac{\partial^2 y_o}{\partial t^* \partial \tau} \right) \tag{9a}$$

$$\phi \frac{\partial}{\partial t^*} z_1 + \frac{\partial}{\partial \tau} z_o = h(y_o, z_o). \tag{9b}$$

A theorem by Kuzmak (1959) can now be generalized to our case to read: Under broad conditions of smoothness of the functions g and h, the asymptotic series (7) is uniformly valid for $0 \leqq t \leqq \tau_o/\epsilon$, provided Equations (8) and (9) admit solutions periodic in t^* of the same period P, which must be independent of τ for $0 \leqq t \leqq \tau_o/\epsilon$.

Equation (8b) immediately shows that z_o is independent of t^* and therefore non-oscillatory, whereas, according to our assumptions about the nature of the potential V following Equation (2), Equation (8a) has a periodic solution at fixed τ which we supposed known, $y_o(t^*, \tau) = y_o(t^*/\phi; a(\tau), z_o(\tau))$, since it corresponds to the adiabatic solution of the original system when $\epsilon = 0$ and z is therefore fixed. Because all the different elements oscillate in phase, this nonlinear mode can be described by a single amplitude $a(\tau)$. We opt to fix the phase in such a way that the t^*-origin (modulo P) coincides with one of the two rest-points. According to the theorem, the function $\phi(\tau)$ in Equation (6) is now chosen such that P becomes independent of τ and we determine $a(\tau)$ and $z_o(\tau)$ by requiring that the next order solutions, $y_1(t^*, \tau)$ and $z_1(t^*, \tau)$, be periodic in t^* with the same period P. Equation (9a) when integrated yields the result:

$$z_1(t^*, \tau) = z_1(o, \tau)$$

$$+ \frac{1}{\phi} \left\{ \int_0^{t^*} h(y_o(t^*, \tau), z_o(\tau)) dt^* - \frac{t^*}{P} \frac{dz_o}{d\tau} \right\}, \tag{10}$$

and periodicity of z_1 requires then that

$$\frac{dz_o}{d\tau} = \frac{1}{P} \oint h(y_o(t^*, \tau), z_o(\tau)) dt^*. \tag{11}$$

Equation (9a), on the other hand, is linear, inhomogeneous, with periodic coefficients ($(\partial g/\partial y_o)(y_o, z_o)$ has period P). The condition for the existence of periodic solutions of period P is that

$$\oint \chi_i \cdot \text{RHS}(\text{Eq. (9a)})dt^* = 0, \tag{12}$$

where χ_i are all the solutions of the homogeneous system adjoint to Equation (9a) which have the same period P. Because of Equation (2), the system is self-adjoint and it appears immediately that $\chi_1 = (\partial y_o/\partial t^*)$ is a solution; we also note that χ_1 is odd in t^*. There always exists another solution, χ_2, also with period P, but which is even in t^*.

Condition (12) for χ_2 shows that the integration constant $z_1(0, \tau)$ must vanish. On the other hand, the use of χ_1 yields the condition

$$\frac{d}{d\tau}\left\{\phi \oint \left(\frac{\partial y_o}{\partial t^*}\right)^2 dt^*\right\} = -\oint \frac{\partial y_o}{\partial t^*} \cdot \frac{\partial g}{\partial z_o} \cdot z_1 dt^*. \tag{13}$$

In view of our assumptions about the nondegeneracy of the adiabatic solution, there are no other χ's to consider in relation (12). We note that these periodicity conditions (11) and (13) generalize the well-known Poincaré condition of absence of resonances.

The solution of the original differential system (1) of order $(2n+s)$ with time variations on the short time P has now been reduced to that of a system of only order $(s+1)$ in the slow time variable (i.e., stretched by a factor of $1/\epsilon$), which is given by Equations (11) and (13). It is also of importance to note that even the zeroth order terms y_o and z_o already include all the significant coupling between pulsation and evolution. It should be stressed that, to this lowest order, the solution for the slow evolution is in general not identical with the solution of the same equation under the suppression of the pulsations. It takes into account the influence of the evolution on the oscillations as well as the feedback of the oscillations on the evolution. In the next section the nature of these feedbacks will be made explicit.

3. THE HARMONIC APPROXIMATION AND VIBRATIONAL STABILITY ANALYSIS

We now turn to a special case of Equation (1a) which, while still of practical interest, greatly simplifies the amount of work required in an actual application of the method. This happens when the oscillators are harmonic, i.e., when g is of the form

$$g(y, z) = A(z) \cdot [y - x(z)], \tag{14}$$

where A is a matrix containing the spring constants and where x(z) is the equilibrium position. This case corresponds to a linearization (in y only!) of Equation (1a). We stress that we do not make any such approximation, however, about the z dependence of either A, x of h, nor do we restrict the y dependence of h. In a stellar application, h is usually a strongly y-dependent function because of the opacity law, while the oscillations are often close to harmonic; actually, asymptotic methods are available (Minorsky, 1962) which allow a straightforward analytical solution of (8a) even in the presence of a small nonlinearity in g. In the harmonic case y_o can immediately be written down:

$$y_o(t^*, \tau) = x(z_o(\tau)) + a(\tau)e(z_o(\tau))\cos[\omega(z_o(\tau))t^*/\phi(\tau)], \tag{15}$$

where $a(\tau)$ is the amplitude and where $e(z_o)$ and $\omega(z_o)$ are the eigenvector and eigenvalue of the normal mode under consideration. We impose the condition of τ-independent periodicity by requiring that $\omega(z_o(\tau)) = \phi(\tau)$. Defining a Fourier expansion of the function h, which is always possible,

$$h\left(y_o(t^*, \tau), z_o(\tau)\right) = \sum_{n=0}^{\infty} b_n(a, z_o)\cos nt^*, \tag{16}$$

we obtain the equations corresponding to (11) and (13), respectively,

$$\frac{dz_o}{d\tau} = b_o(a, z_o) \tag{11'}$$

$$\frac{d}{d\tau} a^2\omega = \frac{1}{4}\omega a^2 \frac{d\ell n\omega^2}{dz_o} \cdot b_2 - a\omega e \cdot \frac{dx}{dz_o} \cdot b_1 . \tag{13'}$$

We note that only the first three Fourier coefficients are needed. Equation (13') can also be transformed with the help of (11') into

$$\frac{da}{d\tau} = -\tfrac{1}{2}a\,\frac{d\ell n\omega^2}{dz_o}\cdot\left(b_o - \tfrac{1}{2}b_2\right) - \tfrac{1}{2}e\cdot\frac{dx}{dz_o}\cdot b_1, \qquad (17)$$

or, by introducing the energy E_o in the pulsational mode,

$$E_o = \frac{1}{P}\oint\left(\frac{\partial y_o}{\partial t^*}\right)^2 dt^* = \tfrac{1}{2}a^2\omega^2, \qquad (18)$$

we can also write it as

$$\frac{dE_o}{d\tau} = \tfrac{1}{2}E_o\,\frac{d\ell n\omega^2}{dz_o}\cdot\left(b_o + \tfrac{1}{2}b_2\right) - \sqrt{\frac{E_o}{2}}\,\omega e\cdot\frac{dx}{dz_o}\cdot b_1. \qquad (19)$$

The analysis of the vibrational (linear) stability of stars in thermal imbalance, that is to say stars for which $dS/dt \neq 0$ and which are thus evolving on a t_{KH} timescale, has been controversial. Basically, the problem has been attacked along two lines of approach, the first concerning the growth of the amplitude of an applied small perturbation, while the second has been based on energy considerations [for further details we refer the reader to, e.g., Aizenman and Cox (1975), Simon (1977), Demaret (1976), and Buchler (1977)]. The two methods have led to divergent results in some applications. Demaret (1976) has recently shown the ineluctable difference between the two approaches. The two-time method in the harmonic approximation is, however, simpler and more transparent for the discussion of such stability coefficients. To that effect, we now expand the Fourier coefficients $b_n(a, z_o)$ of h in powers of the amplitude a:

$$b_o(a, z_o) = b_{oo}(z_o) + b_{o2}(z_o)a^2 + \cdots \qquad (20a)$$

$$b_1(a, z_o) = b_{11}(z_o)a + b_{13}(z_o)a^3 + \cdots \qquad (20b)$$

$$b_2(a, z_o) = b_{22}(z_o)a^2 + b_{24}(z_o)a^4 + \cdots. \qquad (20c)$$

We first note that the evolution to lowest order in a is <u>independent</u> of the amplitude of the pulsation, since

$$\frac{dz_o}{d\tau} = b_{oo}(z_o) + O(a^2), \tag{21}$$

and $z_o(\tau)$ can readily be obtained through integration.

Equation (13') can be written in the form

$$\frac{da}{d\tau} = -\sigma_a a + O(a^3), \tag{22}$$

where the linear stability coefficient for the amplitude σ_a is defined as

$$\sigma_a = \frac{1}{4}\frac{d\ln\omega^2}{dz_o} \cdot b_{oo}(z_o) + \frac{1}{2} e \cdot \frac{dx}{dz_o} \cdot b_{11}(z_o); \tag{23}$$

on the other hand, Equation (18) gives a similar expression for the energy

$$\frac{dE_o}{d\tau} = -2\sigma_E E_o + O(E_o{}^2) \tag{24}$$

where the corresponding linear stability coefficient is given by

$$\sigma_E = -\frac{1}{4}\frac{d\ln\omega^2}{dz_o} \cdot b_{oo}(z_o) + \frac{1}{2} e \cdot \frac{dx}{dz_o} \cdot b_{11}(z_o). \tag{25}$$

There thus appears an irreconcilable difference between amplitude and energy considerations as was first shown by Demaret (1976) with a different procedure:

$$\sigma_E = \sigma_a - \frac{1}{2}\frac{d\ln\omega^2}{dz_o} \cdot b_{oo}(z_o). \tag{26}$$

In the case of thermal balance dS/dt (for $a = 0$) = 0, $b_{oo}(z_o) = 0$ and the energy and amplitude criteria coincide.

This difference can easily be understood physically. As mentioned previously, a star can be regarded as an assembly of coupled oscillators. It is therefore of interest to consider the case of just one harmonic oscillator with a variable spring constant ω^2, but with constant length. The <u>action variable</u> $I_o = a^2\omega$ is then an <u>adiabatic invariant</u> [c.f. Eq. (13')], so that $E_o \propto \omega$ and $a \propto 1/\sqrt{\omega}$. The two stability criteria σ_a and σ_E lead to a contradictory conclusion about stability: A slow decrease in ω is accompanied by an exponential growth in the amplitude ($\sigma_a < 0$), but to an exponential decay in the energy ($\sigma_E > 0$), and vice versa for an increase in ω. Physically, however, neither case would be considered a proper instability, but rather an adiabatic adjustment. It would thus seem more appropriate to define stability neither with respect to amplitude, energy, nor any other combination of a and ω, but rather with respect to the action variable I_o. Similar considerations for the n-dimensional stellar oscillator lead us to define the stability coefficient σ_I for the action variable:

$$\frac{dI_o}{d\tau} = -2\sigma_I I_o + 0(I_o^2) \tag{27}$$

$$\sigma_I = e \cdot \frac{dx}{dz_o} \cdot b_{11}(z_o), \tag{28}$$

where the troublesome term has now disappeared from σ_I.

Finally, we briefly turn to a discussion of stability with respect to hard oscillations. A study of such oscillations is clearly outside the reach of the usual linear analysis, but can be nicely handled with our formalism. Although a general discussion is not possible without a prior specification of the functional behaviour of the Fourier coefficients (b_o, b_1 and b_2 in the harmonic case). If the system is linearly stable then the minimum size perturbation for a given z_o profile, required to excite hard oscillations, can be obtained from the knowledge of the first zero of odd order of the derivative $da/d\tau$ in Equation (17). This amplitude corresponds to what we may call the first unstable limit cycle. The dynamic structure of the system can thus be explored in the phase-space (a, z_o) with the functional knowledge of b_o, b_1 and b_2. It is interesting to note that nonlinearity in the force (g) has no direct bearing on the existence of limit oscillations. The latter result directly from the nonlinear coupling of the oscillation and the evolution.

When the oscillations are so large that the harmonic approximation (14) is physically not acceptable, a similar analysis can still be carried through; however, y_o is no longer sinusoidal and expressions (11') and (13') involve Fourier components b_n of h with $n > 2$.

4. SUMMARY

We conclude by enumerating the main features of the method: First, we have developed a practical scheme to integrate the dynamic evolution Equations (1) in the case when $\epsilon = \text{Period}/t_{evol}$ is very small and when direct integration methods fail. Second, this scheme is particularly simple when the oscillators are harmonic (or even when there is a small nonlinearity). Third, we have rederived linear stability coefficients in a more elegant and transparent way and, finally, we have given a useful tool for the exploration of the phase space of an evolving dynamical system, which allows in particular the study of limit cycles and hard oscillations.

5. ACKNOWLEDGEMENTS

A large part of these results was obtained in collaboration with Dr. Jean Perdang. We wish to thank Professors Barut and Calogero for organizing a very fruitful meeting. We also gratefully acknowledge a grant from the Government of Luxembourg. Finally, we thank the National Science Foundation for supporting this work under grant number AST75-05012.

REFERENCES

1. M. Aizenman and J. P. Cox, Ap. J. 195, 175 (1975).
2. J. R. Buchler, Ap. J., submitted, 1977.
3. J. R. Buchler and J. Perdang, Arch. Rat. Mech. Anal., submitted, 1977.
4. J. R. Buchler, W. R. Yueh, and J. Perdang, Ap. J. 214, 510 (1977).
5. J. D. Cole, Perturbation Methods in Applied Mathematics, Blaisdell Publishers, Waltham, Massachusetts, 1968.
6. J. P. Cox and R. T. Giuli, Principles of Stellar Structure, Vol. II, Gordon and Breach, New York, 1968.
7. J. Demaret, Ap. and Space Sci. 45, 31 (1976).
8. G. E. Kuzmak, J. Appl. Math. Mech. 23, 730 (1959).
9. N. Minorsky, Nonlinear Oscillations, D. Van Nostrand, Princeton, 1962.
10. N. R. Simon, preprint, 1977.

PART II

SOLITONS

SOLITONS IN PHYSICS †

R. K. Bullough

Department of Mathematics, U.M.I.S.T.
P. O. Box 88, Manchester M60 1QD, U.K.

1. INTRODUCTORY MATHEMATICS

These lectures are intended as an introduction to soliton physics. In Section 1, I indicate in a simple way some of the mathematics underlying our present understanding of soliton theory. In Section 2, I summarize some current applications of this to nonlinear physics. The range of these applications is already vast: I cannot hope to survey them all, and in Section 3, I mention a number of areas where solitons plainly have significance—for example, where they can actually be observed—and then select four of these for more detailed study. In Section 4, I look at the problem of quantization and two possible applications.

Solitons arise as solutions of nonlinear evolution equations (NEEs) $u_t = K[u]$ where $K[u]$ is some nonlinear functional of u involving u, u_x, u_{xx}, etc.: in one-space dimension two examples are

$$u_t = -6uu_x - u_{xxx} \tag{1.1}$$

and

$$u_t = \int_{-\infty}^{x} \sin u(x',t)dx'. \tag{1.2}$$

Both of these are nonlinear wave equations: Equation (1.1) is the

† Lectures given at NATO Advanced Study Institute on Nonlinear Equations in Physics and Mathematics, Istanbul, August 1977.

A. O. Barut (ed.), Nonlinear Equations in Physics and Mathematics, 99–141.

Korteweg-de Vries equation [1] (KdV) with one conventional scaling, and (1.2) is the sine-Gordon equation [2] (s-G) which for NEE form is necessarily expressed in light-cone coordinates. Both equations have a large number of applications in mathematical physics [3], [4].

It is convenient to classify solitons as classical solitons or quantal (that is, quantized) solitons. Currently more is known about the classical solitons than about the quantal solitons and their range of application is the greater. But this situation could be changing because, for example, the quantized s-G (1.2) has applications in many-body theory [5] and statistical mechanics [5] outside the area of particle physics for which it might seem to be designed.

As a first working definition of a classical soliton, I take the following: it is a pulsed (e.g., bell-shaped) "solitary wave" solution of a c-number nonlinear wave equation which shows essentially complete stability in collision with other solitons. This may not be the most comprehensive useful definition, but certainly it covers a majority of those "multisoliton" solutions of particular nonlinear wave equations which have now been found. These have been found by strictly analytical means, and any useful definition will be one which will further an extension of the range of nonlinear equations which can also be exactly solved.

A "solitary wave" in one-space dimension is $u(x-Vt)$, a wave of permanent profile moving up x with constant speed V: to be solitary $u(x,t) \to 0$, $|x| \to \infty$. The "boomerons" reported by Degasperis [6] elsewhere in these Proceedings have permanent profile but variable speed $V(t)$. Thus our definition is already too restricted. For the present it will serve to keep V constant, however.

The linear nondispersive wave equation $Vu_x + u_t = 0$ has a solution $u(x-Vt)$ for any function u; the linear dispersive wave equation $-Vu_{xxx} + u_t = 0$ has only the harmonic wave solutions of permanent profile. If $u(x) \to 0$, $|x| \to \infty$, this dispersive wave equation has no solitary wave solutions.

The "simple wave"

$$u_t + uu_x = 0 \qquad (1.3)$$

is indeed one of the simplest nonlinear wave equations. It can be integrated by a hodagraph transformation: if u, t, x are each functions of the two other variables $u_t = -u_x x_t$ so that (since $u_x = 0$ implies $u = \text{constant}$ which is excluded) (1.3) means $-x_t + u = 0$ and $x = ut + g(u)$ where g is arbitrary. Then $u = f(x-ut)$ is the solution of (1.3) where $f = g^{-1}$ and is arbitrary.

Notice that at points (x,t) where u is large, the disturbance u travels at a large speed u; hence, u steepens and shocks. That u shocks also follows from the observation that if u and x are both parametrized by t

$$\frac{du}{dt} = u_t + \frac{dx}{dt} u_x; \tag{1.4}$$

and on the curves for which $dx/dt = u$, $du/dt = 0$ from (1.3) so that $u = \text{constant}$ on these curves and they are the straight lines $x = ut + \text{constant}$. These are the characteristic curves: they intersect for different constant values of u so that u is multivalued at these points of intersection.

We obtain a solitary wave solution by balancing nonlinearity against dispersion. Take, for example, $u_t + u_x + u_{xxx} = 0$ (note the change of sign of the dispersive term compared with that in the linear dispersive wave equation†). This equation is easily scaled to the KdV (1.1) and that form has the solitary wave solution with parameter ξ

$$u(x,t) = 2\xi^2 \operatorname{sech}^2 \xi (x - 4\xi^2 t). \tag{1.5}$$

However, the Burgers equation

$$u_t + 6uu_x - bu_{xx} = 0, \qquad b > 0 \tag{1.6}$$

also balances nonlinearity against dispersion. A solution with permanent profile is

$$u(x,t) = u_1 + (u_2 - u_1)\left[1 + \exp\left\{\frac{3(u_2-u_1)}{b}(x - Vt)\right\}\right]^{-1}. \tag{1.7}$$

This has asymptotes $u(-\infty,t) = u_2$, $u(+\infty,t) = u_1 < u_2$ and $V = \frac{1}{2}(u_1+u_2)$. This step-like function is a shock rather than a solitary wave: it has some of the character of the "kinks" we shall find we must also include amongst the classical solitons. However, these particular kinks are not solitons: the two shocks (u_1,u_2) and (u_2,u_3) with $u_1 > u_2 > u_3$ necessarily collide if (u_1,u_2) is to the left at $t=0$; but asymptotically they form the confluent shock (u_1,u_3) with speed $\frac{1}{2}(u_1+u_3)$ [7]. Thus they do not have the soliton collision property.

---

† This is because the equation refers to a frame moving at the sound speed: the solitons move at speeds exceeding the sound speed, the 'radiation' (see below) moves at speeds below the sound speed.

In contrast, the solitary wave solutions (1.5) of the KdV are solitons. This equation was the first to be solved by the inverse scattering method [8] and it has multisoliton solutions. Asymptotically these multisoliton solutions break up into a train of single solitons (1.5) travelling at different speeds. Notice that the speeds $4\xi^2$ depend on the amplitudes $2\xi^2$. A sequence of such solitons ordered according to decreasing amplitudes $2\xi_1^2 > 2\xi_2^2 > 2\xi_3^2 > \cdots$ and well separated at $x = -\infty$ will collide; but the remarkable result is that at $x = +\infty$ the solution is simply the same set of solitary waves in reversed order. The only change is that each soliton receives a small shift of argument

$$\Delta_i = \sum_{j \neq i} \Delta_{ij}$$

—the argument of the ith soliton becomes $x - 4\xi_i^2 t - \Delta_i$. Moreover,

$$\sum_i \Delta_i = \sum_i \sum_{j \neq i} \Delta_{ij} = 0$$

and total phase shift is conserved. Only pairwise shifts Δ_{ij} are involved. This is the collision property of many of the classical solitons and illustrates the stability in collision demanded for our first working definition of a soliton.

This definition must now be extended to include certain kinks. The linearized Burgers equation is a diffusion equation: it is perhaps not surprising therefore that its kinks do not have the soliton collision property. A wave equation with kink solutions is the "ϕ-four equation"

$$\phi_{xx} - \phi_{tt} = -\phi + \phi^3. \tag{1.8}$$

Its kink (antikink) solutions are

$$\phi = \pm \tanh\left[1/\sqrt{2}\left\{(x-Vt)/(1-V^2)^{\frac{1}{2}}\right\}\right] \tag{1.9}$$

(the antikink takes the -ve sign). The ϕ-four is a generalized Klein-Gordon equation $\phi_{xx} - \phi_{tt} = F(\phi)$. It is Lorentz covariant. In the rest frame with dependent variable u it becomes the ordinary differential equation

$$u_{xx} = -u + u^3. \tag{1.10}$$

This has unstable equilibrium points at $u = \pm 1$, the two zeros of the right side. The kink takes u from u=-1 to u=+1, the anti-kink from +1 to -1. They correspond to the two trajectories satisfying

$$\frac{1}{2}u_x^2 + \frac{1}{2}u^2 - \frac{1}{4}u^4 = \frac{1}{4} \qquad (-1 \leqq u \leqq +1) \tag{1.11}$$

in the phase space u_x against u (the "phase plane").

It might appear curious that these supposedly stable kinks connect unstable equilibrium points. In fact, these points are energetically stable. The Hamiltonian for the ϕ-four is

$$H = \int \mathcal{H} dx \equiv \gamma^{-1} \int \left\{ \tfrac{1}{2}u_t^2 + \tfrac{1}{2}u_x^2 + (\tfrac{1}{4} - \tfrac{1}{2}u^2 + \tfrac{1}{4}u^4) \right\} dx \tag{1.12}$$

(in which γ is a coupling constant) and the potential energy has minima at $u = \pm 1$. There are features here of the one-dimensional classical "instanton." The classical instanton is being used as an indicator for quantum mechanical tunnelling in a quantized field theory [9], [10]. For tunnelling, one can replace t by $\tau = it$ and look for a solution in Euclidean rather than Minkowski space. In our discussion we could Lorentz transform to the time frame: $-u_{tt} = -u + u^3$. Then $t \to it$ yields (1.10) with x=t.

Much more physics can be found in (1.8): Landau-Ginsberg expansions, the laser phase transition [11], displacive 'phase transitions' [12] on a linear lattice [13], [14], and central peak phenomena [13], [14], for example. The quantization of (1.8) provides an instructive model field theory [9], [15], [16]: at low enough order in γ, kink-like fields (1.9) are associated with 1-soliton simultaneous eigenstates of momentum p and energy $\sqrt{M^2+p^2}$: M has the value $\frac{2}{3}\gamma^{-1}$, just that one finds for the rest energy of a kink (1.9) inserted into (1.12) with V=0;

$$\langle p' | u | p \rangle = \int e^{i(p'-p)x'} u(x', o) dx'$$

where u(x,t) is the kink (1.9). These are examples of quantal solitons. The definition of these is looser than that we have adopted for the classical solitons. The criterion is one of finite self-energy and no special collision property is required.

The classical kinks (1.9), solutions of the c-number ϕ-four equations (1.8), are not classical solitons within the working definition. They are not stable in collision but typically emit a small amount of 'radiation.' To understand this a little more, compare the ϕ-four with the s-G. In covariant form the s-G is

$$u_{xx} - u_{tt} = \sin u. \tag{1.13}$$

It has kink (antikink) solutions

$$u = 4\tan^{-1}\exp\left\{\pm(x-Vt)/(1-V^2)^{\frac{1}{2}}\right\} \tag{1.14}$$

which take u from $u=0$ to $u=2\pi$ ($u=2\pi$ to 0) as x goes from $x=-\infty$ to $x=+\infty$. These are the 2π-kinks (antikinks). Their drivatives are sech's and are true solitary waves. The kinks themselves connect equilibrium points of equal energy. One point of difference from the ϕ-four is that there is an infinity of these equilibrium points $u=0, \pm 2\pi, \pm 4\pi, \ldots$. A consequence is that any kink or antikink of the s-G can be followed by any number of kinks or antikinks (for appropriate boundary conditions). The ϕ-four has only two equilibrium points so that a kink (antikink) must be followed by an antikink (kink). If a kink-antikink pair of the ϕ-four collides, the kink and antikink must have the same order before and after the collision: they can only bump (or annihilate).

The kinks and antikinks of the s-G do not need to maintain any order, however, and in collision they appear to pass through each other with only a phase shift. In the case of the kinks alone or the antikinks alone, this is a subjective interpretation: two kinks may be said to bump exchanging energy and momentum as particles (they also exchange steepness of profile). In this interpretation, kinks and antikinks need to exchange kink character on collision however, whilst the collision of a kink or antikink with the "breather" solution (see below) needs further interpretation. In practice many of the classical solitons can be interpreted as passing through each other without change of shape and speed and with no more than a phase shift. Notice that for arbitrary Klein-Gordon equations $u_{xx}=F(u)$, it is then necessary that F(u) is periodic in u.

The important point, however, is that the kink solutions of the s-G do not radiate in collision: they are stable entities. At the present time there are no analytical methods for finding multikink-like solutions of nonlinear wave equations in which the component kinks radiate off each other in collision [17]. No analytical expression for multikink solutions of the ϕ-four has been found: in contrast the s-G has a multisoliton solution which is a multikink solution [18] and can be solved for given initial data by the inverse scattering scheme in the 2×2 Zakharov-Shabat AKNS scheme [19], [20]. NEEs which fit into the ZS-AKNS scheme are examples of infinite dimensional completely integrable Hamiltonian systems [21]. We can exhibit a doubly infinite number of constants of the motion. The ϕ-four does not have this property.

To see this complete integrability of the s-G, note first of all that two conservation laws, for energy and momentum respectively, derive from the ϕ-four equation (1.10) written in terms of u:

$$\gamma\mathcal{H}_t - (u_x u_t)_x = 0$$

$$(u_x u_t)_t - \left(\tfrac{1}{2}u_t^2 + \tfrac{1}{2}u_x^2 - U(u)\right)_x = 0. \qquad (1.15)$$

$\mathcal{H}$ is defined by (1.12) and $U(u) = \frac{1}{4} - \frac{1}{2}u^2 + \frac{1}{4}u^4$. There is a conservation law obtainable by Noether's theorem from invariance of the ϕ-four under infinitesimal Lorentz transformations [22]. Other conserved quantities have not been found. The same three conservation laws can be derived for any generalized Klein-Gordon equation $u_{xx} - u_{tt} = U'(u)$.

In light-cone coordinates energy and momentum symmetrize to

$$\gamma\mathcal{H}_t - (\tfrac{1}{2}u_t^2)_x = 0$$
$$(\tfrac{1}{2}u_x^2)_t - \gamma\mathcal{H}_x = 0 \qquad (1.16)$$

with $\mathcal{H} \equiv 0$, and a third conservation law follows from invariance under the infinitesimal Lie transformation $x \to x(1+\epsilon)$, $t \to t(1-\epsilon)$ [22]. The s-G is exceptional in that the following infinite sequence of <u>polynomial</u> conserved densities can be found [22]:

$$\frac{1}{2}u_x^2,\ \frac{1}{2}u_{xx}^2 - \frac{1}{8}u_x^4,\ \frac{1}{2}u_{xxx}^2 - \frac{5}{4}u_{xx}^2u_x^2 + \frac{1}{16}u_x^6,$$

$$\frac{1}{2}u_{xxxx}^2 - \frac{7}{4}u_{xxx}^2u_x^2 + \frac{7}{8}u_{xx}^4 + \frac{35}{16}u_{xx}^2u_x^4 + \frac{5}{128}u_x^8,\ \ldots. \qquad (1.17)$$

A second infinite set can then be found from the corresponding fluxes with (x,t) interchanged [23]. The conserved densities are

$$\gamma^{-1}(1-\cos u),\ -\gamma^{-1}(\tfrac{1}{2}u_t^2\cos u),\ \ldots \qquad (1.18)$$

and are not polynomial. Evidently the sequence (1.17) is a sequence of conserved momentum densities; the sequence (1.18) is a sequence of energy densities. To see that the s-G is an infinite dimensional completely integrable Hamiltonian system, it is now sufficient to see that the s-G in light-cone coordinates is Hamiltonian. From (1.18) we take the symmetrized energy

$$H = -\frac{1}{4\gamma}\int_{-\infty}^{\infty}\left\{2 - \cos\left(-2\int_{-\infty}^{x} q(x')dx'\right) - \cos(2\gamma p)\right\}dx \quad (1.19a)$$

with

$$q = -\tfrac{1}{2}u_x, \quad p = -\gamma^{-1}\int_{-\infty}^{x} q(x')dx' = \tfrac{1}{2}\gamma^{-1}u. \quad (1.19b)$$

Then

$$-q_t(=\tfrac{1}{2}u_{xt}) = -\frac{\delta H}{\delta p} = \tfrac{1}{2}\sin u$$

$$p_t(=\tfrac{1}{2}\gamma^{-1}u_t) = -\frac{\delta H}{\delta q} = \frac{1}{2\gamma}\int_{-\infty}^{x}\sin u\, dx'. \quad (1.20)$$

It is now natural to take the second member of (1.18) as a Hamiltonian density. The problem surrounding the presence of u_t in this density is surmounted by working with Hamilton's principal function

$$\int_{-\infty}^{\infty} dt \int_{-\infty}^{\infty} dx \{ pq - \mathcal{H} \}.$$

The result equivalent to (1.20) is

$$u_{xt} = \frac{1}{2}\left\{u_t^2 \sin u - u_{tt}\cos u\right\}. \quad (1.21)$$

This equation is not invariant under the infinitesimal Lie transformation and is not Lorentz covariant in laboratory coordinates. Evidently, of the sequence derived from (1.18), only the s-G itself is Lorentz covariant—a result of special importance to applications in particle physics.

The special character of the s-G follows from a number of other results: first, Derrick's theorem [9], [16] shows that no Lorentz covariant equation $\partial_\mu\partial^\mu u + U'(u) = 0$ in more than one space dimension has immediately comparable properties. We prove that the only time independent solution in more than one space dimension is the ground state. Take

$$H = V_1[u] + V_2[u] : V_1 = \int \tfrac{1}{2}(\vec{\nabla}u)^2 d^D\vec{x}, \quad V_2 = \int U(u)d^D\vec{x}$$

and

$$V_1, V_2 \geqq 0.$$

Take a particular solution u and consider the one-parameter family $u(\vec{x};\lambda) = u(\lambda\,\vec{x})$. Evidently

$$H = \lambda^{2-D} V_1 + \lambda^{-D} V_2 \tag{1.22}$$

and, by Hamilton's principle, this is stationary at $\lambda = 1$. Thus if $D > 2$, $V_1 = V_2 = 0$. If $D = 2$, $V_2 = 0$. Since this is its minimum value, it is stationary. Then V_1 is stationary and $V_1 = 0$. The same argument carries through for coupled fields $u = (u_1, \ldots, u_N)$ in more than one space dimension. This is one reason for extensions of the s-G in one dimension to several coupled fields [25] or more generally for studies in more than one dimension of the classical gauge fields.

Next the s-G is exceptional in having a Bäcklund transformation. A first theorem is [26]:

Theorem 1: The generalized Klein-Gordon equation

$$\partial_\mu \partial^\mu u + U'(u) = 0$$

in $q = (q-1) + 1$ dimensions has no Bäcklund transformation if $q > 2$. The theorem is proved in Reference 26 for Monge-Ampère type equations in $q = 3$ dimensions

$$a u_{xy} + b u_{yz} + c u_{zx} + d u_{xx} + e u_{yy} + f u_{zz} + g = 0$$

in which $a, b, \ldots, g$ are arbitrary functions of u, u_x, u_y, and u_z. The proof generalizes to any q.

This theorem is consistent with Derrick's theorem for the following reason: the s-G has the Bäcklund transformation (BT)

$$\begin{aligned} u'_x &= u_x + 2k \sin \tfrac{1}{2}(u'+u) \\ u'_t &= u_t + 2k^{-1} \sin \tfrac{1}{2}(u'-u) \end{aligned} \tag{1.23}$$

which relates a solution $u(x,t)$ of $u_{xt} = \sin u$ to a solution $u'(x,t)$ of $u'_{xt}(x,t) = \sin u'$. Because u and u' satisfy the same equation, this BT is an auto-Bäcklund transformation (aBT). From the solution $u = 0$ one finds the kink solution

$$u'(x,t) = 4 \tan^{-1} \exp(kx + k^{-1}t). \tag{1.24}$$

If the s-G had a BT in two space dimensions, one might expect it to have a time independent solution other than the ground state $u \equiv 0$ and analogous to (1.24) (which becomes (1.14)). The aBT

(1.23) was known to Bäcklund in 1882. Bäcklund transformations for other nonlinear evolution equations have been found very much more recently [27]. Degasperis [6] reports generalized BTs for NEEs in his lectures.

A BT for the s-G which is not an aBT is

$$\begin{aligned} u'_x &= k^{-1}\sin(u'-u) \\ u'_t &= u_t + k\sin u. \end{aligned} \tag{1.25}$$

This maps a solution u of $u_{xt} = \sin u$ to a solution u' of

$$u'_{xt} = \left\{1 - k^2 u_x^2\right\}^{\frac{1}{2}} \sin u'. \tag{1.26}$$

If we write

$$u' = \sum_{n=0}^{\infty} f_n(u) k^n,$$

we find from (1.25) that

$$u' = u + ku_x + k^2 u_{xx} + k^3\left(u_{xxx} + \frac{1}{3!}u_x^3\right)$$
$$+ k^4\left(u_{xxxx} + u_x^2 u_{xx}\right) + \dots . \tag{1.27}$$

A conservation law for (1.26) is

$$\left\{\left(1 - k^2 u'^2_x\right)^{\frac{1}{2}}\right\}_t - k^2\left\{\cos u'\right\} = 0. \tag{1.28}$$

If (1.27) is inserted into (1.28) we get, after dropping constants and constant factors, the expression (1.29) for the conserved density in (1.28)

$$\left.\begin{aligned} & u_x^2 + k(2u_x u_{xx}) + k^2\left(u_{xx}^2 + 2u_x u_{xxx} + \frac{1}{4}u_x^4\right) \\ & + k^3\left(2u_x u_{xxxx} + 2u_{xx}u_{xxx} + 2u_x^3 u_{xx}\right) \\ & + k^4\left(u_{xxx}^2 + 2u_{xx}u_{xxxx} + 2u_x u_{xxxxx} + \frac{1}{8}u_x^6 + \right. \end{aligned}\right\} \tag{1.29}$$

$$+ 3u_{xxx}u_x^3 + \frac{13}{2}u_x^2u_{xx}^2\Big) + \ \ldots\ \Big]$$

Since k is arbitrary, the quantities u_x^2, u_{xx}^2, $2u_xu_{xxx} + \frac{1}{4}u_x^4$, ... are conserved separately. This is the sequence (1.17) modulo certain perfect differentials (which are trivially conserved).

A BT is not a property which is peculiar to equations with soliton solutions: the Burgers equation (1.6) in the form $u_t + uu_x - \nu u_{xx}$ $(\nu > 0)$ has the BT

$$\begin{aligned} u'_x &= -(2\nu)^{-1}uu' \\ u'_t &= -(4\nu)^{-1}\left(2\nu u_x - u^2\right)u' \end{aligned} \tag{1.30}$$

which maps solutions u of the Burgers equation into solutions u' of the heat equation $u'_t - \nu u_{xx}' = 0$. The transformation $u = -2\nu(\ell n\ u')_x$ is the Hopf-Cole transformation: it linearizes the Burgers equation as the heat equation and the Burgers equation is exactly solved this way.

Auto BTs linearize in a more subtle way: In the aBT (1.23) set $\Gamma = \tan\frac{1}{4}(u + u')$ to reach

$$\begin{aligned} \Gamma_x &= \tfrac{1}{2}u_x(1 + \Gamma^2) + k\Gamma \\ \Gamma_t &= k^{-1}\Gamma\cos u - (2k)^{-1}(1 - \Gamma^2)\sin u. \end{aligned} \tag{1.31}$$

By the Ricatti transformation $\Gamma = v_2/v_1$, one obtains the linearized equations

$$\begin{aligned} v_{1,x} + \tfrac{1}{2}u_xv_2 &= -\tfrac{1}{2}kv_1 \\ -\tfrac{1}{2}u_xv_1 + v_{2,x} &= +\tfrac{1}{2}kv_2 \end{aligned} \tag{1.32a}$$

$$\begin{aligned} v_{1,t} &= -(v_1/2k)\cos u - (v_2/2k)\sin u \\ v_{2,t} &= -(v_1/2k)\sin u + (v_2/2k)\cos u. \end{aligned} \tag{1.32b}$$

Equation (1.32a) is a linear scattering problem in which $-\frac{1}{2}u_x$ acts as a scattering potential and $-\frac{1}{2}ik$ is an eigenvalue. This can be seen as follows: The more general problem

$$\hat{L}v = \zeta v$$

$$\hat{L} = i\begin{bmatrix} \partial/\partial x & -q(x,t) \\ r(x,t) & -\partial/\partial x \end{bmatrix}, \quad v = \begin{bmatrix} v_1(x,t) \\ v_2(x,t) \end{bmatrix} \tag{1.33}$$

with ζ a complex eigenvalue maps into (1.32a) by $\zeta = -\frac{1}{2}ik$, $q = -\frac{1}{2}u_x = -r$. On the other hand, for $r = -1$ in (1.33) v_2 satisfies the Schrödinger eigenvalue problem $-v_{2,xx} - qv_2 = \zeta^2 v_2$ in which the scattering potential is $-q(x,t)$. Notice that this scattering potential depends on t so that in general the eigenvalues ζ^2 do so also. The same is true of the eigenvalues ζ in (1.33).

We generalize the pair of equations (1.32b) to

$$\hat{A}v = v_t, \quad \hat{A} = \begin{bmatrix} A(x,t) & B(x,t) \\ C(x,t) & -A(x,t) \end{bmatrix}. \tag{1.34}$$

A, B and C are functionals of u(x,t). Under the condition ζ is independent of t, the relation

$$[\hat{L},\hat{A}]v \equiv (\hat{L}\hat{A} - \hat{A}\hat{L})v = \hat{L}v_t - \zeta v_t = -\hat{L}_t v$$

or

$$\hat{L}_t = [\hat{A},\hat{L}] \tag{1.35}$$

holds. The operator relation (1.35) is actually a pair of NEEs. In the case of (1.32) we find

$$i\begin{bmatrix} 0 & \frac{1}{2}u_{xt} \\ \frac{1}{2}u_{xt} & 0 \end{bmatrix} = -\frac{\sin u}{2\zeta}\begin{bmatrix} 0 & 1 \\ 1 & 0 \end{bmatrix}\begin{bmatrix} \partial/\partial x & -\frac{1}{2}u_x \\ \frac{1}{2}u_x & -\partial/\partial x \end{bmatrix} = i\begin{bmatrix} 0 & \frac{1}{2}\sin u \\ \frac{1}{2}\sin u & 0 \end{bmatrix} \tag{1.36}$$

which is the original NEE. We now have an important result: the system (1.33) and (1.34) together with $\zeta_t = 0$ is equivalent to the NEEs (1.35). The form (1.35) is known as the Lax form† [28].

† The result (1.35) assumes that v span the space, but in (1.36) the matrix $\hat{A}$ defined as in (1.32b) must be chosen separately

Of course, the equivalence of (1.33) and (1.34) with $\zeta_t = 0$ to the NEEs (1.35) is useless unless we can solve these coupled systems. Ultimately this must require discovery of the potentials $q(x,t)$ and $r(x,t)$. Thus (1.33) is to be solved inversely. This is the origin of the description "the inverse scattering method." The 2×2 scheme (1.33) with (1.34) is the ZS-AKNS scheme [19], [20]. We return to the formulation of the inverse scattering transform later in these lectures.

Two other theorems indicate the rather special character of the sine-Gordon equation. These are [26], [22]:

<u>Theorem 2</u>: The equations $u_{xt} = F(u)$, $u'_{xt} = G(u')$ have an invertible Bäcklund transformation $u \leftrightarrow u'$ iff F, G satisfy $\ddot{F} + \alpha^2 F = 0$, $\ddot{G} + \alpha^2 h^{-2} G = 0$ for some complex valued constant α (not excluding $\alpha = 0$) and for some real $h^2 \neq 0$.

<u>Theorem 3</u>: A necessary and sufficient condition for the existence of an infinity of polynomial conserved densities for the equation $u_{xt} = F(u)$ is that $F(u) = Ae^{\alpha u} + Be^{-\alpha u}$ where α is a nonzero complex valued constant (A, B are s.t. $F(u) = \text{real}$). A corollary from Theorem 1 is that $u_{xt} = F(u)$ has an aBT if and only if $\ddot{F} + \alpha^2 F = 0$. The s-G is the case $\alpha = i$ and F is then periodic. The other non-periodic cases prove not to have soliton solutions so that this result is consistent with the need for periodic $F(u)$ if solitons must pass through each other in collision.

These results also show that the ϕ-four (1.8) does not have an aBT and that it does not have an infinity of polynomial conserved densities. The analysis to (1.32) rather shows that there is no 2×2 inverse scattering transform for the ϕ-four equation. These results are all consistent with numerical work which shows that the kinks (1.9) scatter off each other in collisions.

The multikink solutions of the s-G can be found by direct methods [29] (Hirota's method [30]) or by the inverse method (which, however, also solves the initial value problem). One form of the multikink solutions of (1.13) (laboratory coordinates) is

2π-kink $$u(x,t) = 4\tan^{-1}\theta_1 \tag{1.37}$$

(continued from preceding page)

for each eigenvalue $\zeta = \frac{1}{2}ik$ and thereby for each eigenvector belonging to it. The case $\zeta = 0$ is necessarily excluded. Two operators $\hat{L}$ and $\hat{A}$ can be chosen to represent the s-G as the operator equation $[\hat{L}, \hat{A} + \partial/\partial t] = 0$ (which is (1.35)) in terms of 4×4 matrices (see, e.g., Reference 59), but I do not know a 2×2 description.

4π-kink
$$u(x,t) = 4\tan^{-1}\frac{\sinh\frac{1}{2}(\theta_1+\theta_2)}{(a_{12})^{\frac{1}{2}}\cosh\frac{1}{2}(\theta_1+\theta_2)}$$
$$a_{12} = (a_1-a_2)^2(a_1+a_2)^{-2}; \qquad (1.38)$$

0π-kink (or "breather")

$$r = a_R a_I^{-1}$$
$$\Theta_R = \frac{1}{2}a_R\left[\left(1+\frac{1}{a_R^2+a_I^2}\right)x+\left(1-\frac{1}{a_R^2+a_I^2}\right)t+x_R\right] \qquad (1.39)$$
$$\Theta_I = \frac{1}{2}a_I\left[\left(1-\frac{1}{a_R^2+a_I^2}\right)x+\left(1+\frac{1}{a_R^2+a_I^2}\right)t+x_I\right].$$

The 2Nπ-kink is given by

$$\cos u(x,t) = 1 - 2\left(\frac{\partial^2}{\partial x^2}-\frac{\partial^2}{\partial t^2}\right)\ln f(x,t)$$
$$f(x,t) = \det\|M\|. \qquad (1.40)$$

The NxM matrix $\|M\|$ has elements

$$M_{ij} = 2(a_i+a_j)^{-1}\cosh\left\{\frac{1}{2}(\Theta_i+\Theta_j)\right\},$$
$$\Theta_i = \pm\gamma_i(x-V_it+x_i) \qquad (1.41)$$
$$a_i^2 = (1-V_i)(1+V_i)^{-1}, \quad \gamma_i^2 = (1-V_i^2)^{-1}$$

in (1.40). In (1.37) Θ_1 is given by Θ_i as in (1.41). In (1.38) Θ_1 and Θ_2 are similar: a_1 and a_2 are the two (real) parameters of this two-parameter solution and determine two velocities v_1 and v_2 as in (1.41). Asymptotically, (1.38) is

$$u(x,t) = 4\tan^{-1}\exp\left[\Theta_1+\eta_1^{\pm}\right] + 4\tan^{-1}\left[\Theta_2+\eta_2^{\pm}\right]$$

as $x\to\pm\infty$:

$$\eta_1^{\pm} = -\eta_2^{\pm} = \pm \ln a_{12}. \tag{1.42}$$

This illustrates the breakup of a two-kink solution into 2π-kinks: it indicates both the form of the phase shifts and their conservation.

The 0π or "breather" solution of (1.39) is of a type not discussed before. It is a two-parameter solution which does not break up. Instead it has an internal oscillation which progresses and is modulated by a moving external envelope. It is obtained from (1.40) by choosing a_1 and a_2 complex $a_1 = a_2^* = a_R + ia_I$ and $x_1 = x_2 = x_R + ix_I$. Its energy is

$$16\gamma_R\left\{r^2(1+r^2)^{-1}\right\}^{\frac{1}{2}} = 16\gamma_R \sin\mu$$

for

$$a_1 \equiv ae^{i\mu} : \gamma_R^2 = \left(1-V_R^2\right)^{-1}$$

and

$$V_R = \frac{1-a_R^2-a_I^2}{1+a_R^2-a_I^2} .$$

The corresponding energy of the 2π-kinks is $8\gamma_i$. The breather acts like a soliton: it can collide with 2π-kinks, 2π-antikinks and other breathers and passes through them with only a phase shift. It can be thought of as a bound kink-antikink pair in which u moves from 0 to some angle $< 2\pi$ and back to 0 as x moves from $-\infty$ to $+\infty$. It therefore has a total kink angle of zero. Its rest energy $16\sin\mu \leqq 2\times 8$. When $\mu = 1$ the breather is unstable to break up into two 2π-kinks. Notice that the breather is time dependent in the rest frame in which $V_R = 0$. Indeed in this frame, changing x_I and x_R slightly,

$$u(x,t) = 4\tan^{-1}\left[\cot\mu \sin\left\{(\sin\mu)t + x_I\right\} \operatorname{sech}\left\{(\cos\mu)x + x_R\right\}\right]. \tag{1.43}$$

It has an internal oscillation frequency $\sin\mu$ which appears in the energy $16\gamma_R \sin\mu$. It proves to be discretely quantized in the quantized case (see Section 4). The solution (1.43) is of a type excluded in the considerations leading to Derrick's theorem.

We now turn to some current applications of solitons in physics.

2. APPLICATIONS OF SOLITONS TO NONLINEAR PHYSICS

A less than definitive list of applications of solitons in physics already includes

i) Gravity waves in deep and shallow water.
ii) The theory of plasmas and the interaction of radiation with plasmas.
iii) Superconductivity: the theory of Josephson junctions.
iv) Fermi liquid theory: spin waves in the A- and B-phases of liquid ^{3}He below 2.6mK.
v) Ferromagnetics: Bloch wall motion.
vi) Resonant and nonresonant nonlinear optics and laser physics.
vii) Nonlinear crystal physics: theory of dislocations, anharmonic crystals; recurrence phenomena in thermal transport and nonergodic behaviour (Fermi-Pasta-Ulam problem [3], [31]); displasive and other phase transitions and central peak phenomena; linear conductors (like TTF-TCNQ).
viii) Theory of fundamental particles.
ix) Astrophysics: solitons in the solar corona have been suggested (the Great Red Spot in Jupiter has been called a soliton also!). Examples which embrace almost all of this list are treated in Reference 3. A large number of illustrative diagrams is given there.

The key equations are:

The KdV
$$u_t + 6uu_x + u_{xxx} = 0. \qquad (2.1)$$

The modified KdV
$$u_t + 6u^2u_x + u_{xxx} = 0. \qquad (2.2)$$

The nonlinear Schrödinger equation (NLS)
$$iu_t + 2u|u|^2 + u_{xx} = 0. \qquad (2.3)$$

The s-G
$$u_{xt} - \sin u = 0. \qquad (2.4)$$

Other important equations are:

The Hirota equation
$$iu_t + 3i\alpha|u|^2u_x + \rho u_{xx} + i\sigma u_{xxx} + \delta|u|^2u = 0 \quad (\alpha\rho=\sigma\delta) \qquad (2.5)$$

The Reduced Maxwell-Bloch equations

$$
\begin{aligned}
u_x &= -\mu s \\
v_x &= Ew + \mu u \\
w_x &= -Ev \\
E_t &= v \qquad (\mu = \text{const.}).
\end{aligned}
\tag{2.6}
$$

The Boussinesq equation

$$u_{tt} - (12uu_x + u_{xxx})_x = 0. \tag{2.7}$$

The three-wave interaction (decay type)

$$
\begin{aligned}
u_{1,x} + c_1 u_{1,t} &= iqu_2 u_3^* \\
u_{2,x} + c_2 u_{2,t} &= iqu_1 u_3 \\
u_{3,x} + c_3 u_{3,t} &= iqu_1^* u_2 .
\end{aligned}
\tag{2.8}
$$

The Toda lattice

$$u_{n,tt} = e^{-(u_n - u_{n+1})} - e^{-(u_{n-1} - u_n)}. \tag{2.9}$$

The Kadomtsev-Petviashvili equation

$$\frac{3}{4}\beta^2 u_{yy} + \left\{\alpha u_t + \lambda u_x + \frac{1}{4}(u_{xxx} + 6uu_x)\right\}_x = 0. \tag{2.10}$$

All of these equations have multisoliton solutions. The majority are in the form of systems of NEEs and only the initial data $u(x,0)$ is needed to determine the motion. It is interesting to note that the KdV type equation $u_t + 6u^p u_x + u_{xxx} = 0$ appears to have soliton solutions if and only if $p=1$ or 2. The Boussinesq equation appears to be a particular case of the Kadomtsev-Petviashvili equation. But the latter has multisoliton solutions in two space dimensions which are localized and which do not have the multisoliton solution of the Boussinesq equation as limit. Zakharov and Shabat [32] gave a multisoliton solution of the K-P equation which is not localized. Recently Manakov, Zakharov, Bordag, Its and Matveev found the localized solitons as a limit of this earlier result. We quote it here: without loss of generality we take $\lambda=0$, $\alpha=\frac{1}{4}$ and $\beta^2=\pm 1$. For $\beta^2=-1$ the multisoliton solution is

$$u = 2\frac{\partial^2}{\partial x^2}\ln\det\|V\|$$

$$V_{mn} = \delta_{mn}\left(x - i\nu_n y - \xi_n - 3\nu_n^2 t\right) + (1-\delta_{mn})2(\nu_m - \nu_n)^{-1}. \tag{2.11}$$

The one-soliton is: $N = 2$, $\nu_2 = -\bar{\nu}_1$, $\xi_2 = \bar{\xi}_1$

$$\det\|V\| = 4(\nu_1 + \bar{\nu}_1)^{-2} + |x - i\nu_1 y - \xi_1 - 3\nu_1^2 t|^2 > 0. \tag{2.12a}$$

The resultant u is real, localized and nonsingular with the asymptotic behaviour

$$u \sim \left[x^2 + 2(\mathrm{Im}\,\nu_1)xy + |\nu_1|^2 y^2\right]^{-1}. \tag{2.12b}$$

For $\beta^2 = +1$, the 'two-dimensional KdV equation' for water, $y \to iy$ (whilst $\bar{\nu}_1 \to -\bar{\nu}_1$ in (2.12a) which becomes singular). The most remarkable feature of these multisoliton solutions is that single solitons like that described in Equations (2.12) collide with other similar solitons with absolutely no resultant interaction asymptotically—not even a phase shift.

I quote a few physical applications: the KdV to shallow water waves, lattice recurrences, plasma ion acoustic waves; the modified KdV to Alfven waves in a cold collisionless plasma; the NLS to gravity waves in deep water, to one-dimensional self-focusing, self-phase modulation, Langmuir turbulence in plasmas [33] (Zakharov's 'caverns' [34]), laser plasma interactions (optical filament formation), hydrodynamical vortices [35], the one-dimensional Heisenberg ferromagnet [36]; the s-G to spin waves, Josephson junctions, nonlinear optics (self-induced transparency) [37], [38], lattice theory and particle physics [9], [16]; the RMB to self-induced transparency [3], [38]; the Boussinesq to hydrodynamics and plasmas; the three-wave to stimulated Raman back scattering in plasmas; the Toda lattice as a soluble lattice with hard-core and harmonic limits; the Kadomtsev-Petviashvili equation as a two-dimensional nonstationary problem in a weakly dispersive medium [32], [39]. Refer to Reference 3 for more comprehensive references on the other topics.

3. SOME PARTICULAR APPLICATIONS OF SOLITONS IN PHYSICS

I develop particular examples in more detail. These are the application of the multisoliton solutions of the NLS to optical filament formation in laser irradiated neutral dielectrics and the application of the s-G to spin waves, optical pulses and Josephson junctions. Spin waves in a paramagnetic gas can be taken in one jump with optical pulses and Josephson junctions. I treat spin waves in the A- and B-phases of liquid ^{3}He separately after these.

Consider the neutral dielectric: in a scalar field $E(\underset{\sim}{x},t)$, the dipole $P(\underset{\sim}{x},t)$ is

$$P(x,t) = \alpha E(x,t) + \alpha_{NL} E(\underset{\sim}{x},t)^3 + \cdots , \tag{3.1}$$

and α and α_{NL} are the constant linear and first nonlinear susceptibilities. Maxwell's equation is linear and is

$$\nabla^2 E(\underset{\sim}{x},t) - c^{-2}\partial^2 E(\underset{\sim}{x},t)/\partial t^2 = 4\pi n c^{-2}\partial^2 P(\underset{\sim}{x},t)/\partial t^2 \tag{3.2}$$

where n is the atomic number density. We look for complex envelope solutions $\epsilon(x,y,z,t)$ modulating carrier waves $\exp\{i(\omega t - kx)\}$:

$$E(\underset{\sim}{x},t) = \epsilon(x,y,z,t)e^{i(\omega t-kz)} + \text{c.c.} \tag{3.3}$$

We impose the linear dispersion relation $\omega^2 = c^2k^2 - 4\pi n\alpha\omega^2$. We equate coefficients of all terms in $\exp\{i(\omega t-kz)\}$ (in principle one should go on and obtain a coupled sequence of equations for the complex envelopes of the harmonics). We find

$$\epsilon_{xx} + \epsilon_{yy} + \omega^2 c^{-2} 12\pi n\alpha_{NL}|\epsilon|^2\epsilon + 2ic^{-2}\left[\omega\epsilon_t(1+4\pi n\alpha) - c^2k\epsilon_z\right] = 0. \tag{3.4}$$

We look for steady-state solutions (ϵ does not depend on t). By suitable scaling

$$\epsilon_{xx} + \epsilon_{yy} + 2|\epsilon|^2\epsilon - i\epsilon_z = 0 \tag{3.5}$$

which is the two-dimensional NLS in (x,y) and a 'time' (z).

In one space dimension (x) the NLS (3.5) has the one-soliton solution

$$\epsilon(x,z) = \frac{2\eta\exp\{4i(\xi^2-\eta^2)z - 2i\xi x + i\delta\}}{\cosh[2\eta(x-x_0) - 8\eta\xi z]} . \tag{3.6}$$

It contains a carrier which corrects the linearized carrier in (3.4). Solitons (3.6) are typical of the envelope solitons. This particular one can be seen in experiments on deep water [40]. As an electric field, it carries intensity $4\eta^2 \mathrm{sech}^2[2\eta(x-x_o) - 8\eta\xi z]$. This is the intensity across a wave-guide like channel induced in the medium by the intense field making an angle $\tan^{-1} 4\xi$ to the z-axis. An arbitrary laser profile $\epsilon(x,0)$ breaks up into a number of such soliton channels as z increases.

The second application is more complicated. We consider first a collection of spin-$\frac{1}{2}$ systems ($n\ \mathrm{cc}^{-1}$) in a constant magnetic field B_o along $-z$. We use the gyromagnetic ratio $\gamma\hbar = e\hbar m_e^{-1} c^{-1}$ (g-factor = 2). The Larmor frequency is $\omega_L = \gamma B_o$. The $m_s = \pm\frac{1}{2}$ states have energies $\pm\frac{1}{2}\hbar\omega_L$. An inhomogeneous transverse R.F. field $B_x(\underset{\sim}{x},t)$ flips the spins and these flips propagate as spin waves. From the Hamiltonian density

$$H(\underset{\sim}{x}) = \tfrac{1}{2}\hbar\omega_L \sigma_z(\underset{\sim}{x}) - \tfrac{1}{2}\gamma\hbar\sigma_x(\underset{\sim}{x})B_x, \tag{3.7}$$

one easily finds

$$\dot{\underset{\sim}{\sigma}} = \underset{\sim}{\omega} \times \underset{\sim}{\sigma}, \quad \underset{\sim}{\omega} = (-\gamma B_x, 0, \omega_L). \tag{3.8}$$

We define the Bloch vector density $\underset{\sim}{r}(\underset{\sim}{x},t)$ by the expectation value

$$\underset{\sim}{r}(\underset{\sim}{x},t) = \langle \sigma(\underset{\sim}{x},t) \rangle .$$

Then

$$\dot{\underset{\sim}{r}} = \underset{\sim}{\omega} \times \underset{\sim}{r}. \tag{3.9}$$

The transverse magnetic dipole is $\frac{1}{2}\gamma\hbar r_1(\underset{\sim}{x},t) = \frac{1}{2}\gamma\langle\sigma_x(\underset{\sim}{x},t)\rangle$ and Maxwell's equations are

$$\nabla^2 B_x - c^{-2} B_{x,tt} = 4\pi c^{-2}\tfrac{1}{2}\gamma r_{1,tt}. \tag{3.10}$$

With the choice of ω in (3.8), Equations (3.9) and (3.10) form a system of coupled nonlinear partial differential equations governing the nonlinear propagation of spin waves. We call this the Bloch-Maxwell (BM) system. It is linearized by noting that $r_{1,t} = -\omega_L r_2$, $r_{2,t} = \omega_L r_1 + \gamma B_x r_3$, and, for constant inversion, r_3, $r_{1,tt} = -\omega_L^2 - \omega_L \gamma B_x r_3$—the usual pseudo-Bose system.

The astute reader will notice that as σ_y moves in time it too generates a magnetic field: the result is a circularly polarized transverse field, but as the theory is not changed conceptually this way we consider only the plane polarized case. We also ignore the longitudinal field generated as σ_z moves in time. These approximations are made for comparison with a second then totally equivalent problem which arises in the propagation of plane polarized 10^{-9} sec optical pulses through media with resonant nondegenerate atomic transitions. We take n two-level atoms cc^{-1} each with resonant frequency ω_s. Pulses are envelopes modulating carriers of frequency $\omega \approx \omega_s$. This justifies the two-level atom approximation. This atom is a two-state system with a spin representation: spin up (down) is occupation of the upper (lower) state. The equations take precisely the BM form with [3], [38]

$$\underset{\sim}{\omega} = (-2p\hbar^{-1}E(\underset{\sim}{x},t), 0, \omega_s). \tag{3.11}$$

The electric field replaces the magnetic field in the Maxwell equation

$$\nabla^2 E - c^{-2}E_{tt} = 4\pi c^{-2} pr_{1,tt} \tag{3.12}$$

(the electric dipole proves to be $pr_1 = p\langle\sigma_x\rangle$ with p the dipole matrix element: the transition supposedly does not change magnetic quantum numbers; the light is plane polarized and, in contradistinction to the magnetic case, σ_y and σ_z are not sources).

A third equivalent problem arises in the Josephson junction of large area (such a junction is typically $\lesssim$ 1 mm. long). The junction is a two-state system (the two sides of the junction). If a voltage V is applied between these two sides, the two states differ in energy by 2eV (2e is the charge of the Cooper pair). The sides will couple by some parameter K (say) characteristic of the junction. The junction equation will be (3.9) [3] with

$$\underset{\sim}{\omega} = (2K\hbar^{-1}, 0, 2eV\hbar^{-1}). \tag{3.13}$$

Consider the plane Josephson "sandwich" with superconductors top and bottom separated by a thin uniform layer of oxide of effective thickness (including penetration depth) d. If V is applied across the oxide, a field $E = d^{-1}V$ exists there in the direction normal (called z) to the plane of the junction (the x-y plane). Consider plane E-waves propagating along x and carrying a magnetic field B along y.

The quantity $r_3 \equiv \langle\sigma_z\rangle$ is the difference in occupation number densities either side of the junction. The current density component j_z along z is thus $j_z = \beta r_{3,t}$ (for some constant β). This drives the (transverse) wave equation in the usual way:

$$E_{xx} - \bar{c}^{-2}E_{tt} = 4\pi\beta\bar{c}^{-1}r_{3,tt}. \qquad (3.14)$$

We are in a material medium so $\bar{c}=ck^{-\frac{1}{2}}$. This arises through the displacement current CV_t: $C=k/4\pi d$ is the capacitance of the junction per unit area.

There are two differences from the two BM systems considered previously: r_3 rather than r_1 drives the wave equation, and V not K will depend on t in $\underset{\sim}{\omega}$ (K is probably more complicated but we do not investigate this here). We thus reach the BM problem with the simple switch of components $1 \leftrightarrow 3$, $2 \leftrightarrow 2$!

The BM system

$$E_{xx} - \bar{c}^{-2}E_{tt} = 4\pi\bar{c}^{-2}\beta k^{-1}r_{3,tt} \qquad (3.15a)$$

$$\underset{\sim}{r}_{1,t} = \underset{\sim}{\omega}\times\underset{\sim}{r} \quad (\underset{\sim}{\omega} = (2K\hbar^{-1}, 0, 2ed\,E\hbar^{-1})) \qquad (3.15b)$$

does not have multisoliton solutions [3]. The reduced Maxwell-Bloch system which replaces (3.15a) by the one-way going (single characteristic) wave equation $E_x+\bar{c}^{-1}E_t=2\pi\bar{c}^{-1}\beta k^{-1}r_{3,t}$ can be scaled to (2.6) [38], [41] and does have multisoliton solutions. The elimination of the backward-going characteristic is admis sible in the short optical pulse problem because typically one is concerned with the very low density of metal vapors ($n \sim 10^{11}cc^{-1}$). For the large area junction, the numbers scarcely permit this approximation (although the multisoliton RMB behaviour will come through approximately).

In the Josephson junction problem one usually assumes instead that $r_3 \approx 0$ everywhere (little charge imbalance on the two sides). Then the Bloch equations (3.15b) become

$$r_{2,t} = 2ed\hbar^{-1}Er_1, \quad r_{1,t} = -2ed\hbar^{-1}Er_2 \qquad (3.16)$$

with solution

$$r_2 = -\sin\sigma, \quad r_1 = -\cos\sigma$$
$$\sigma = 2ed\hbar^{-1}\int_{-\infty}^{t} E(x,t')dt'. \qquad (3.17)$$

Thus

$$\sigma_t = 2ed\hbar^{-1}E, \quad \text{or} \quad \sigma_t = 2eV\hbar^{-1}$$
$$j_z = -2\beta K\hbar^{-1}r_2, \quad \text{or} \quad j_z = j_{zo}\sin\sigma. \qquad (3.18)$$

The forms on the right are exactly Josephson's two equations: σ is now identified as the Josephson phase.

We have still to satisfy Maxwell's equation (3.15a). After one integration all through by time this becomes

$$\sigma_{xx} - \bar{c}^{-2}\sigma_{tt} = \lambda_o^{-2}\sin\sigma. \tag{3.19}$$

This is the s-G scaled by the natural length

$$\lambda_o = \{k\hbar\bar{c}^2/8\pi\beta K\}^{\frac{1}{2}}. \tag{3.20}$$

It is also possible to obtain the s-G from the optical problem. We look for plane wave envelope solutions modulating resonant carrier waves by setting

$$\begin{aligned} E(x,t) &= \hbar p^{-1}\epsilon(x,t)\cos\left\{\omega_s(t-c^{-1}x)\right\} \\ r_1(x,t) &= Q(x,t)\cos\left\{\omega_s(t-c^{-1}x)\right\} + P(x,t)\sin\left\{\omega_s(t-c^{-1}x)\right\}. \end{aligned} \tag{3.21}$$

One finds by using the fact that P,Q vary on a 10^{-9} sec time scale and $\omega_s \sim 10^{15}$ Hz that $Q=0$ and

$$P_t = \epsilon N, \quad N_t = -\epsilon P \tag{3.22}$$

(where $N \equiv r_3(x,t)$) whilst

$$\epsilon_x + c^{-1}\epsilon_t = \alpha P \qquad (\alpha = 2\pi p^2 n\omega_s \hbar^{-1} c^{-1}). \tag{3.23}$$

An "attenuator" is an initially unexcited medium. For this (3.22) is solved by

$$\begin{aligned} P &= -\sin\sigma, \quad N = -\cos\sigma \\ \sigma(x,t) &= \int_{-\infty}^{t} \epsilon(x,t')dt'. \end{aligned} \tag{3.24}$$

From (3.24), (3.23) is

$$\sigma_{xt} + c^{-1}\sigma_{tt} = -\alpha\sin\sigma. \tag{3.25}$$

This is the s-G in unusual independent variables. Set

$$\sqrt{c}\,\xi = \sqrt{\alpha}\,(ct-2x), \quad \eta = \sqrt{\alpha c}\ t. \tag{3.26}$$

Then

$$\sigma_{\xi\xi} - \sigma_{\eta\eta} = \sin\sigma . \tag{3.27}$$

The crucial idea here is to exploit the resonance condition, and the approximation is very different from the assumption $r_3 \approx 0$ for the junction problem (r_3 is, of course, r_1 in the optical problem and this goes over to the out-of-phase component P of the dipole).

Both the s-G (3.27) for optical pulses and the s-G (3.19) for the large area junction have the multisoliton solutions (1.37)-(1.41). In the optical problem, an arbitrary intensity envelope ($\propto \epsilon^2$) breaks up in general into a train of sech^2 pulses. Each of these is the squared time derivative in the x,t coordinate system of a 2π-kink. The sech^2 pulses are solitons: they persist and the medium is transparent to them. The phenomenon is therefore called self-induced transparency (SIT) which explains this phenomenon introduced in Section 2. In the (x,t) system the one-soliton is

$$\epsilon(x,t) = pE_0\hbar^{-1}\,\mathrm{sech}\,\tfrac{1}{2}pE_0\hbar^{-1}(t-xv^{-1}). \tag{3.28}$$

It is a soliton solution of the so-called SIT equations [38]. The "area" $\Theta(x,t) \equiv \sigma(x,\infty)$ is independent of x:

$$\Theta(x,t) = \int_{-\infty}^{\infty} \epsilon(x,t')dt' = \Big[4\tan^{-1} e^{t}\Big]_{-\infty}^{\infty} = 2\pi. \tag{3.29}$$

A pulse of arbitrary area Θ entering the resonant medium undergoes a jump in Θ at the boundary of the medium to the value $2\nu\pi$ where ν is an integer. The pulse then reshapes at this constant area inside the medium to ν_1 2π-sech pulses (ν_1 kinks in σ), ν_2 2π-sechs of opposite sign (ν_2 antikinks in σ) and ν_3 breathers. We have $\nu_1 - \nu_2 = \nu$. Details are given in References 3 and 38 and the references therein.

The theory of the Josephson junction displays similar features in a slightly different way. In the rest frame the kink solution of (3.19) is

$$\sigma = 4\tan^{-1}\exp[(x-x_0)/\lambda_0]. \tag{3.30}$$

Use $\sigma_{tx} = \sigma_{xt} = 2e\hbar^{-1}V_x = 2ed\hbar^{-1}c\,B_t$ to reach

$$\sigma_x = 2ed\hbar^{-1}c^{-1}B \tag{3.31}$$

(B is the y-component of the magnetic field). Then

$$2\pi = \sigma(\infty,t) - \sigma(-\infty,t) = 2ed\hbar^{-1}c^{-1}\int Bdx = 2e\hbar^{-1}c^{-1}\int BdS. \tag{3.32}$$

Thus the kink carries one unit of flux hc/2e—the single "fluxon." The boundary conditions on a large junction will be $\sigma(\infty,t) = 2\nu_1\pi$, $\sigma(-\infty,t) = 2\nu_2\pi$ in general, and the total flux is $(\nu_2-\nu_1)$ hc/2e. The sum of the numbers of kinks and antikinks is therefore $(\nu_2-\nu_1)$. The breathers carry no net flux.

The breakup of optical pulses into their constituent solitons has been observed [42], [3]. Evidence for the existence of kinks in large area Josephson junctions has been obtained by looking at changes in the voltage/current characteristics [43]. Extra current spikes have appeared with particular voltage spacings [43]. In frequency units this spacing is just the fundamental even mode frequency of the junctions equivalent cavity. Reference 43 argues that a kink which reaches the end of an open ended cavity is reflected as an antikink. There is a natural cavity "mode" consisting of a kink with its antikink. There is therefore one node and the mode corresponds in that respect to a harmonic cavity mode with twice the fundamental frequency. Successive kink-antikink pairs induce current spikes (or current steps) much as radiation induces current steps in small area junctions. The observations [43] are in extraordinarily good agreement with the results of this argument. However, the s-G has not been solved for open ended or close ended boundary conditions on finite support $-\frac{1}{2}L \leqq x \leqq \frac{1}{2}L$. All the solutions considered in these lectures are for the real line $-\infty < x < \infty$. Certain equations, notably the KdV (1.1), have been solved for periodic boundary conditions $u(0,t) = u(L,t)$, $u_x(0,t) = u_x(L,t)$, etc. [44] and the method developed by Novikov [44] seems rather general.

Finally we turn to a further application of spin wave theory. This is to spin waves in the Fermi liquid ^{3}He below 2.6mK. Above 2.6mK ^{3}He is a normal Fermi liquid. At 2.6mK and 34 atmospheres it undergoes a second order phase transition to the so-called A-phase and at lower temperatures and pressures it becomes the B-phase. The change A to B is signalled by a change in magnetic properties.

Both phases are supposed to be the result of a superconducting type phase transition [45], [46], [47]. ^{3}He atoms have an unpaired nuclear spin. Pairing is between atoms, and, to avoid their close approach, pairs take on angular momentum $\ell = 1$. (The possibility of $\ell = 3$ has also been investigated.) Accordingly, spins are paired in the triplet (S = 1) configuration, not S = 0. The A-phase is exceptional in that spin fluctuations appear to suppress the symmetric spin state $S_z = 0$. The A-phase is thus a two-state system and can be thought of as two interpenetrating superfluids labelled by "spin up" and "spin down." These two superfluids couple through the weak spin dipole interaction. This "Josephson junction" satisfies the s-G equation even though the excitations are longitudinal.

In contrast, the B-phase admits all three spin states. Spin waves prove to satisfy the 'double sine-Gordon equation' (double s-G)

$$u_{xx} - u_{tt} = -(\sin u + \tfrac{1}{2}\sin\tfrac{1}{2}u) \tag{3.33}$$

in this case. The negative sign on the right side is important. The double s-G with negative sign replaced by positive occurs in the theory of SIT when the resonant transitions are degenerate [49]. With either sign the corollary from Theorem 2 shows that the equation does not have an aBT: Theorem 3 shows that (in light-cone coordinates) it does not have an infinity of polynomial conserved densities. Hirota's direct method (see Dr. Caudrey's lectures [30]) applied to the double s-G terminates only for the solitary wave solutions. The solutions are soliton-like, however, and remarkably interesting.

We first take as Hamiltonian (compare Kleinert's lectures [50])

$$H = \int d\underset{\sim}{x}\,\psi_\alpha^\dagger(\underset{\sim}{x})\left[-(2m)^{-1}\nabla^2 - \mu\right]\psi_\alpha(\underset{\sim}{x}) + \frac{1}{2}\int\sum_{\alpha\beta}\psi_\alpha^\dagger(\underset{\sim}{x})\psi_\beta^\dagger(\underset{\sim}{x}')V(\underset{\sim}{x},\underset{\sim}{x}')\psi_\beta(\underset{\sim}{x}')\psi_\alpha(\underset{\sim}{x}')\,d\underset{\sim}{x}\,d\underset{\sim}{x}' \tag{3.34}$$

[μ is the chemical potential and $\hbar = k_B = 1$ (k_B is Boltzmann's constant); $\psi_\alpha^\dagger$ creates a fermion of spin α]. The potential term is factorized [46] to have integrand

$$\Delta^*_{\alpha\beta}(\underset{\sim}{x},\underset{\sim}{x}')\psi_\beta(\underset{\sim}{x}')\psi_\alpha(\underset{\sim}{x}) + \Delta_{\alpha\beta}(\underset{\sim}{x},\underset{\sim}{x}')\psi_\alpha^\dagger(\underset{\sim}{x})\psi_\beta^\dagger(\underset{\sim}{x}') \tag{3.35}$$

in which the components $\Delta_{\alpha\beta}$ of the 'gap matrix' Δ are defined by

$$\Delta_{\alpha\beta}(\underset{\sim}{x},\underset{\sim}{x}') = V(\underset{\sim}{x},\underset{\sim}{x}')\langle\psi_\beta(\underset{\sim}{x}')\psi_\alpha(\underset{\sim}{x})\rangle = -\Delta_{\beta\alpha}(\underset{\sim}{x}',\underset{\sim}{x}) \tag{3.36}$$

with complex conjugate $\Delta^*_{\alpha\beta}$. The symbol $\langle\ldots\rangle$ is thermal average. Through Gorkov's formalism for thermal Green's functions, one finds [46] the BCS type 'gap equation'

$$\underset{\approx}{\Delta}^{\dagger}(p) = \int d\underset{\sim}{p}' V(\underset{\sim}{p},\underset{\sim}{p}')\underset{\approx}{\Delta}^{\dagger}(p')2E(p')\tanh\{E(p')/2T\} \qquad (3.37)$$

where $E^2 = \{(2m)^{-1}p^2 - \mu\} + |\Delta|^2$ and we have imposed the unitary condition $\underset{\approx}{\Delta}\underset{\approx}{\Delta}^{\dagger} = \underset{\approx}{1}|\Delta|^2$ with 1 the unit matrix: $\underset{\approx}{\Delta}^{\dagger}$ is the Hermitian conjugate of $\underset{\approx}{\Delta}$. We are working in momentum space and, e.g., $\underset{\approx}{\Delta}^{*}(-\underset{\sim}{p}) = \underset{\approx}{\Delta}^{\dagger}(\underset{\sim}{p})$ where $\Delta^{\dagger}_{\alpha\beta} = \Delta^{*}_{\beta\alpha}$. We expand $V(\underset{\sim}{p},\underset{\sim}{p}')$ in spherical harmonics and retain only $V_1\,\underset{\sim}{n}\cdot\underset{\sim}{n}'$: $\underset{\sim}{n}$ is the normal to the Fermi surface. We assume $V_1 > 0$ is a constant within a narrow energy shell about the Fermi surface and vanishes outside this. Then the matrix $\underset{\approx}{\Delta}$ depends only on $\underset{\sim}{n}$ and is zero outside the energy shell. The gap equation is now

$$\underset{\approx}{\Delta}^{\dagger}(\underset{\sim}{n}) = \int d\underset{\sim}{p}' V_1 \underset{\sim}{n}\cdot\underset{\sim}{n}'\underset{\approx}{\Delta}^{\dagger}(\underset{\sim}{n})2E(p')\tanh\{E(p')/2T\}. \qquad (3.38)$$

The transition temperature T_c is determined in terms of the density of states at the Fermi surface by letting $\underset{\approx}{\Delta} \to 0$ in (3.38).

From this gap equation (3.38), $\underset{\approx}{\Delta}^{\dagger}(\underset{\sim}{n}) = -\underset{\approx}{\Delta}^{\dagger}(-\underset{\sim}{n})$. From the fermion property $\Delta_{\alpha\beta}(\underset{\sim}{x},\underset{\sim}{x}) = -\Delta_{\beta\alpha}(\underset{\sim}{x},\underset{\sim}{x})$ of (3.36) now follows $\Delta_{\alpha\beta}(\underset{\sim}{n}) = \Delta_{\beta\alpha}(\underset{\sim}{n})$. We can therefore set

$$\underset{\approx}{\Delta} = \begin{bmatrix} \Delta_{\uparrow\uparrow} & \Delta_{\uparrow\downarrow} \\ \Delta_{\downarrow\uparrow} & \Delta_{\downarrow\downarrow} \end{bmatrix} = \underset{\sim}{d}(\underset{\sim}{n})\cdot\underset{\sim}{\sigma}\,\sigma_v = \begin{bmatrix} id_u + d_v & -id_w \\ -id_w & -id_u + d_v \end{bmatrix}. \qquad (3.39)$$

From the transformation properties of $\underset{\approx}{\Delta}$, $\underset{\sim}{d}(\underset{\sim}{n}) = (d_u, d_v, d_w)$ behaves as a vector under rotations in spin space (u,v,w): $d_{\alpha}(\underset{\sim}{n})$ gives the amplitude for pairing at $\underset{\sim}{n}$ to be in the state with $M_S = 0$ along α; $\underset{\sim}{\sigma} = (\sigma_u, \sigma_v, \sigma_w)$ is the usual set of Pauli matrices.

The gap equation (3.38) shows that solutions for the gap matrix are linear combinations of $\ell = 1$ spherical harmonics. The equation is linear for $T = T_c$, where $|\Delta|$ vanishes, but is nonlinear for $T < T_c$; although this nonlinearity mixes the ℓ values in (3.37), it is supposed to be a good approximation to ignore this mixing. For $\ell = 1$ solutions we introduce the bivector $\underset{\approx}{D}$ by

$$\underset{\sim}{d}(\underset{\sim}{n}) = \underset{\approx}{D}\cdot\underset{\sim}{n}. \qquad (3.40)$$

Then $\underset{\approx}{D}$ transforms as a vector in the second suffix under rotations in orbit space and as a vector in the first suffix under rotations in spin space. The energy of a solution of (3.38) is invariant under separate rotations of spin and orbit space and these solutions form a degenerate manifold. This manifold is split by the weak dipole interactions.

It is believed that the A-phase is described by the ABM state [45]

$$\underset{\approx}{D} = i\sqrt{3}\,\Delta_o \hat{\underset{\sim}{v}}\,\hat{\underset{\sim}{\ell}}^{(+)} \tag{3.41}$$

plus rotations: $\hat{\underset{\sim}{v}}$ is a real unit vector in spin space and $\hat{\underset{\sim}{\ell}}^{(+)}$ is a (complex) spherical vector, $\hat{\underset{\sim}{\ell}}^{(+)} \equiv (1/\sqrt{2})\,(\hat{\ell}_x + i\hat{\ell}_y) \equiv (1/\sqrt{2})(\hat{\ell}_1 + i\hat{\ell}_2)$ with axis along $\hat{\underset{\sim}{\ell}} = \hat{\underset{\sim}{\ell}}^{(+)} \times \hat{\underset{\sim}{\ell}}^{(-)} = \hat{\ell}_3$ in orbit space; $\hat{\underset{\sim}{\ell}}$ is along z, Δ_o is a complex number, and rotation of $\Delta_o\hat{\underset{\sim}{\ell}}^{(+)}$ about $\hat{\underset{\sim}{\ell}}$ (rotation in the x,y plane) changes the phase of $\underset{\approx}{\Delta}$ by $-\phi$. Thus ϕ is the analogue of the phase in the $\ell = 0$ case. The state defined by (3.41) is intrinsically anisotropic with both $\ell = 1$ and expectation value $\underset{\sim}{\ell} \neq \underset{\sim}{0}$.

The B-phase is believed to be described by the BW state [45]

$$\underset{\approx}{D} = \Delta_o \sum_{i=1}^{3} \hat{\underset{\sim}{v}}_i\, \hat{\underset{\sim}{\ell}}_i \tag{3.42}$$

where the $\hat{\underset{\sim}{v}}_i$, $\hat{\underset{\sim}{\ell}}_i$, define respectively sets of orthogonal axis in spin and orbit space. The state is macroscopically isotropic with $\ell = 1$ $(\ell^2 = 2)$ but the expectation value $\underset{\sim}{\ell} = \underset{\sim}{0}$.

For the A-phase we consider a magnetic field B_o along y: the orbital direction $\hat{\underset{\sim}{\ell}}$ is a unique direction for this system and the role of B_o along y is to align $\hat{\underset{\sim}{\ell}}$ perpendicular to it (along z). The direction $\hat{\underset{\sim}{v}}$ is a second unique direction—in spin space. States with $\hat{\underset{\sim}{v}}$ in an arbitrary direction with respect to $\hat{\underset{\sim}{\ell}}$ are degenerate in the BCS approximation (3.34) with (3.37). But the weak spin dipole interaction energies are minimized with $\hat{\underset{\sim}{v}}$ parallel to $\hat{\underset{\sim}{\ell}}$. The rotations of $\hat{\underset{\sim}{v}}$ in the plane perpendicular to $\hat{\underset{\sim}{v}} \times \hat{\underset{\sim}{\ell}}$ propagate as spin waves. We consider here rotations in the x-, z-plane propagating up the y direction, the direction of B_o. We define $\theta = \cos^{-1} \hat{\underset{\sim}{v}} \cdot \hat{\underset{\sim}{\ell}}$ so that

$$\underset{\approx}{\Delta} = \frac{\sqrt{3}}{2}\,\Delta_o (n_x + i n_y) \begin{bmatrix} i\cos\theta + \sin\theta & 0 \\ 0 & -i\cos\theta + \sin\theta \end{bmatrix} . \tag{3.43}$$

Total spin up (down) lies parallel (antiparallel) to B_o. The totally symmetric spin state evidently has amplitude d_w always zero. As indicated earlier, we can think of the A-phase as two interpenetrating superfluids with spins up and down respectively (the spin fluctuations suppress the symmetric states by changing the effective pair interaction [48]). From this view, the system is a two-state system and will propagate spin waves governed by the s-G [3], [45], [51], [52]. From (3.43) the phase difference between the two states is -2θ whilst the phase sum is -2ϕ. Spin currents are therefore determined by $\nabla\theta$ whilst (as in a normal superconductor) number currents are determined by $\nabla\phi$.

The B-phase is simpler: any rotation $R(\theta)$ through angle θ of spin axes against orbital axes takes (3.42) to another solution of (3.38). The axis of $R(\theta)$ is a unique direction for the B-phase. Leggett [45] argues that this direction will tend to lie along B_o. Spin waves are again rotations of θ and can in particular propagate along y. We treat only this case here.

The dipole interaction energies are very weak; they lead to NMR frequencies $\sim 10^5$ Hz (compare atomic hyperfine frequencies $\sim 10^8$ Hz). They are also fast compared with the motions of $\hat{\underset{\sim}{\ell}}$ in the A-phase (orbital waves [46], [50]) and we can keep $\hat{\underset{\sim}{\ell}}$ fixed in the A-phase case whilst in both cases consider <u>adiabatic</u> motions of the spin vector $\underset{\sim}{d}(\underset{\sim}{n})$ such that all other degrees of freedom react instantaneously to these motions. We therefore introduce the adiabatic Hamiltonian density [52]

$$\mathcal{H}(x) = \tfrac{1}{2}\gamma^2\chi^{-1}\sigma_w^2 - \gamma\sigma_w B_o + \tfrac{1}{4}\chi\gamma^{-2}\bar{c}^2\nabla\theta\cdot\rho\cdot\nabla\theta + H_D(\theta). \quad (3.44)$$

[$\gamma = e/m_p c$ is the gyromagnetic ratio, χ is the magnetic susceptibility, $\bar{c}$ is a velocity (~ 10 cm sec^{-1}) and $H_D(\theta)$ is the dipole interaction: $\underset{\sim}{\rho}$ = diag (2,2,1) for the B-phase and diag (1,2,1) for the A-phase because of the different preferred directions z and y, and the fraction "$\tfrac{1}{4}$" is a dummy taken to ensure $\bar{c}$ is the spin-wave speed in each case [52]; $\mathcal{H}$ needs scaling by $\rho_s\hbar$; ρ_s is the superfluid number density.]

We define the frequency $\Omega_\chi \equiv 4\gamma^2\chi^{-1}$, the longitudinal NMR frequency Ω_ℓ, and the coupling constant $\gamma_o \equiv \Omega_\chi\Omega_\ell^{-1}$ (for technical reasons [52] $\Omega_\ell = \Omega_{\ell A}$ for the A-phase but $\Omega_\ell = \sqrt{16/15}\,\Omega_{\ell B}$ for the B-phase). We introduce the natural commutation relations

$$[\sigma_w(\underset{\sim}{x},t), \theta(\underset{\sim}{x}',t)] = -i\delta(\underset{\sim}{x}-\underset{\sim}{x}'). \quad (3.45)$$

Heisenberg's equations are

$$\theta_t = +\left(\gamma^2\chi^{-1}\sigma_w - \gamma B_o\right) \quad (3.46)$$

$$\sigma_{w,t} = -\left((\delta H_D/\delta\theta) - \tfrac{1}{2}\chi\gamma^{-2}\bar{c}^2\nabla\cdot\underset{\approx}{\rho}\,\nabla\theta\right) \quad (3.47)$$

and

$$\theta_{tt} = \tfrac{1}{2}\bar{c}^2\nabla\cdot\underset{\approx}{\rho}\cdot\nabla\theta - \gamma^2\chi^{-1}(\delta H_D/\delta\theta) \tag{3.48a}$$

with

$$[\theta,\theta_t] = \gamma^2\chi^{-1}i\delta(\underset{\sim}{x}-\underset{\sim}{x}'). \tag{3.48b}$$

Results for the A-phase are

$$H_D(\theta) = -\frac{3}{5}g_D(T)\cos^2\theta \tag{3.49}$$

and for the B-phase

$$\begin{aligned} H_D(\theta) &= \frac{1}{5}g_D(T)\left[\{\mathrm{Tr}\,R(\theta)\}^2 + \mathrm{Tr}\,R^2(\theta)\right] + \text{constant} \\ &= \frac{2}{5}g_D(T)\left[\mathrm{Tr}\,R(\theta) + \mathrm{Tr}\,R^2(\theta)\right] + \text{constant} \\ &= \frac{4}{5}g_D(T)\ (\cos\theta + \cos 2\theta) + \text{constant}. \end{aligned} \tag{3.50}$$

In both cases, $g_D(T)$ is a temperature-dependent parameter which will be absorbed in Ω_ℓ. The second form in (3.50) illustrates the double spin description which underlies the double sine-Gordon equation [3], [49], [52], [53], [54].

We consider plane waves propagating up B_0 the direction of y. It is convenient to relabel this direction as x. For the A-phase we find

$$\theta_{tt} - \bar{c}_A^2\theta_{xx} = -\Omega_{\ell A}^2\sin\theta\cos\theta \tag{3.51}$$

$(\Omega_{\ell A}^2 = (6/5)g_D(T)\gamma^2\chi^{-1})$. For the B-phase

$$\theta_{tt} - \bar{c}_B^2\theta_{xx} = \frac{4}{15}\Omega_{\ell B}^2\{\sin\theta + 2\sin 2\theta\} \tag{3.52}$$

$(\Omega_{\ell B}^2 = 3g_D(T)\gamma^2\chi^{-1})$. We set $u=-2\theta$ (A-phase) and $u=2\theta$ (B-phase). We use t for $\Omega_\ell t$. Then we easily reach the <u>operator</u> equations

$$u_{xx} - u_{tt} = \sin u \tag{3.53}$$

(A-phase) and

$$u_{xx} - u_{tt} = -(\sin u + \tfrac{1}{2}\sin\tfrac{1}{2}u) \tag{3.54}$$

(B-phase) with the commutation relations

$$[u, u_t] = i\gamma_0 \delta(\underset{\sim}{x}-\underset{\sim}{x}') \qquad (3.55)$$

in both cases.

We note that the gap matrix defined by (3.36) is a c-number; hence, θ is a c-number. Nevertheless, it decorrelates through the Hartree approximation (3.35), the quantized problem (3.34). It therefore seems worth investigating the quantized forms of both Equations (3.53) and (3.54), and this we do in Section 4. Here we look at the c-number forms. The essentially new case is the operator double s-G (3.54) which in c-number form is exactly the double s-G (3.33).

From the Hamiltonian density for this equation

$$\mathcal{H} = \gamma_0^{-1}\left\{\tfrac{1}{2}u_x^{\,2} + \tfrac{1}{2}u_t^{\,2} + 2(\cos\tfrac{1}{2}u + \tfrac{1}{4})^2\right\}; \qquad (3.56)$$

(note again that $\gamma_0 \equiv \Omega_\chi \Omega_\ell^{-1}$ is now the coupling constant and $\gamma \equiv e/m_p c$) the dipole interaction energy minimum occurs at $u = \delta \equiv 2\cos^{-1}(-\frac{1}{4})$ and at $4\pi-\delta$. There are two solitary wave kinks: one, the 2δ-kink, rotates u from $-\delta(\text{mod } 4\pi)$ to $+\delta$; the other, the $4\pi-2\delta$, rotates u from $+\delta$ to $4\pi-\delta$. These kinks can only bump (since $2\delta \neq 4\pi-2\delta$). Computer results confirm this prediction [52], [53], [54].

The analytical forms for the kinks prove to be

$$\begin{aligned} u &= 2\pi + 4\tan^{-1}\left\{\sqrt{\frac{3}{5}}\tanh\tfrac{1}{2}\theta\right\} \\ u &= 4\tan^{-1}\left\{\sqrt{\frac{5}{3}}\tanh\tfrac{1}{2}\theta\right\} \qquad (3.57) \\ \theta &= \kappa(x-Vt), \quad \kappa = \sqrt{\frac{15}{16}}(1-V^2)^{-\frac{1}{2}}. \end{aligned}$$

The form for the $(4\pi-2\delta)$-kink is given first. The energies are

$$\int_{-\infty}^{\infty}\mathcal{H}dx = \gamma_0^{-1}\, 8(1-V^2)^{-\frac{1}{2}}\left\{\sqrt{\frac{15}{16}} + \frac{1}{4}\left\{\tan^{-1}\frac{1}{\sqrt{15}} \pm \frac{\pi}{2}\right\}\right\} \qquad (3.58)$$

and the "masses" are 5.1097 γ_0^{-1} and 11.3929 γ_0^{-1} units.

Notice now that a 2δ-kink-antikink pair takes u from $u = -\delta$ (the boundary condition) to $+\delta$ to $-\delta$. Also a $-(4\ -2\delta)$-antikink-kink pair takes $-\delta$ to $\delta-4\pi (= \delta\ (\text{mod } 4\pi))$ to $-\delta$. It is therefore possible for the 2δ-kink-antikink pair to collide and re-emerge as the $-(4\pi-2\delta)$-antikink-kink pair. Since the 2δ has rest mass $11.3929\,\gamma_0^{-1}$ units, it can do this without additional kinetic energy. Computer results [52], [53], [54] show that this conversion actually takes place.

On the other hand, for the boundary conditions $u \to +\delta$ for $|x| \to \infty$, the $(4\pi-2\delta)$-kink-antikink pair has an energy threshold for conversion to the -2δ-antikink-kink pair (the critical velocity is 0.8938). Figure 1a shows that just above threshold it still finally bumps because of the loss of energy by radiation (the velocity is 0.91) but notice the trough where the -2δ-antikink-kink pair is formed and is <u>then</u> returned to a (4 -2δ)-kink-antikink pair by the continuing loss of energy by radiation. Figure 1b shows part of the very long trough created when the velocity is increased to 0.925. The reader is referred to References 52, 53 and 54 cited already for other computer results both above and below threshold. Reference 53 shows in particular the case when the velocity is 0.92—a case which compares with that shown in Figure 1b.

This is not the place to discuss the prospects for experiments based on these various results. The point has been to show the rather remarkable properties of the double s-G equation (3.33). As indicated near (3.33), the double s-G also occurs in the theory of SIT: the resonant transition must be degenerate and specifically of Q(2) symmetry [49]. In the case of the attenuator, the negative sign on the right side of (3.33) is replaced by a positive one although either sign can arise in principle [49]. For the +ve sign the solitary wave is a 4π-kink. Since the electric field ϵ is obtained by differentiation of the field variable [taken with respect to time in laboratory coordinates—compare Section 3, Equations (3.24)-(3.26)], this is a double-humped 4π-pulse. It proves to be a <u>bound pair</u> of 2π-sech pulses each of the form (3.28) with a characteristic separation between them [49]. The bound pair has an internal degree of freedom and can wobble about this equilibrium separation [3], [49], [54]. This wobbler has been observed in SIT experiments on hyperfine degenerate $F=2 \to F'=2$ D_1 transitions in sodium vapor [55].

In the case of Q(J) degenerate transitions the J-tple sine-Gordon equation arises. For example, for $J=3$

$$u_{xx} - u_{tt} = \pm\left(\sin u + \tfrac{1}{3}\sin\tfrac{1}{3}u + \tfrac{2}{3}\sin\tfrac{2}{3}u\right). \qquad (3.59)$$

In the case of the +ve sign this equation has 6π triple-peaked wobbling solutions and similar properties seem to be displayed by all the multiple s-Gs. Phase plane analysis of (3.59) indicates

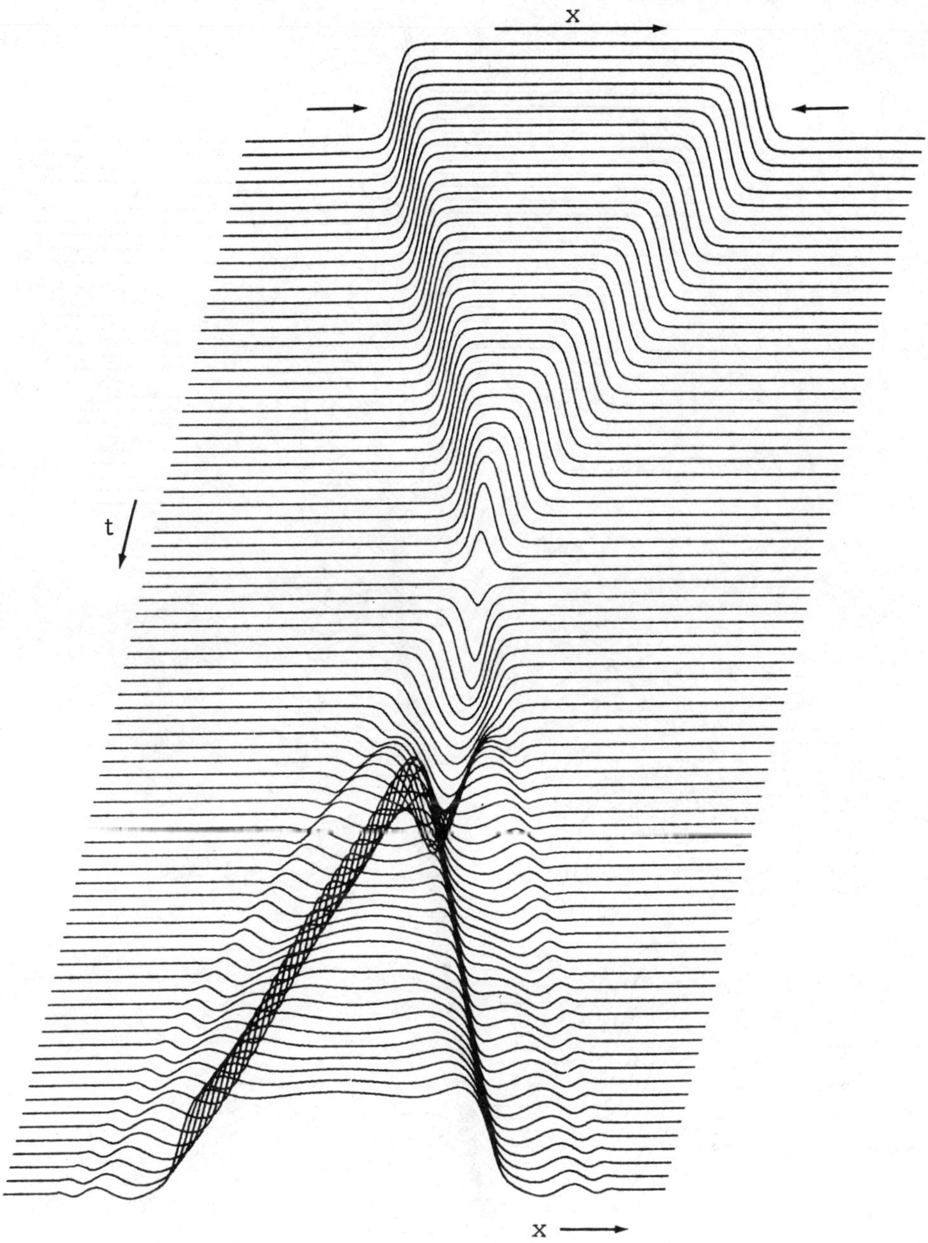

Figure 1a. Collision of the $(4\pi-2\delta)$-kink-antikink pair taken in the center-of-mass frame with velocities 0.91, greater than the threshold velocities. The pair converts to the 2δ-antikink-kink pair but reverts to the $(4\pi-2\delta)$-pair after losing energy by radiation.

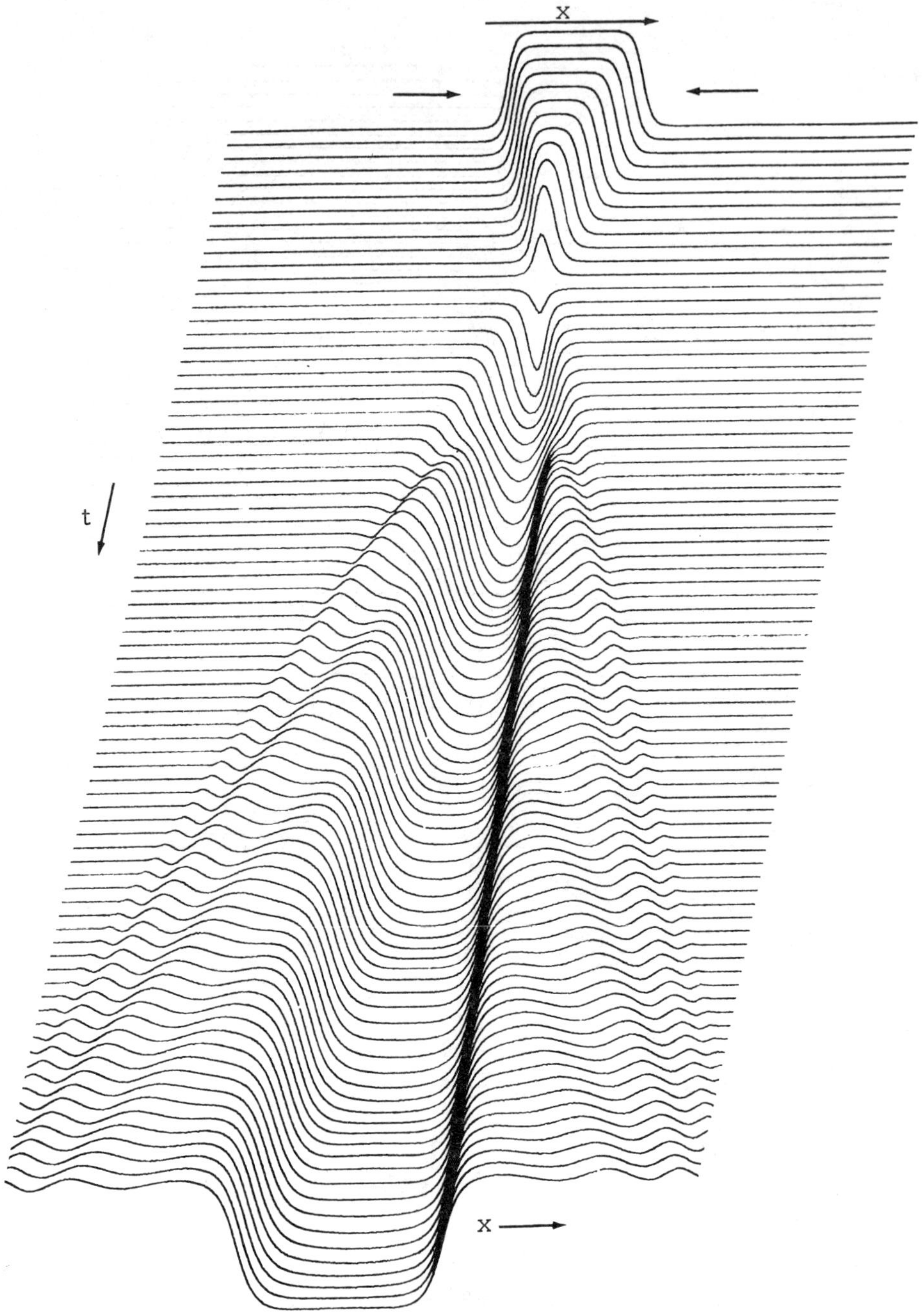

Figure 1b. Collision of the $(4\pi-2\delta)$-kink-antikink pair with velocity 0.925. The pair is expected to revert eventually to the $(4\pi-2\delta)$ pair.

the existence of eight different solitary waves for ϵ which are variously triple, double, or single humped; all the multiple-peaked pulses wobble. Only the four solitary wave solutions of the double s-G are analytically particularly simple: these include the 4π, the 2δ and the $4\pi-2\delta$ as well as a fourth 0π solution satisfying the boundary condition $u \to 2\pi$ as $|x| \to \infty$ [49]. The 4π wobbler solution of the double s-G has been described successfully by perturbation theory about the two-kink solution (1.38) of the simple s-G [56]. Dr. Mason in his lecture [57] describes perturbation theory for the 0π solution of the double s-G taken about the breather solution (1.43) of the simple s-G: this 0π solution can either oscillate or break up. Perturbation theory for the kinks (3.57) has not yet been done.

The radiative collision of kink-antikink pair solutions of (3.33) suggests that the solutions (3.57) are not solitons. Numerical results for collisions of 4π wobblers show that wobble can be transferred in collision. The evidence is that the solitary wave solutions of the double s-G with either sign are not solitons. These results are consistent with the conclusions from Theorems 2 and 3. No inverse scattering method has been found for the multiple s-Gs and perturbation theory is likely to remain the only mode of analysis. It is therefore remarkable that Dr. Burt in his lectures [58] shows that the double s-G has soliton solutions in more than one-space dimension ($q+1$ dimensions, $q \geqq 2$). Theorem 1 shows there are no BTs there.

4. QUANTIZED SOLITONS

I restrict attention to quantization of the s-G as in (3.53) and apply the results to the two cases of SIT and of ^{3}He A spin waves discussed in Section 3. The discussion is necessarily brief and more details appear in Reference 24. The canonical formalism we need is treated in References 21. I add a brief comment on the quantized double s-G at the end of this part.

From (1.19a) we can take

$$H(p,q) = \frac{m^2}{4\gamma_0}\int_{-\infty}^{\infty}\left\{2-\cos\left(-2\int_{-\infty}^{x} q(x')dx'\right)-\cos(2\gamma_0 p)\right\}dx \qquad (4.1)$$

(γ_0 is the coupling constant and I have introduced a 'meson' mass m). From the momentum density $\frac{1}{2}u_x^2$ in (1.17), we form

$$P(p,q) = \frac{1}{2\gamma_o}\int_{-\infty}^{\infty}\left(q^2 + \gamma_o^2 p_x^2\right)dx. \tag{4.2}$$

As in (1.32a) we choose $r=-q$ in the scattering problem (1.33). Since [cf. (1.19b)] $q=-\frac{1}{2}u_x$ and is real $r=-q=-q^*$. This symmetry requires that the bound state eigenvalues ζ of (1.33) in the upper half ζ-plane arise at $\zeta=i\eta\,(\eta>0)$ or in pairs at ζ and $-\zeta^*$. The scattering problem (1.33) can be solved for the initial data $q(x,0)$ on q. One obtains [19], [24] the so-called scattering data $\{\zeta_i, k_j, a(\xi), w(\xi)\}$, $\xi\in R$; $a^{-1}(\xi)$ and $w(\xi)=b(\xi)a^{-1}(\xi)$ are transmission and reflection coefficients associated with the eigenvalue ξ of the continuous spectrum; ζ_j is a bound state eigenvalue and $b(\zeta_j)=ik_j\,da/d\zeta\big|_{\zeta=\zeta_j}$ defines k_j. The inverse scattering method follows the time evolution of the scattering data through (1.34) and then solves inversely for $q(x,t)$ for any $t>0$.

The transform of q to scattering data is a canonical transformation [21], [24]. The inverse transform is also a canonical transformation. A semiclassical quantization procedure finds canonical coordinates in terms of the scattering data. The Poisson brackets are read as commutators and because of the invariance of the Poisson brackets under canonical transformations, this quantization is equivalent to quantizing the Poisson brackets in terms of $q(x,t)$.

In terms of scattering data (4.1) and (4.2) become [24]

$$P = \frac{2i}{\gamma_o}\left\{-2\sum_{j=1}^{K}(\zeta_j-\zeta_j^*) - 2i\sum_{j=1}^{M}\eta_j - i\gamma_o\int_{-\infty}^{\infty}d\xi\,\xi\,P(\xi)\right\}$$

$$H = \frac{-m^2 i}{2\gamma_o}\left\{-2\sum_{j=1}^{K}\left[\zeta_j^{-1}-\zeta_j^{*-1}\right] + 2i\sum_{j=1}^{N}\eta_j^{-1}\right.$$

$$\left. + i\gamma_o\int_{-\infty}^{\infty}d\xi\,\xi^{-1}P(\xi)\right\}. \tag{4.3}$$

These are expressible in the forms

$$P = \sum_{j=1}^{K}\hat{p}_j + \sum_{j=1}^{M}p_j + \int_{-\infty}^{\infty}d\xi\,p(\xi)\,P(\xi)$$

$$H = \sum_{j=1}^{K}\hat{h}_j + \sum_{j=1}^{M}h_j + \int_{-\infty}^{\infty}d\xi\,h(\xi)\,P(\xi)$$

with ($\theta_j = \arg \zeta_j$)

$$h_j p_j = 4m^2\gamma_o^{-2}, \quad \hat{h}_j\hat{p}_j = 16m^2\gamma_o^{-2}\sin^2\theta_j$$
$$h(\xi)p(\xi) = m^2. \tag{4.4}$$

All these dispersion relations become those for the linearized s-G $u_{xt} = u$ by appropriate scaling. Notice the separate contributions of M 2π kinks, K breathers, and of the continuous spectrum associated with the meson field.

Define

$$h(\xi) = \tfrac{1}{2}\left\{h'(\xi) + p'(\xi)\right\}$$

$$p(\xi) = \tfrac{1}{2}\left\{h'(\xi) - p'(\xi)\right\}, \text{ etc.}$$

Set $m' = 2m$, drop primes and find

$$h(\xi) = \left\{m^2 + p^2(\xi)\right\}^{\frac{1}{2}}, \quad h_j = \left\{4m^2\gamma_o^{-2} + p_j^2\right\}^{\frac{1}{2}}$$

$$\hat{h}_j = \left\{16m^2\gamma_o^{-2}\sin^2\theta_j + \hat{p}_j^2\right\}^{\frac{1}{2}} \tag{4.5}$$

A check through the Poisson brackets [24] shows $\gamma_o \to \gamma_o' = 4\gamma_o$ and $2m\gamma_o^{-1} \to 8m\gamma_o'^{-1}$ (called then $8m\gamma_o^{-1}$). Thus now

$$H = \sum_{j=1}^{K}\left[256m^2\gamma_o^{-2}\sin^2\theta_j + \hat{p}_j^2\right]^{\frac{1}{2}} + \sum_{j=1}^{M}\left[64m^2\gamma_o^{-2} + p_j^2\right]^{\frac{1}{2}}$$
$$+ \int_{-\infty}^{\infty}\left[m^2 + p^2(\xi)\right]^2 P(\xi)d\xi, \tag{4.6}$$

the result obtained by Fadeev [59]. The spectra have thresholds at m, $8m\gamma_o^{-1}$ and $16m\gamma_o^{-1}$ and $16m\gamma_o^{-1}\sin\theta_j$: the lowest possible value of the last is zero and the breather thresholds form a continuous band with energies between 0 and $16m\gamma_o^{-1}$ when the breather breaks up into a kink-antikink pair.

The coordinates q_j canonical to the p_j are found through the canonical transformation to scattering data. The system is then quantized by setting $[q_j, p_k] = i\delta_{jk}$, etc. Through (4.4), the phase space is R^2 for all except the internal coordinates of the breathers. These coordinates are ($\gamma_o^{-1}\theta_j$, $4\arg b_j$) and this internal phase space is compact ($0 < \theta_j < (\pi/2)$, $0 < 4\arg b_j < 8\pi$). The internal momenta therefore have discrete eigenvalues. We can drop the label j of the particular breather. The discrete eigenvalues now called θ_n of θ are

$$\theta_n \sim n(\pi/2N), \quad N \sim 8\pi\gamma_o^{-1}. \tag{4.7}$$

The number N is the number of discrete eigenvalues and is equal to the phase space volume $4\pi^2\gamma_o^{-1}$ divided by 2π (for $\hbar = 1$). Then, as $\gamma_o \to 4\gamma_o$, $N = 8\pi\gamma_o^{-1}$. The breather masses are therefore quantized at values

$$M_n = \frac{16m}{\gamma_o} \sin \frac{n\gamma_o}{16}, \quad n = 1, \ldots, N. \tag{4.8}$$

This result was first found by Fadeev [60] following a route through the canonical formalizm. Dashen et al. [61] found the same result through path integral techniques. There is a normal ordering problem associated with the canonical formalism and the result [60], [61] is a renormalization which everywhere replaces γ_o by $\gamma_o'' = \gamma_o[1 - (\gamma_o/8\pi)]^{-1}$.

Since the SIT problem is initially a quantized problem (the e.m. field is quantized), it is desirable to study the theory in that form. This has not yet been successfully done but in the nondegenerate case it is interesting to connect this quantized theory with that of the s-G [53]. There are problems associated with the collective commutation relations on which Kleinert's methods [50] may shed some light.

Ignoring these problems we find in the nondegenerate case [53], [54] a quantized s-G theory with γ_o'' (the observed coupling constant) $\sim 10^{-6}$. It follows that N is large ($\sim 10^7$) and

$$M_n = \frac{16m}{\gamma_o''} \sin \frac{n\gamma_o''}{16} \sim nm \tag{4.9}$$

for small $n = 1, 2, \ldots$. The energy spacing $m (= h^{-1}mc^2$ in frequency units) is [53, [54] $\tau_C \sim 10^9$ Hz: this is comparable with atomic radiative level shifts and so seems to be of the right order of magnitude. Until a theory of quantized collective operators can be derived for SIT [50], conclusions are speculative. But from the arguments of Reference 53 and the results of this Section 4, one finds that the SIT breather spectrum would be discretely quantized with frequency spacings $\sim 10^9$ Hz—a value, however, small enough to make the spectrum quasicontinuous and hard to distinguish from the continuous c-number spectrum. This may be no more than one way of saying that the c-number semiclassical approximation to SIT is very good and the usual "many photon" ideas are certainly applicable. The masses of the kinks $8m\gamma_o^{-1}$ are large (~ 10^{-33} gm) compared with m(~ 10^{-39} gm). They contain many "mesons" each carrying the mass of the field m: this mass is in some sense the mass of an equivalent photon contributing to the e.m. wave pulse traversing the resonant medium. Unfortunately, e.m. breather pulses have not been identified in SIT (antikink-kink pairs have been shown, however, to have enhanced transparency [3]) so that any quantized features, which ultimately must derive from the quantization of the e.m. field, cannot so far be determined experimentally.

The ^{3}He A-phase is very different: here γ_o'' (the observed value) proves to be 10^5! This is therefore a very strong quantum theory if it is quantized at all. It has been assumed that $\gamma_o'' << 1$ [60], [61]. However, Dashen et al. [61] give the condition $\gamma_o'' < 8\pi$ as a condition for one discrete breather state. The value of the bare coupling constant γ_o corresponding to $\gamma_o'' \sim 10^5$ is $\gamma_o = 8\pi - \epsilon$ where $\epsilon \sim 10^{-4}$. The value $\gamma_o'' \sim 10^5$ seems to mean that the number of quantized breather states is zero. The breathers therefore split into pairs of 2π kinks and the breather spectrum is eliminated. In principle, one could look for the zero energy threshold of the c-number breathers to test whether the A-phase spin waves are quantized or not: unfortunately the kink masses $8m\gamma_o''^{-1}$ are low (~ 1 Hz in frequency units). This is the frequency of orbital waves in the A-phase so that in the threshold region spin waves cannot be dissociated from orbital waves and the simple s-G theory is not obviously applicable. In Reference 52 we speculate on the spin wave spectrum in the B-phase. No c-number breather has yet been found for the double s-G. We have not yet been able to quantize this equation via any canonical formalism since no exact canonical transformation to scattering data is available for it. Perturbation theory [57] is likely to be the only applicable method but this has not yet been developed for this purpose.

ACKNOWLEDGMENTS

I am grateful to P. W. Kitchenside for Figures 1a and 1b, and to P. J. Caudrey, R. K. Dodd and other colleagues of the Manchester Group who have shared in much of the work here described.

REFERENCES

1. D. J. Korteweg and G. de Vries, Phil. Mag. 39, 422 (1895).
2. J. Rubinstein, J. Math. Phys. 11, 258 (1970).
3. R. K. Bullough, Solitons, in Interaction of Radiation with Condensed Matter, Vol. 1, IAEA-SMR-20/51, International Atomic Energy Agency, Vienna, 1977, pages 381-469.
4. A. Barone, F. Esposito, C. J. Magee, and A. C. Scott, Rivista del Nuovo Cimento 1, 227 (1971).
5. A. Luther, Phys. Rev. B 14, 2153 (1976).
6. A. Degasperis, these Proceedings.
7. G. B. Whitham, Linear and Nonlinear Waves, John Wiley & Sons, New York, 1974, Chapter 4.
8. M. D. Kruskal, these Proceedings.
9. R. Jackiw, Rev. Mod. Phys. 49, 681 (1977).
10. S. Coleman, these Proceedings. E. Corrigan and D. B. Fairlie, Phys. Lett. B 67, 69 (1977) establish a connection between a c/4 ϕ-four theory and an SU_2 Yang-Mills instanton.
11. H. Haken, Synergetics, in Phys. Bulletin, September 1977, page 413.
12. No finite temperature phase transition occurs in one dimension—a result due to Landau.
13. T. Schneider and E. Stoll, Phys. Rev. Lett. 35, 296 (1975). Also see T. Schneider, these Proceedings.
14. J. A. Krumhansl and J. R. Schrieffer, Phys. Rev. B 11, 3535 (1975).
15. R. F. Dashen, B. Hasslacher, and A. Neveu, Phys. Rev. D 10, 4130 (1974).
16. S. Coleman, Classical Lumps and Their Quantum Descendants, in Lectures at the 1975 International School of Subnuclear Physics, Ettore Majorana, 1975.
17. Like all such sweeping remarks, this one may need qualification. See V. E. Zakharov, The Inverse Scattering Method, in Solitons, R. K. Bullough and P. J. Caudrey (eds.), Springer Topics in Modern Physics, Springer-Verlag, Heidelberg, 1978, to appear.
18. R. Hirota, J. Phys. Soc. Japan 33, 1459 (1972); P. J. Caudrey, J. D. Gibbon, J. C. Eilbeck, and R. K. Bullough, Phys. Rev. Lett. 30, 237 (1973); P. J. Caudrey, J. C. Eilbeck and J. D. Gibbon, J. Inst. Maths. Applics. 14, 375 (1974)
19. M. J. Ablowitz, D. J. Kaup, A. C. Newell, and H. Segur, Phys. Rev. Lett. 31, 125 (1973); V. E. Zakharov and A. B.

Shabat, Zh. Eksp. Teor. Fiz. 61, 118 (1971) (Sov. Phys. J.E.T.P. 34, 62 (1972)).
M. J. Ablowitz, D. J. Kaup, A. C. Newell, and H. Segur, Phys. Rev. Lett. 30, 1262 (1973); M. J. Ablowitz, D. J. Kaup, A. C. Newell, and H. Segur, Studies in Applied Math. 53, 249 (1974).

20. J. D. Gibbon, P. J. Caudrey, R. K. Bullough, and J. C. Eilbeck, Lett. al Nuovo Cimento 8, 775 (1973).

21. V. E. Zakharov and L. D. Fadeev, Funkt. Anal. i Ego Prilozh. 5, 18 (1971); H. Flaschka and A. C. Newell, Integrable Systems of Nonlinear Evolution Equations, in Dynamical Systems. Theory and Applications, J. Moser (ed.), Springer Notes in Physics, Springer-Verlag, Heidelberg, 1975; R. K. Dodd and R. K. Bullough, The Generalised Marchenko Equation and the Canonical Structure of the AKNS-ZS Method. To appear.

22. R. K. Dodd and R. K. Bullough, Proc. Roy. Soc. A 352, 481 (1977).

23. Alternatively the set (1.17) is found by expanding log $a(\zeta)$ about infinity; the set (1.18) is found by expanding about $\zeta = 0$. The quantity $a(\zeta)$ is the reciprocal of the transmission coefficient associated with the continuous part of the spectrum in the inverse scattering method. See, e.g., Reference 24 and Section 4 below.

24. R. K. Bullough and R. K. Dodd, Solitons II. Mathematical Structures, in Synergetics, H. Haken (ed.), Springer-Verlag, Heidelberg, 1977, pages 104-119.

25. K. Pohlmeyer, Commun. Math. Phys. 46, 207 (1976); F. Lund, Phys. Rev. Lett. 38, 1175 (1977) and these Proceedings; A. Budazov and L. Tachtadjan, Dokl. Akad. Nauk. SSR, 1977.

26. R. K. Dodd and R. K. Bullough, Proc. Roy. Soc. London A 351, 499 (1976).

27. H. D. Wahlquist and F. B. Estabrook, Phys. Rev. Lett. 31, 1386 (1973); H. Chen, Phys. Rev. Lett. 33, 925 (1974).

28. P. D. Lax, Commun. Pure Appl. Math. 21, 467 (1968).

29. P. J. Caudrey, J. C. Eilbeck, and J. D. Gibbon, J. Inst. Maths. Applics. 14, 375 (1974); R. Hirota, Direct Methods in Soliton Theory, in Solitons, R. K. Bullough and P. J. Caudrey (eds.), Springer Topics in Modern Physics, Springer-Verlag, Heidelberg, 1978. To appear.

30. P. J. Caudrey, these Proceedings.

31. E. Fermi, J. R. Pasta, and S. M. Ulam, Studies of Nonlinear Problems, I, Los Alamos Rept. LA-1940, May 1955,

and Collected Works of E. Fermi, Vol. II, Univ. of Chicago Press, 1965, pages 978-88.
32. V. E. Zakharov and A. B. Shabat, Funkt. Anal. i Ego Prilozh. 8, 43 (1974).
33. J. Gibbons, S. G. Thornhill, M. J. Wardrop, and D. ter Haar, On the Theory of Langmuir Solitons, preprint Univ. of Oxford, Department of Theoretical Physics, Ref. 36/76 (1976).
34. V. E. Zakharov, Zh. Eksp. Teor. Fiz. 62, 1745 (1972) (Soviet Physics J.E.T.P. 35, 908 (1972).
35. H. Hasimoto, J. Fluid Mech. 51, 477 (1972).
36. M. Lakshamanan, Phys. Lett., preprint 1977. L. A. Taktadjan, communicated at the Warsaw Solitons Meeting, September 1977.
37. G. L. Lamb, Rev. Mod. Phys. 43, 99 (1971).
38. J. C. Eilbeck, P. J. Caudrey, J. D. Gibbon, and R. K. Bullough, J. Phys. A: Math. Nucl. Gen. 6, 1337 (1973).
39. B. B. Kadomtsev and V. I. Petviashvili, Dokl, Akad. Nauk. SSR, 192, 753 (1970).
40. H. C. Yuen and S. M. Lake, Phys. Fluids 18, 956 (1975).
41. J. D. Gibbon, P. J. Caudrey, R. K. Bullough, and J. C. Eilbeck, Lett. al Nuovo Cimento 8, 775 (1973).
42. H. M. Gibbs and R. E. Slusher, Phys. Rev. A 6, 2326 (1972).
43. T. A. Fulton and R. C. Dynes, Solid State Comm. 12, 57 (1973).
44. S. P. Novikov, Funkt. Anal. i. Ego Prilozh. 8, 54 (1974).
45. A. J. Leggett, Rev. Mod. Phys. 47, 331 (1975).
46. M. Cross, dissertation, Corpus Christi College, Cambridge, 1974.
47. A. J. Leggett, Ann. Phys. 85, 11 (1974).
48. W. F. Brinkman, J. W. Serens, and P. W. Anderson, Phys. Rev. A 10, 2386 (1974).
49. S. Duckworth, R. K. Bullough, P. J. Caudrey, and J. D. Gibbon, Phys. Lett. 57 A, 19 (1976).
50. H. Kleinert, these Proceedings.
51. K. Maki and T. Tsuneto, Phys. Rev. B 11, 2539 (1975).
52. R. K. Bullough and P. J. Caudrey, Bumping Spin Waves in the B-phase of Liquid ^{3}He, in J. Phys. C: Solid State Physics. To appear in 1978. For the double s-G in the B-phase, see also work by K. Maki and P. Kumar cited there.
53. R. K. Bullough and P. J. Caudrey, Proc. Symposium on Nonlinear Evolution Equations Solvable by the Inverse

Spectral Transform, Rome, June 1977. Proceedings, F. Calogero (ed.), to be published by Pitman in 1978.

54. R. K. Bullough and P. J. Caudrey, Optical Solitons and Their Spin Wave Analogues in ^{3}He, in Proc. Fourth Rochester Conference on Coherence and Quantum Optics, L. Mandel and E. Wolf (eds.), Plenum, New York. To be published in 1978.

55. R. K. Bullough, P. J. Caudrey, J. D. Gibbon, S. Duckworth, H. M. Gibbs, B. Bölger, and L. Baede, Optics Commun. 18, 200 (1976).

56. A. C. Newell, J. Math. Phys. 18, 922 (1977).

57. A. L. Mason, these Proceedings.

58. P. B. Burt, Exact, Multiple Soliton Solutions of the Double Sine-Gordon Equation, preprint 1977 and these Proceedings.

59. L. A. Taktadjan and L. D. Fadeev, Teor. Mat. Fiz. 21, 160 (1974).

60. V. E. Korepin and L. D. Fadeev, Teor. Mat. Fiz. 25, 47 (1975).

61. R. F. Dashen, B. Hasslacher, and A. Neveu, Phys. Rev. D11, 3424 (1975).

SOLITONS AND GEOMETRY†‡

Fernando Lund

Institute for Advanced Study
Princeton, New Jersey

1. INTRODUCTION

The biggest advance in the theory of nonlinear equations in recent years has been the invention of the so-called inverse scattering method by Gardner et al. [1] in 1967, who used it to solve the initial value problem for the Korteweg-de Vries [2] equation. The method was systematized by Lax [3] in 1968 and it was given definitive respectability by Zacharov and Shabat [4] in 1971 when they solved the Schrödinger equation with a cubic nonlinearity. By now, a large number of equations can be solved this way [5]. This method, in the language of Lax [3], is the following: Given an evolution equation, say, of the form

$$\frac{\partial u}{\partial t} = K(u), \tag{1.1}$$

one finds firstly operators L and B such that Equation (1.1) is equivalent to the relation

$$\frac{\partial L}{\partial t} = i[L,B], \tag{1.2}$$

In this case the spectrum of L does not depend on time and one says that the flow of L is isospectral. Secondly, one tries to associate a scattering problem with the operator L (which acts,

† Lectures given at NATO Advanced Study Institute on Nonlinear Equations in Physics and Mathematics, Istanbul, August 1977.
‡ Supported in part by the NSF Grant No. GP-40768X.

A. O. Barut (ed.), Nonlinear Equations in Physics and Mathematics, 143–175.

for instance, on a Hilbert space). This typically restricts the type of asymptotic behavior allowed on u. Finally, the direct and inverse scattering problems associated with L must be solved. When these three steps can be taken, the nonlinear initial value problem for Equation (1.1) can be reduced to a series of linear problems, and it is said to be solved. Although explicit solutions can be found only in special cases, the equation is considered solved as standard linear techniques may now be used.

In this lecture I want to address the first of the three problems referred to in the preceding paragraph: Given a nonlinear evolution equation, how is one to go about finding operators L and B that will allow writing it in the form (1.2)? This problem has received close attention lately and progress has been made in a series of articles by using Cartan's exterior form calculus in the manner pioneered by Estabrook and Wahlquist [6], in the study of the motion of helical curves by Lamb [7], and in the study of surfaces embedded in spaces of higher dimension [8]. It is the last approach that will be described here. They all have in common the fact that they rely on ideas of differential geometry, and they add one more surprising ingredient to the Cauchy problem for nonlinear equations where scattering theory, a problem of analysis, also came unexpectedly to the rescue. One is tempted to think that this indicates the existence of a deep underlying structure whose properties are only beginning to be understood.

The first time a nonlinear evolution equation appeared in geometry seems to have been at the beginning of this century in the description of surfaces of constant curvature through the sine-Gordon equation [9] (then called "the fundamental equation of differential geometry"). Already then this equation was in the form (1.2), as it is the integrability condition for the <u>linear</u> equations (of Gauss-Weingarten) obeyed by the tangent frame to the surface. What geometers at the time did not have were two things: one, an extra parameter interpretable as the eigenvalue of the operator L, and two, the scattering theory that was developed with the impulse of quantum mechanics. On the other hand, it was realized only recently [8] that the linear system used to solve sine-Gordon [10] was a rewriting of the equations of surface theory. One sees then, that the embedding of surfaces in higher dimensional manifolds provides a framework in which nonlinear equations and linear ones appear intimately related. This is the main idea that will be exploited in the present work.

In Section 2 the necessary concepts from geometry will be developed [11]. It will be shown that under certain conditions the embedding equations describe a Lagrangian field theory with a naturally associated linear problem of which sine-Gordon theory is a particular case. This theory arises in a number of physical situations which will be described in Section 3. Section 4 contains the solution of the initial value problem using the inverse scattering method.

2. GEOMETRY

In this section the necessary results from geometry will be derived. They are all standard but in general unfamiliar to physicists.

Consider then an n-dimensional Riemannian manifold V_n embedded in an (n+1)-dimensional Euclidean space E. It is possible to consider the more general case of the surrounding space having an arbitrary dimension larger than n and also an arbitrary geometry, not necessarily a Euclidean one. However, it is sufficient to take this special case to understand what is going on.

The space V_n has coordinates y^μ ($\mu = 1, \ldots, n$) and a metric

$$ds^2 = g_{\mu\nu} dy^\mu dy^\nu . \qquad (2.1)$$

The space E has coordinates x^i ($i = 1, \ldots, n+1$) and a metric

$$d\bar{s}^2 = dx^i dx^i .$$

The embedding of V_n in E is said to be isometric if V_n can be defined through the relations

$$x^i = x^i (y^1, \ldots, y^n) \qquad (2.2)$$

such that the metric $g_{\mu\nu}$ satisfies

$$g_{\mu\nu} = \frac{\partial x^i}{\partial y^\mu} \frac{\partial x^i}{\partial y^\nu} . \qquad (2.3)$$

("The metric of V_n is induced by that of E".) This already shows that not every V_n can be isometrically embedded in E, as (2.3) represents a set of $\frac{1}{2}n(n+1)$ equations in (n+1) unknowns (the x^i) which overdetermines the system except for $n = 2$, and the fundamental question in this context is to identify those conditions on V_n that will insure its embeddability in E.

Call now

$$X_\mu \equiv \frac{\partial \vec{x}}{\partial y^\mu}, \tag{2.4}$$

the vectors tangent to V_n. They are a family of n linearly independent vectors. A basis in E (the "moving frame") is completed with the unit normal to V_n,

$$\hat{X}_{n+1}, \quad ||\hat{X}_{n+1}|| = 1. \tag{2.5}$$

Having a basis, any vector may be expressed as a linear combination of its elements. In particular, there exist coefficients $\Gamma^\mu{}_{\nu\lambda}$, $L_{\mu\nu}$, $a_\mu{}^\nu$ such that

$$\frac{\partial X_\mu}{\partial y^\nu} = \Gamma^\lambda{}_{\mu\nu} X_\lambda + L_{\mu\nu} \hat{X}_{n+1} \tag{2.6}$$

$$\frac{\partial \hat{X}_{n+1}}{\partial y^\nu} = a_\nu{}^\mu X_\mu . \tag{2.7}$$

These are the equations of Gauss-Weingarten. One has, by definition, that

$$X_\lambda \cdot \hat{X}_{n+1} = 0, \quad \lambda = 1,\ldots,n. \tag{2.8}$$

Consequently, taking the scalar product of (2.6) with $\hat{X}_{n+1}$ gives

$$L_{\mu\nu} = \frac{\partial X_\mu}{\partial y^\nu} \cdot \hat{X}_{n+1} = -X_\mu \cdot \frac{\hat{X}_{n+1}}{\partial y^\nu}. \tag{2.9}$$

These quantities $L_{\mu\nu}$ are the components of a symmetric, rank-2 tensor called the extrinsic curvature or second fundamental form of V_n—the first fundamental form being the metric tensor $g_{\mu\nu}$. This tensor describes properties of V_n relative to the ambient space E, while its intrinsic properties (i.e., independent of E) are described by $g_{\mu\nu}$. For example, a cylinder and a plane have the same metric but their extrinsic curvatures differ, reflecting the fact that all normals to a plane are parallel while those of a cylinder are not.

Taking the scalar product of (2.7) with X_ρ gives

$$X_\mu \cdot X_\rho a_\nu{}^\mu = \frac{\partial \hat{X}_{n+1}}{\partial y^\nu} \cdot X_\rho . \tag{2.10}$$

Using (2.3), (2.4) and (2.9), this implies

$$a^{\mu}_{\ \nu} = -g^{\mu\lambda}L_{\lambda\nu}. \tag{2.11}$$

Finally, taking the scalar product of (2.6) with X_ρ yields

$$g_{\lambda\rho}\Gamma^{\lambda}_{\ \mu\nu} = \frac{\partial X_\mu}{\partial y^\nu}\cdot X_\rho .$$

Using (2.3), (2.4), it is immediate to show that

$$\Gamma^{\lambda}_{\ \mu\nu} = \tfrac{1}{2}g^{\lambda\rho}\left(-g_{\mu\nu,\rho} + g_{\nu\rho,\mu} + g_{\rho\mu,\nu}\right). \tag{2.12}$$

They are the Christoffel symbols. The Gauss-Weingarten equations are then a set of linear equations for the tangent and normal vector X_μ, $\hat{X}_{n+1}$ whose coefficients depend on the metric $g_{\mu\nu}$ and extrinsic curvature $L_{\mu\nu}$. This system consists of $(\frac{1}{2}n^2(n^2+1) + n(n+1)-1)$ independent equations for $(n+1)^2$ scalar unknowns and is overdetermined so it has a solution only if integrability conditions are satisfied; crossed second derivatives must be equal in pairs:

$$\frac{\partial^2 X}{\partial y^\mu \partial y^\nu} = \frac{\partial^2 X}{\partial y^\nu \partial y^\mu}. \tag{2.13}$$

Imposing these conditions on (2.6)-(2.7) leads to the following restrictions on the coefficients $g_{\mu\nu}$ and $L_{\mu\nu}$:

$$R_{\mu\nu\lambda\rho} = L_{\mu\lambda}L_{\nu\rho} - L_{\mu\rho}L_{\nu\lambda} \tag{2.14}$$

$$L_{\mu\nu;\lambda} = L_{\mu\lambda;\nu}, \tag{2.15}$$

(Gauss-Codazzi equations)

where

$$R^{\mu}_{\ \nu\lambda\rho} = \frac{\partial\Gamma^{\mu}_{\nu\rho}}{\partial y^\lambda} - \frac{\partial\Gamma^{\mu}_{\nu\lambda}}{\partial y^\rho} + \Gamma^{\mu}_{\ \sigma\lambda}\Gamma^{\sigma}_{\ \nu\rho} - \Gamma^{\mu}_{\ \sigma\rho}\Gamma^{\sigma}_{\ \nu\lambda} \tag{2.16}$$

is the Riemann tensor of V_n and the semicolon denotes covariant differentiation on V_n.

The equations of Gauss-Codazzi (2.14)-(2.15) are the conditions that must be satisfied by V_n for it to be embeddable in E. More precisely, the result of geometry is the following: Given two tensors $g_{\mu\nu}$ and $L_{\mu\nu}$, the equations of Gauss-Weingarten (2.6)-(2.7) have a unique solution for X_μ, $\hat{X}_{n+1}$ once their initial value has been prescribed at a point if and only if the equations of Gauss-Codazzi (2.14)-(2.15) are satisfied. Moreover, this solution determines uniquely (up to rigid motions) a surface $\vec{x}(y^\mu)$ such that $\partial\vec{x}/\partial y^\mu = X_\mu$, X_{n+1} is the unit normal, $X_\mu \cdot X_\nu = g_{\mu\nu}$ and $\partial X_\mu/\partial y^\nu \cdot X_{n+1} = L_{\mu\nu}$.

Here we are less interested in this existence theorem than in the fact that the embedding of surfaces gives nonlinear equations as integrability conditions of linear ones. To further explore this situation we take the simplest possible case, $n=2$.

Calling $y^1 = \sigma$, $y^2 = \tau$, the metric induced on a surface by a three-dimensional Euclidean space is

$$ds^2 = \left(\frac{\partial\vec{x}}{\partial\sigma}\right)^2 d\sigma^2 + 2\left(\frac{\partial\vec{x}}{\partial\sigma}\cdot\frac{\partial\vec{x}}{\partial\tau}\right)d\sigma d\tau + \left(\frac{\partial\vec{x}}{\partial\tau}\right)^2 d\tau^2, \tag{2.17}$$

and it is always possible to choose coordinates σ, τ such that

$$\frac{\partial\vec{x}}{\partial\sigma}\cdot\frac{\partial\vec{x}}{\partial\tau} = 0 \tag{2.18}$$

$$\left(\frac{\partial\vec{x}}{\partial\sigma}\right)^2 + \left(\frac{\partial\vec{x}}{\partial\tau}\right)^2 = 1. \tag{2.19}$$

A reason to make this choice, apart from the fact that it gives good results, will be given in Section 3. There exists then a variable $\theta(\sigma,\tau)$ such that

$$\left(\frac{\partial\vec{x}}{\partial\sigma}\right)^2 = \cos^2\theta, \quad \left(\frac{\partial\vec{x}}{\partial\tau}\right)^2 = \sin^2\theta \tag{2.20}$$

and the line element (2.17) reads

$$ds^2 = \cos^2\theta d\sigma^2 + \sin^2\theta d\tau^2. \tag{2.21}$$

A second condition is imposed on the extrinsic curvature:

$$L_{22} - L_{11} = 2\sin\theta\cos\theta. \tag{2.22}$$

This, as opposed to the previous condition on the metric, does restrict the type of surface under consideration. Again, a reason for this restriction will be given in Section 3.

Introducing the unit tangents

$$\hat{X}_1 = \frac{1}{\cos\theta}\frac{\partial\vec{x}}{\partial\sigma}, \quad \hat{X}_2 = \frac{1}{\sin\theta}\frac{\partial\vec{x}}{\partial\tau} \tag{2.23}$$

and substituting into (2.6)-(2.7) using the restrictions (2.21)-(2.22) gives the following equations of Gauss-Weingarten

$$\frac{\partial}{\partial\sigma}\begin{bmatrix}\hat{X}_1\\ \hat{X}_2\\ \hat{X}_3\end{bmatrix} = \begin{bmatrix} 0 & \frac{\partial\theta}{\partial\tau} & \frac{L_{11}}{\cos\theta} \\ -\frac{\partial\theta}{\partial\tau} & 0 & \frac{L_{12}}{\sin\theta} \\ -\frac{L_{11}}{\cos\theta} & -\frac{L_{12}}{\sin\theta} & 0 \end{bmatrix}\begin{bmatrix}\hat{X}_1\\ \hat{X}_2\\ \hat{X}_3\end{bmatrix} \tag{2.24}$$

$$\frac{\partial}{\partial\tau}\begin{bmatrix}\hat{X}_1\\ \hat{X}_2\\ \hat{X}_3\end{bmatrix} = \begin{bmatrix} 0 & \frac{\partial\theta}{\partial\sigma} & \frac{L_{12}}{\cos\theta} \\ -\frac{\partial\theta}{\partial\sigma} & 0 & \frac{L_{22}}{\sin\theta} \\ -\frac{L_{12}}{\cos\theta} & -\frac{L_{22}}{\sin\theta} & 0 \end{bmatrix}\begin{bmatrix}\hat{X}_1\\ \hat{X}_2\\ \hat{X}_3\end{bmatrix} . \tag{2.25}$$

Similarly, substituting into the equations of Gauss-Codazzi (2.14)-(2.15) gives

$$\frac{\partial B}{\partial\tau} - \frac{L_{12}}{\partial\sigma} + \frac{\partial\theta}{\partial\tau}\,\frac{B}{\sin\theta\cos\theta} - \frac{\partial\theta}{\partial\sigma}\,\frac{L_{12}}{\sin\theta\cos} = 0 \tag{2.26}$$

$$\frac{\partial B}{\partial\sigma} - \frac{L_{12}}{\partial\tau} + \frac{\partial\theta}{\partial\tau}\,\frac{L_{12}}{\sin\theta\cos\theta} - \frac{\partial\theta}{\partial\sigma}\,\frac{B}{\sin\theta\cos\theta} = 0 \tag{2.27}$$

$$B^2 - (L_{12})^2 = \sin^2\theta\cos^2\theta + \sin\theta\cos\theta\,\frac{\partial^2\theta}{\partial\tau^2} - \frac{\partial^2\theta}{\partial\sigma^2} \tag{2.28}$$

where

$$B \equiv \tfrac{1}{2}(L_{11} + L_{12}) .$$

The main point to be made now is that the Gauss-Weingarten equations (2.24)-(2.25) have an underlying group structure. In fact, they are of the form

$$\frac{\partial X}{\partial\sigma} = AX, \quad \frac{\partial X}{\partial\tau} = BX \tag{2.29}$$

where X is not really a column vector but a three-by-three orthogonal matrix while A and B are antisymmetric 3×3 matrices. In other words, X is an element of the rotation group in three dimensions SO(3) while A and B are elements of the Lie algebra of SO(3). In still other words, A and B are linear combinations of the anti-hermitian generators of the rotation group, and their explicit form (2.24)-(2.25) is obtained in the spin one representation (remember that the irreducible representations of the rotation group are labeled by a number S, the spin, which can take (positive) integer and half-integer values so that the Casimir operator of the group takes the value S(S+1)). What Equations (2.29) are saying then, is that the frame tangent to the surface moves about by rotating itself which is, of course, the only possible motion for an orthonormal set of vectors.

The matrices A and B are then of the form

$$A = \vec{w}_\sigma \cdot \vec{L} \qquad B = \vec{w}_\tau \cdot \vec{L} \tag{2.30}$$

where $\vec{L}$ are the three anti-hermitian generators of SO(3) and

$$\vec{w}_\sigma = \left(\frac{L_{12}}{\sin\theta}, -\frac{L_{11}}{\cos\theta}, \frac{\partial\theta}{\partial\tau}\right) \tag{2.31}$$

$$\vec{w}_\tau = \left(\frac{L_{22}}{\sin\theta}, -\frac{L_{12}}{\cos\theta} \quad \frac{\partial\theta}{\partial\sigma}\right). \tag{2.32}$$

It is now possible to go to the spin-$\frac{1}{2}$ representation, where $\vec{L} = (i/2)\vec{\sigma}$, with $\vec{\sigma} = (\sigma_1, \sigma_2, \sigma_3)$ being the Pauli matrices. In this case Equations (2.29) are

$$\frac{\partial v}{\partial\sigma} = \left(\frac{i}{2}\vec{w}_\sigma \cdot \vec{\sigma}\right)v \tag{2.33}$$

$$\frac{\partial v}{\partial\tau} = \left(\frac{i}{2}\vec{w}_\tau \cdot \vec{\sigma}\right)v \tag{2.34}$$

where v is an SU(2) spinor:

$$v = \begin{pmatrix} v_1 & -v_2^* \\ v_2 & v_1^* \end{pmatrix}, \qquad |v_1|^2 + |v_2|^2 = 1$$

related to the $\hat{X}$'s of Equations (2.24)-(2.25) (the spin-1 spinor) through

$$X_{ij} = \tfrac{1}{2}\,\mathrm{tr}(v\sigma_i v^\dagger \sigma_j). \tag{2.35}$$

Notice that although v is a complex quantity, the $\hat{X}_{ij}$ given by this relation are automatically real, as they should be. The important fact now is that the integrability conditions for the spin-$\frac{1}{2}$ equations (2.33)-(2.34) are still (2.26)-(2.28). The first two of these equations can be written

$$\frac{\partial}{\partial\tau}(\mathrm{tg}\theta B) = \frac{\partial}{\partial\sigma}(\mathrm{tg}\theta\, L_{12}) \tag{2.36}$$

$$\frac{\partial}{\partial\sigma}(\cot\theta B) = \frac{\partial}{\partial\tau}(\cot\theta\, L_{12}). \tag{2.37}$$

From (2.36) it follows that there exists an integrating factor such that

$$B = \cot\theta\,\frac{\partial\lambda}{\partial\sigma}, \quad L_{12} = \cot\theta\,\frac{\partial\lambda}{\partial\tau}. \tag{2.38}$$

Substituting back into (2.37) and (2.38), we obtain that the integrability conditions for (2.33)-(2.34) are

$$\frac{\partial}{\partial\tau}\left(\cot^2\theta\,\frac{\partial\lambda}{\partial\tau}\right) - \frac{\partial}{\partial\sigma}\left(\cot^2\theta\,\frac{\partial\lambda}{\partial\sigma}\right) = 0 \tag{2.39}$$

$$\frac{\partial^2\theta}{\partial\tau^2} - \frac{\partial^2\theta}{\partial\sigma^2} + \sin\theta\cos\theta + \frac{\cos\theta}{\sin^3\theta}\left[\left(\frac{\partial\lambda}{\partial\tau}\right)^2 - \left(\frac{\partial\lambda}{\partial\sigma}\right)^2\right] = 0. \tag{2.40}$$

We have then a system of two coupled, nonlinear differential equations in two dependent and two independent variables as integrability conditions of a system of two linear equations. We now want to interpret one of these equations as an eigenvalue problem. To do this an extra parameter, that can be interpreted as eigenvalue, is needed. The simplest way to introduce such an eigenvalue is to use the Lorentz invariance of Equations (2.39)-(2.40), taking the following steps: First, go to light cone coordinates

$$\begin{aligned}\xi &= \tfrac{1}{2}(\sigma+\tau)\\ \eta &= \tfrac{1}{2}(\sigma-\tau).\end{aligned} \tag{2.41}$$

Second, perform a spinor rotation on the linear equations:

$$v \longrightarrow e^{(i/2)\sigma_3\theta}v \tag{2.42}$$

and change components

$$v_1 \longrightarrow v_1 + iv_2$$

$$v_2 \longrightarrow v_1 - iv_2 . \tag{2.43}$$

Finally, make a Lorentz transformation

$$\xi \longrightarrow \frac{1}{\zeta}\,\xi$$
$$\eta \longrightarrow \zeta\,\eta . \tag{2.44}$$

The end product is that the equations

$$\frac{\partial}{\partial\xi}\left(\cot^2\theta\,\frac{\partial\lambda}{\partial\eta}\right) + \frac{\partial}{\partial\eta}\left(\cot^2\theta\,\frac{\partial\lambda}{\partial\xi}\right) = 0 \tag{2.45}$$

$$\frac{\partial^2\theta}{\partial\xi\partial\eta} - \tfrac{1}{2}\sin 2\theta + \frac{\cos\theta}{\sin^3\theta}\,\frac{\partial\lambda}{\partial\xi}\,\frac{\partial\lambda}{\partial\eta} = 0 \tag{2.46}$$

are the integrability conditions for the linear system

$$\frac{\partial v}{\partial\xi} = \begin{pmatrix} -i\zeta + ip & q \\ -q^* & i\zeta - ip \end{pmatrix} v \tag{2.47}$$

$$\frac{\partial v}{\partial\eta} = i\begin{pmatrix} r & s \\ s & -r \end{pmatrix} v \tag{2.48}$$

where

$$p = \frac{\cos 2\theta}{2\sin^2\theta}\,\frac{\partial\lambda}{\partial\xi} \tag{2.49}$$

$$q = -\frac{\partial\theta}{\partial\xi} + i\cot\theta\,\frac{\partial\lambda}{\partial\xi} \tag{2.50}$$

$$r = \frac{1}{4\zeta}\cos 2\theta - \frac{1}{2\sin^2\theta}\,\frac{\partial\lambda}{\partial\eta} \tag{2.51}$$

$$s = \frac{1}{4\zeta}\sin 2\theta . \tag{2.52}$$

Equation (2.47) is then an eigenvalue problem whose spectrum is left invariant when θ and λ evolve according to Equations (2.45)-(2.46). Equation (2.48) gives the "time" (η) evolution of the corresponding eigenfunctions. When $\lambda = 0$ one recovers the sine-Gordon theory. In this case, $B = L_{12} = 0$ and the corresponding surface has constant intrinsic curvature $R = g^{\mu\nu}R^{\lambda}{}_{\mu\lambda\nu}$. The converse is not true in general, as a surface of constant curvature has only that $B^2 = L_{12}^2$.

The system (2.45)-(2.46) or, in laboratory coordinates, (2.39)-(2.40) is Lagrangian,

$$\mathcal{L} = \frac{1}{2}\left\{\left(\frac{\partial\theta}{\partial\tau}\right)^2 - \left(\frac{\partial\theta}{\partial\sigma}\right)^2 - \sin^2\theta + \cot^2\theta\left[\left(\frac{\partial\lambda}{\partial\tau}\right)^2 - \left(\frac{\partial\lambda}{\partial\sigma}\right)^2\right]\right\} \tag{2.53}$$

and arises in a number of physical systems. These will be described in the next section.

3. PHYSICS

3.1 Strings in interaction through a scalar field [8]

A system of one-dimensional relativistic objects $x^{\mu}(\sigma,\tau)$ moving in four-dimensional space time and in interaction with a field $A_{\mu\nu} = -A_{\nu\mu}$ may be described by the following action integral:

$$S = -N\int\sqrt{-g}\,d\sigma d\tau + f\int A_{\mu\nu}\frac{\partial x^{\mu}}{\partial\sigma}\frac{\partial x^{\nu}}{\partial\tau}\,d\sigma d\tau - \tfrac{1}{4}\int F_{\mu}F^{\mu}, \tag{3.1}$$

where N, f are constants, $F^{\mu} = \epsilon^{\mu\nu\lambda\rho}\partial_{\nu}A_{\lambda\rho}$ and $\sqrt{-g}$ is the square root of the metric tensor for the surface described by the string $x^{\mu}(\sigma,\tau)$ as it moves in Minkowski space. The first term in the action (3.1) is then the area of the surface swept by the string in space time. The second term is an interaction term and f is a coupling constant. The last term is the free field contribution describing the evolution of $A_{\mu\nu}$. It was shown in Reference 8 that in spite of the apparent tensor nature of this field, most of its degrees of freedom are gauge ones and that it has only one true dynamical degree of freedom per space point: It is a scalar field. This system is a natural extension of classical electrodynamics in which one has relativistic point particles in interaction through a vector field A_{μ}. The action in that case is

$$S = m\int ds + e\int A_{\mu}\frac{\partial x^{\mu}}{\partial\tau}\,d\tau - \tfrac{1}{4}\int F_{\mu\nu}F^{\mu\nu} \tag{3.2}$$

where m is the mass of the particle, e the charge, and

$F^{\mu\nu} = \epsilon^{\mu\nu\lambda\rho}\partial_\lambda A_\rho$. The first term is the length of the curve described by the particle in space time; the second term is the interaction with e as a coupling constant, and the last term is the free field contribution.

The problem now is to find the motion of a string in a prescribed external field in the same way as one studies the motion of an electron in a fixed magnetic field. The equations of motion obtained by minimizing the action (3.1) with respect to x^μ are

$$N\left(\frac{\partial^2 x^\mu}{\partial\sigma^2} - \frac{\partial^2 x^\mu}{\partial\tau^2}\right) = f\epsilon^{\mu\nu\lambda\rho} F_\nu \frac{\partial x_\rho}{\partial\sigma}\frac{\partial x_\lambda}{\partial\tau} \tag{3.3}$$

with the constraints

$$\left(\frac{\partial x}{\partial\sigma}\right)^2 + \left(\frac{\partial x}{\partial\tau}\right)^2 = 0 \tag{3.4}$$

$$\frac{\partial x^\mu}{\partial\sigma}\frac{\partial x_\mu}{\partial\tau} = 0. \tag{3.5}$$

The prescribed external field will be

$$F^o = \tilde{F}, \quad F^i = 0 \tag{3.6}$$

and the motion of the string will be studied in the reference frame where

$$\tau = x^o. \tag{3.7}$$

With these choices the $\mu = 0$ of Equations (3.3) is identically satisfied and the other three can be written

$$\frac{\partial^2\vec{x}}{\partial\tau^2} - \frac{\partial^2\vec{x}}{\partial\sigma^2} = -2c\left(\frac{\partial\vec{x}}{\partial\tau}\times\frac{\partial\vec{x}}{\partial\sigma}\right) \tag{3.8}$$

where $c = f\tilde{F}/2N$, and the constraints (3.4)-(3.5) become

$$\left(\frac{\partial\vec{x}}{\partial\tau}\right)^2 + \left(\frac{\partial\vec{x}}{\partial\sigma}\right)^2 = 1 \tag{3.9}$$

$$\frac{\partial\vec{x}}{\partial\sigma}^2 \cdot \frac{\partial\vec{x}}{\partial\tau} = 0. \tag{3.10}$$

What one notices now is that solving Equations (3.8)-(3.10) consists in finding three functions $x^i(\sigma,\tau)$ that satisfy them. But finding these three functions is exactly the problem of embedding a two-dimensional surface in a three-dimensional Euclidean space, and the above equations of motion can be regarded as restrictions on the geometry of the surface. In particular, (3.9) and (3.10) are the restrictions (2.18)-(2.19) imposed in the previous section on the metric tensor. To see what (3.8) leads to, take its scalar product with

$$X_3 = \left(\frac{\partial\vec{x}}{\partial\sigma} \times \frac{\partial\vec{x}}{\partial\tau}\right)\Bigg/\left(\left\|\frac{\partial\vec{x}}{\partial\sigma} \times \frac{\partial\vec{x}}{\partial\tau}\right\|\right). \tag{3.11}$$

Using (3.10), (2.20) and (2.9), one gets

$$L_{22} - L_{11} = 2\sin\theta\cos\theta \tag{3.12}$$

which is the restriction (2.22) of Section 2. Taking the scalar product with X_1 and X_2 gives identities.

The problem of finding the motion of a string in a uniform external field is then seen to be equivalent to the problem of embedding a surface in Euclidean three-space with the restrictions (3.9), (3.10) and (3.12) on its geometry, which is the problem that led to the nonlinear equations associated with the Lagrangian (2.53) and with the linear system (2.47)-(2.48).

3.2 Massless fermions with scalar contact interactions [12]

A Lagrangian describing a theory of N classical (non-quantum), massless fermions with a scalar contact interaction is the following:

$$\mathcal{L} = i\sum_{k=1}^{N}\bar{\psi}_k \partial\!\!\!/ \psi_k + \frac{g^2}{2}\left\{\left(\sum_{k=1}^{N}\bar{\psi}_k\psi_k\right)^2 - \left(\sum_{k=1}^{N}\bar{\psi}_k\gamma_5\psi_k\right)^2\right\}. \tag{3.13}$$

Theories of this type were first introduced by Nambu and Jona-Lasinio [13] as model field theories for superconductors. The system (3.13) without the γ_5 term is the so-called Gross-Neveu model [14]. It was shown to have a particle spectrum by Dashen, Hasslacher and Neveu [15] using semiclassical techniques. Analogous results for the full Lagrangian (3.13) were established by Shei [16]. These results were obtained when space time is two-dimensional and we shall also use this restriction here. The derivation that follows is taken from the work of Neveu and Papanicolaou [12]. Notation is as follows: Light cone coordinates

$$\xi = \tfrac{1}{2}(t-x), \quad \eta = \tfrac{1}{2}(t+x) \tag{3.14}$$

are used, and the Dirac matrices are

$$\gamma^o = \begin{pmatrix} 0 & 1 \\ 1 & 0 \end{pmatrix}, \quad \gamma^1 = \begin{pmatrix} 0 & 1 \\ -1 & 0 \end{pmatrix}, \quad \gamma^5 = \gamma^o\gamma^1 . \tag{3.15}$$

The equations of motion following from (3.13) are

$$\left[i\partial\!\!\!/ - (\sigma + i\pi\gamma^5)\right]\psi_k = 0 \tag{3.16}$$

where

$$\begin{aligned} \sigma &\equiv \frac{1}{2}\sum_{k=1}^{N} \bar{\psi}_k \psi_k \\ \pi &\equiv \frac{i}{2}\sum_{k=1}^{N} \bar{\psi}_k \gamma^5 \psi_k . \end{aligned} \tag{3.17}$$

Denoting

$$\psi_k = \begin{pmatrix} u_k \\ v_k \end{pmatrix},$$

one has

$$\phi \equiv \sigma - i\pi = \sum_{k=1}^{N} u_k^* v_k \equiv u^* v \tag{3.18}$$

and the equations of motion (3.16) take the form

$$\begin{aligned} iu,_\xi &= \phi^* v \\ iv,_\eta &= \phi u \end{aligned} \tag{3.19}$$

where the comma followed by the subscript denotes partial differentiation.

A number of results will now be stated in the form of Lemmas:

Lemma 1: The equations of motion (3.19) give rise to four conservation laws:

$$(u^*u),_\xi = 0 = (v^*v),_\eta \tag{3.20}$$

$$(u^*u,_\eta - uu^*,_\eta),_\xi = 0 = (v^*v,_\xi - vv^*,_\xi),_\eta . \tag{3.21}$$

Proof: A direct computation using (3.18) and (3.19). As a consequence, one has that the following is true:

$$\begin{aligned} g_1 &\equiv u^*u = g_1(\eta) \\ g_2 &\equiv v^*v = g_2(\xi) \\ h_1 &\equiv \frac{i}{2}(u^*u,_\eta - uu^*,_\eta) = h_1(\eta) \\ h_2 &\equiv \frac{i}{2}(v^*v,_\xi - vv^*,_\xi) = h_2(\xi). \end{aligned} \tag{3.22}$$

Lemma 2: For arbitrary complex functions $f_1 = f_1(\eta)$ and $f_2 = f_2(\xi)$, the transformation

$$\begin{aligned} u &= f_1 u' \\ v &= f_2 v' \\ d\xi' &= |f_2|^2 d\xi \\ d\eta' &= |f_1|^2 d\eta \end{aligned} \tag{3.23}$$

leaves the equations of motion (3.19) invariant.

Proof: By direct substitution.

Lemma 3:

$$\psi,_{\eta\xi} = u^*,_\eta v,_\xi + i(g_2\phi,_\eta - g_1\phi,_\xi) - |\phi|^2\phi. \tag{3.24}$$

Proof: Again by direct calculation using (3.18)-(3.19).

Lemma 4: With the notation of Lemma 2, it is possible to choose f_1 and f_2 such that

$$\begin{aligned} g_1 &= \text{const} \\ g_2 &= \text{const} \\ h_1 &= 0 = h_2. \end{aligned} \tag{3.25}$$

Proof: The transformation (3.23) introduces the following changes in the conserved quantities (3.22):

$$g_1 = |f_1|^2 g_1'$$
$$h_1' = \frac{h_1}{|f_1|^2} + g_1' (\arg f_1)_{,\eta} . \tag{3.26}$$

It is clear then that there is no loss of generality in considering g,h of the form (3.25).

Up to now, results have been valid for arbitrary N. The considerations that follow will hold only for $N=2$.

Lemma 5: For $N=2$,

$$u^*_{,\eta} = \frac{\phi_{,\eta} + ig_1\phi}{g_1g_2 - |\phi|^2}(-\phi^* u^* + g_1 v^*),$$
$$v_{,\xi} = \frac{\phi_{,\xi} - ig_2\phi}{g_1g_2 - |\phi|^2}(g_2 u - \phi^* v). \tag{3.27}$$

Proof: u and v form a basis in a two-dimensional complex vector space, hence all vectors can be written as a linear combination of them. Use of (3.25) then leads to (3.27).

Lemma 6 [17]:

$$\left(g_1g_2 - |\phi|^2\right)\phi_{,\eta\xi} = -\phi^*\phi_{,\eta}\phi_{,\xi} + \phi\left(g_1g_2 - |\phi|^2\right)^2. \tag{3.28}$$

Proof: Substitute (3.27) into (3.24) and make the change of variables

$$u \longrightarrow u e^{-ig_1\eta}$$
$$v \longrightarrow v e^{-ig_2\xi}, \tag{3.29}$$

noticing that this induces a change from $h_1 = 0 = h_2$ to $h_1 = g_1^2$, $h_2 = g_2^2$.

After all these preliminaries we are ready for a

Proposition: There exist scalars θ and λ such that

$$\phi = \sqrt{g_1g_2}\,\sin\frac{\theta}{2}\,e^{i(\lambda/2)}. \tag{3.30}$$

In this case, (3.28) reduces to

$$\theta,_{\eta\xi} - \sin\theta + \frac{tg^2(\theta/2)}{\sin} \lambda,_{\eta} \lambda,_{\xi} = 0 \qquad (3.31)$$

$$(\sin\theta)\, \lambda,_{\eta\xi} + \theta,_{\xi} \lambda,_{\eta} + \theta,_{\eta} \lambda,_{\xi} = 0. \qquad (3.32)$$

Proof: By Schwartz's inequality, one has

$$|\phi| = |u^*v| \leqq |u|\,|v| = \sqrt{g_1 g_2}$$

which guarantees the existence of θ,λ, satisfying (3.30). Equations (3.31)-(3.32) are obtained by substitution into (3.28).

Equations (3.31)-(3.32) are precisely Equations (2.45)-(2.46) with the change $\theta \to (\theta/2)$, $\lambda \to (\lambda/2)$. In laboratory coordinates, they are derivable from the Lagrangian (2.53). It is clear that a solution u,v to (3.19) will give solutions θ,λ to (3.31)-(3.32) through (3.30). Vice versa, a solution to (3.31)-(3.32) gives (u,v), a solution to (3.19), as explained by Neveu and Papanicolaou in Reference 12, who also give the relation between the linear problems (3.19), (3.27) and (2.47)-(2.48).

3.3 Nonlinear σ model with O(4) symmetry [8],[18]

The O(4) nonlinear σ model in two dimensional space time consists of four scalar fields $\phi^i(x^1,x^2)$, $i=1,\ldots,4$ with an interaction given through the constraint

$$\phi^i\phi^i = 1. \qquad (3.33)$$

The Lagrangian density for this system is then

$$\mathcal{L} = \tfrac{1}{2}\partial_\mu\phi^i\partial^\mu\phi^i + \tfrac{1}{2}N(\phi^i\phi^i - 1), \quad \mu = 1,2 \qquad (3.34)$$

where N is a Lagrange multiplier and the equations of motion are (3.33) and

$$\partial_\mu\partial^\mu\phi^i - N\phi^i = 0. \qquad (3.35)$$

This model was introduced originally as an example of a chiral-invariant system (to help understand the results predicted by current algebra [19]) and has received close attention lately in what respects its critical behavior in quantum theory [20] and as a classical system that gives a model for a ferromagnet having "instanton" solutions [21].

Going back to Equations (3.33) and (3.35), one notices that the fields ϕ^i can be interpreted as the components of a vector in four-dimensional Euclidean space. Equation (3.33) says that this vector must lie on the surface of a sphere and a solution to Equation (3.35) describes a surface embedded in this sphere. The problem of solving Equations (3.33)-(3.35) reduces then to the problem of embedding a surface in a three-sphere which is itself embedded in four-dimensional Euclidean space. In Section 1 the embedding of surfaces in flat spaces was described in detail. The modifications introduced when the ambient space is of constant curvature are minor and may be found in Eisenhart's book [22].

The metric of the four-dimensional Euclidean space is

$$ds^2 = d\phi^i d\phi^i \tag{3.36}$$

and induces a metric on the two-dimensional surface $\phi^i(\sigma,\tau)$ given by

$$ds^2 = \left(\frac{\partial\vec{\phi}}{\partial\sigma}\right)^2 d\sigma^2 + 2\,\frac{\partial\vec{\phi}}{\partial\sigma}\cdot\frac{\partial\vec{\phi}}{\partial\tau}\,d\sigma d\tau + \left(\frac{\partial\vec{\phi}}{\partial\tau}\right)^2 d\tau^2 \tag{3.37}$$

where we have written $\sigma = x'$, $\tau = x^2$. Calling $\hat{X}_3$ the unit normal to the surface which is contained in the tangent plane to the sphere, the components of the extrinsic curvature tensor are

$$L_{\mu\nu} = \frac{\partial^2\vec{\phi}}{\partial x^\mu \partial x^\nu}\cdot\hat{X}_3. \tag{3.38}$$

The equation of motion (3.35) restricts these components to satisfy

$$L_{11} = L_{22} \tag{3.39}$$

as is immediately seen taking the scalar product of that equation with $\hat{X}_3$. The scalar product with X_1 and X_2 leads only to identities. Consequently, the only restriction the equations of motion impose on the embedded surface that is to be constructed is given by (3.39), and one is free to use the full freedom of changing the surface coordinates (σ,τ). It was shown by Tchebychev[23] that it is always possible to find coordinates—which we shall still call (σ,τ)—such that the line element (3.37) reads

$$ds^2 = \cos^2\theta\, d\sigma^2 + \sin^2\theta d\tau^2. \tag{3.40}$$

An orthonormal tetrad X tangent to the surface can then be constructed with $\hat{X}_1, \hat{X}_2$, the unit vectors in the direction of the tangents $(\partial\phi/\partial\sigma)$, $(\partial\phi/\partial\tau)$, respectively, with X_3 and with $\hat{X}_4 = \vec{\phi}$. The Gauss-Weingarten equations obeyed by this frame can be worked out to be

$$\frac{\partial X}{\partial \sigma} = AX, \quad \frac{\partial X}{\partial \tau} = BX \tag{3.41}$$

where

$$A = \begin{bmatrix} 0 & \frac{\partial\theta}{\partial\tau} & \frac{L_{11}}{\cos\theta} & -\cos\theta \\ -\frac{\partial\theta}{\partial\tau} & 0 & \frac{L_{12}}{\sin\theta} & 0 \\ -\frac{L_{11}}{\cos\theta} & -\frac{L_{22}}{\sin\theta} & 0 & 0 \\ \cos\theta & 0 & 0 & 0 \end{bmatrix} \tag{3.42}$$

$$B = \begin{bmatrix} 0 & \frac{\partial\theta}{\partial\sigma} & \frac{L_{12}}{\cos\theta} & 0 \\ -\frac{\partial\theta}{\partial\sigma} & 0 & \frac{L_{11}}{\sin\theta} & -\sin\theta \\ -\frac{L_{12}}{\cos\theta} & -\frac{L_{11}}{\sin\theta} & 0 & 0 \\ 0 & \sin\theta & 0 & 0 \end{bmatrix} \tag{3.43}$$

Their integrability (Gauss-Codazzi) conditions are

$$\cos\theta\sin\theta\left(\frac{\partial^2\theta}{\partial\sigma^2} - \frac{\partial^2\theta}{\partial\tau^2}\right) - \cos^2\theta\sin^2\theta - \left(L_{11}^2 - L_{12}^2\right) = 0 \tag{3.44}$$

$$\frac{\partial}{\partial\tau}(\tan\theta L_{11}) = \frac{\partial}{\partial\sigma}(\tan\theta L_{12}) \tag{3.45}$$

$$\frac{\partial}{\partial\sigma}(\cot\theta L_{11}) = \frac{\partial}{\partial\tau}(\cot\theta L_{12}) \tag{3.46}$$

Again, from (3.45) it follows that there exists a field $\lambda(\sigma,\tau)$ such that

$$L_{11} = \cot\theta \frac{\partial\lambda}{\partial\sigma}, \quad L_{22} = \cot\theta \frac{\partial\lambda}{\partial\tau} \tag{3.47}$$

whence the conditions (3.44)-(3.46) change to

$$\frac{\partial^2\theta}{\partial\sigma^2} - \frac{\partial^2\theta}{\partial\tau^2} - \cos\theta\sin\theta + \frac{\cos\theta}{\sin^3\theta}\left[\left(\frac{\partial\lambda}{\partial\sigma}\right)^2 - \left(\frac{\partial\lambda}{\partial\tau}\right)^2\right] = 0 \tag{3.48}$$

$$\frac{\partial}{\partial\tau}\left(\cot^2\theta \frac{\partial\lambda}{\partial\tau}\right) = \frac{\partial}{\partial\sigma}\left(\cot^2\theta \frac{\partial\lambda}{\partial\sigma}\right). \tag{3.49}$$

Thus we have recovered Equations (2.39)-(2.40), this time as integrability conditions for a linear system, Equations (3.41)-(3.43), that look different from the original (2.24)-(2.25). We now show that both systems are in fact the same.

The basis X, being orthonormal, can be understood as an element of the group O(4) and the matrices A and B appearing in (3.41) are linear combinations of its generators. This group being the direct product of two three-dimensional rotation groups and its Lie algebra being the direct sum of the corresponding Lie algebras, it follows that there exist uniquely determined 3×3 matrices C, D, E, F such that

$$\begin{aligned} A &= C \oplus D \\ B &= E \oplus F. \end{aligned} \tag{3.50}$$

It is not difficult to show that

$$C = \begin{bmatrix} 0 & \frac{\partial\theta}{\partial\tau} & -\frac{1}{\sin\theta}\frac{\partial\lambda}{\partial\sigma} \\ -\frac{\partial\theta}{\partial\tau} & 0 & -\cos\theta - \frac{\cos\theta}{\sin^2\theta}\frac{\partial\lambda}{\partial\tau} \\ \frac{1}{\sin\theta}\frac{\partial\lambda}{\partial\sigma} & \cos\theta + \frac{\cos\theta}{\sin^2\theta}\frac{\partial\lambda}{\partial\tau} & 0 \end{bmatrix} \tag{3.51}$$

$$E = \begin{bmatrix} 0 & \frac{\partial\theta}{\partial\sigma} & \sin\theta - \frac{1}{\sin\theta}\frac{\partial\lambda}{\partial\tau} \\ -\frac{\partial\theta}{\partial\sigma} & 0 & -\frac{\cos\theta}{\sin^2\theta}\frac{\partial\lambda}{\partial\sigma} \\ -\sin\theta + \frac{1}{\sin\theta}\frac{\partial\lambda}{\partial\tau} & \frac{\cos\theta}{\sin^2\theta}\frac{\partial\lambda}{\partial\sigma} & 0 \end{bmatrix} \tag{3.52}$$

and that D and F are obtained from C and E by the transformation

$$\begin{aligned} \theta &\to \pi - \theta \\ \lambda &\to -\lambda, \end{aligned} \tag{3.53}$$

which leaves the metric and extrinsic curvature of the surface unchanged. It is then possible to introduce three-dimensional orthonormal frames Y and Z so that Equations (3.41) become

$$\frac{\partial Y}{\partial \sigma} = CY, \quad \frac{\partial Y}{\partial \tau} = EY \tag{3.54}$$

and

$$\frac{\partial Z}{\partial \sigma} = DZ, \quad \frac{\partial Z}{\partial \tau} = FZ. \tag{3.55}$$

It is enough to consider (3.54) as (3.55) contains no new information, its solution being in one-to-one correspondence with those of (3.54) through (3.53). The linear system (3.54) coincides exactly with (2.24)-(2.25) with the change $\theta \to \theta + \pi$, $\sigma \leftrightarrow \tau$.

4. SOLITONS

In Section 2 a set of nonlinear evolution equations with a naturally associated eigenvalue problem was derived in a geometrical setting. It turns out, as was shown in Section 3, that this system also arises in a number of theoretical physics problems. In this section the initial value problem for these equations will be solved. We rewrite them here for reference:

$$\partial_\mu \partial^\mu \theta + \sin\theta \cos\theta + \frac{\cos\theta}{\sin^3\theta} (\partial_\mu \lambda)(\partial^\mu \lambda) = 0 \tag{4.1}$$

$$\partial_\mu \partial^\mu \lambda - \frac{1}{\sin\theta\cos\theta} (\partial_\mu \theta)(\partial^\mu \lambda) = 0. \tag{4.2}$$

In this section we shall use the notation $\sigma = x^1 = x$, $\tau = x^2 = t$. The first question is whether these equations have stationary solutions. The answer is yes, and it is simple to find them in their rest frame, $\partial/\partial t = 0$, using the Lorentz invariance of the theory. In this case then, from (4.2) one has

$$\frac{\partial \lambda}{\partial x} = A \operatorname{tg}^2 \theta \tag{4.3}$$

where A is a constant. Substituting into (4.1), one gets

$$\frac{\partial^2\theta}{\partial x^2} - \sin\theta\cos\theta + A^2\frac{\sin\theta}{\cos^3\theta} = 0 \qquad (4.4)$$

which is easy to integrate by elementary means to get periodic and solitary waves. For the periodic case the field λ is not bounded so we shall not consider it. The solitary wave is

$$\theta(x) = \operatorname{Arcsin}\left[\frac{(1-A^2)^{\frac{1}{2}}}{\cosh(x-x_0)(1-A^2)^{\frac{1}{2}}}\right] \qquad (4.5)$$

$$\lambda(x) = \operatorname{Arctg}\left[(A^{-2}-1)^{\frac{1}{2}}\,\operatorname{th}\left((x-x_0)(1-A^2)^{\frac{1}{2}}\right)\right] + \lambda(0) \qquad (4.6)$$

where $\lambda(0)$ is an irrelevant additive constant that will be omitted in what follows, and a wave travelling with constant speed v is obtained by Lorentz-boosting the solution (4.5)-(4.6)

$$x \longrightarrow \frac{x - vt}{(1-v^2)^{\frac{1}{2}}}\,. \qquad (4.7)$$

The shape of this solitary wave is sketched in Figures 1 and 2. When $|x| \to \infty$, it has the asymptotic behavior

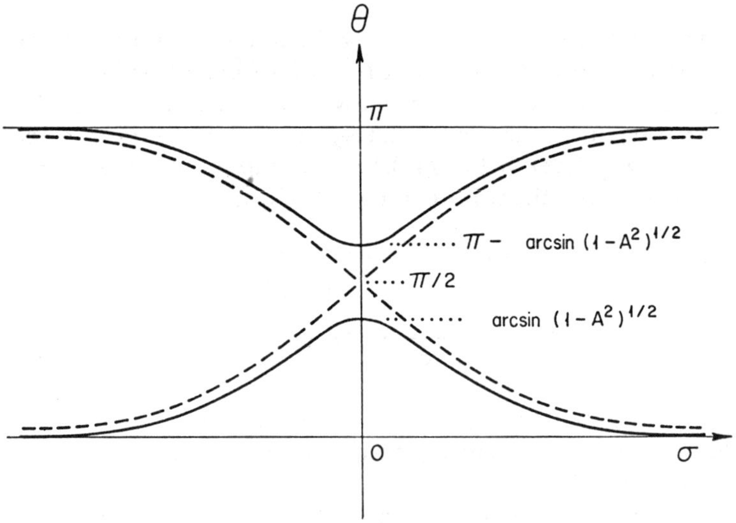

Figure 1. The solid line is the soliton given by (4.5). In the limit A = 0 it goes over to the sine-Gordon soliton (dashed line). Notice that if θ is a solution, so is $\pi-\theta$. Both cases are drawn here.

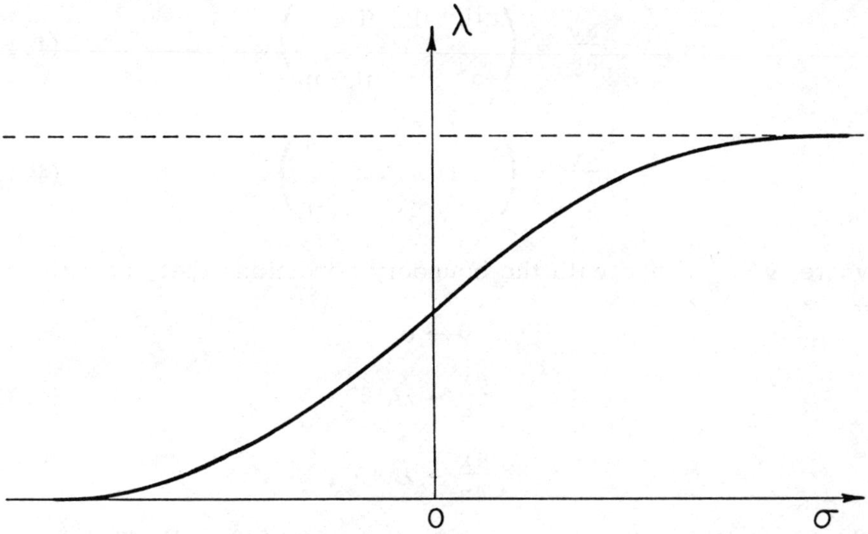

Figure 2. Shape of field λ given by (4.6) when θ is a soliton.

$$\theta \to 0,$$

$$\frac{\partial\lambda}{\partial x}, \frac{\partial\lambda}{\partial t} \sim (\text{constant})\,\theta^2. \qquad (4.8)$$

The goal is now to characterize all solutions to the equations of motion that have this same behavior at infinity. This is done by studying the scattering problem of which (4.1)-(4.2) are the integrability conditions. This scattering problem is not the same when one uses light-cone coordinates as when one used laboratory coordinates, in which case it is more involved. Both cases can be dealt with however [8], and here we shall give some details of the light-cone case. The goal is then to solve the initial value problem for equations

$$\frac{\partial}{\partial\xi}\left(\cot^2\theta\,\frac{\partial\lambda}{\partial\eta}\right) + \frac{\partial}{\partial\eta}\left(\cot^2\theta\,\frac{\partial\lambda}{\partial\xi}\right) = 0 \qquad (4.9)$$

$$\frac{\partial^2\theta}{\partial\xi\partial\eta} - \frac{1}{2}\sin 2\theta + \frac{\cos\theta}{\sin^3\theta}\,\frac{\partial\lambda}{\partial\xi}\,\frac{\partial\lambda}{\partial\eta} = 0 \qquad (4.10)$$

using the linear problem

$$\frac{\partial v}{\partial \xi} = \begin{pmatrix} -i\zeta + ip & q \\ -q^* & i\zeta - ip \end{pmatrix} v \tag{4.11}$$

$$\frac{\partial v}{\partial \eta} = \begin{pmatrix} r & s \\ s & -r \end{pmatrix} v \tag{4.12}$$

where $v = \begin{pmatrix} v_1 \\ v_2 \end{pmatrix}$ and with the boundary conditions that, as $|x| \to \infty$,

$$\begin{aligned} \theta &\to 0 \\ \frac{\partial \lambda}{\partial \xi} &\to 2A'\theta^2 \\ \frac{\partial \lambda}{\partial \eta} &\to 2A''\theta^2 . \end{aligned} \tag{4.13}$$

The quantities r, s, p, q are given in terms of θ, λ by (2.49)-(2.52). The coordinate ξ will be the spatial one while η will be the temporal one. The problem now is to solve the scattering problem for the operator (4.11) [24]. It is very similar to the one of Zacharov and Shabat [4], the main difference being in the asymptotic behavior of the Jost solutions and transmission coefficient in the complex eigenvalue plane. For $\lambda = 0$, it of course reduces to the problem studied by Ablowitz et al. [10] when solving the sine-Gordon equation.

The asymptotic form of (4.11) is

$$\frac{\partial v}{\partial \xi} = \begin{pmatrix} -i\zeta + iA' & 0 \\ 0 & i\zeta - iA' \end{pmatrix} v . \tag{4.14}$$

This suggests the introduction of Jost solutions $\varphi(x,\xi)$ and $\psi(x,\xi)$. They are solutions of (4.11) for $\zeta = \alpha + i\beta$ when $\beta = 0$ with the asymptotic behavior

$$\varphi \to \begin{pmatrix} 1 \\ 0 \end{pmatrix} e^{-i(\alpha - A')\xi}, \quad \xi \to -\infty \tag{4.15}$$

$$\psi \to \begin{pmatrix} 0 \\ 1 \end{pmatrix} e^{i(\alpha - A')x}, \quad \xi \to +\infty . \tag{4.16}$$

They satisfy integral equations—written explicitly in the last paper of Reference 8—which can be used to show that analytic continuation is possible to the upper half plane $\beta > 0$ with the following asymptotic behavior when $\zeta \to \infty$:

$$\varphi_1(\xi,\zeta)e^{i(\zeta-A')} \to e^{i\mu_-(\xi)}\left(1+\frac{1}{2i\zeta}\int_{-\infty}^{\xi} dy|q|^2\right) \quad (4.17a)$$

$$\varphi_2(\xi,\zeta)e^{i(\zeta-A')} \to \frac{-1}{2i\zeta}\, e^{i\mu_-(\xi)} q^* \quad (4.17b)$$

$$\psi_1(\xi,\zeta)e^{-i(\zeta-A')} \to \frac{1}{2i\zeta}\, e^{i\mu_+(\xi)} q \quad (4.17c)$$

$$\psi_2(\xi,\zeta)e^{-i(\zeta-A')} \to e^{i\mu_+(\xi)}\left(1-\frac{1}{2i\zeta}\int_{\xi}^{\infty} dy|q|^2\right) \quad (4.17d)$$

where

$$\mu_-(\xi) = \int_{-\infty}^{\xi} du(p(u)-A') \quad (4.18)$$

$$\mu_+(\xi) = \int_{\xi}^{+\infty} du(p(u)-A'). \quad (4.19)$$

Now, for $\beta=0$,

$$\bar{\psi}(\xi,\alpha) \equiv \begin{pmatrix} \psi_2^*(\xi,\alpha) \\ -\psi_1^*(\xi,\alpha) \end{pmatrix}$$

is also a solution of (4.11) and together with $\psi(\xi,\alpha)$ they form a complete set so there exist coefficients $a(\alpha)$, $b(\alpha)$ (the inverse of the transmission coefficient and the ratio of the reflection to the transmission coefficient respectively) such that

$$\varphi(\xi,\alpha) = a(\alpha)\,\bar{\psi}(\xi,\alpha) + b(\alpha)\,\psi(\xi,\alpha). \quad (4.20)$$

Moreover, in general, if $v(\xi,\zeta_1)$ and $w(\xi,\zeta_2)$ are solutions of (4.11) corresponding to eigenvalues ζ_1 and ζ_2, the following relation holds:

$$\frac{\partial}{\partial x}(v_1 w_2 - w_1 v_2) = -i(\zeta_1-\zeta_2)(w_2 v_1 + w_1 v_2). \quad (4.21)$$

In particular, for $v=\varphi$, $w=\psi$, $\zeta_1=\zeta_2=\alpha$, the right-hand side vanishes and $(\varphi_1\psi_2-\varphi_2\psi_1)$ does not depend on x. Evaluating at

$\xi \to +\infty$, one obtains

$$a(\alpha) = \varphi_1(\alpha)\psi_2(\alpha) - \psi_1(\alpha)\varphi_2(\alpha) \tag{4.22}$$

which shows that also $a(\alpha)$ can be continued to the upper ζ plane, its asymptotic behavior being, according to (4.17),

$$a(\zeta) \xrightarrow[\zeta \to \infty]{} e^{i\mu}(1 + 0(1/\zeta)) \tag{4.23}$$

where $\mu \equiv \mu_- + \mu_+$. A similar behavior occurs in the Dirac equation [25] and in a scattering problem associated with a set of water-wave equations studied by Kaup [26].

From (4.15)-(4.16), one sees that the zeroes of $a(\zeta)$ in the upper half ζ plane give the eigenvalues ζ_k of (4.11). At these points one has

$$\varphi(\xi, \zeta_k) = c_k \psi(\xi, \zeta_k) \tag{4.24}$$

where c_k is a constant related to the normalization of the k-th eigenfunction.

The time evolution of $a(\alpha)$, $b(\alpha)$ and c_k can be obtained by inserting (4.20) and (4.24) into (4.12). It is

$$\begin{aligned} a(\alpha,\eta) &= a(\alpha,0) \\ b(\alpha,\eta) &= b(\alpha,0)e^{-2i((1/4\xi)-A'')\eta} \\ c_k(\eta) &= c_k(0)e^{-2i((1/4\zeta_k)-A'')\eta}. \end{aligned} \tag{4.25}$$

Also, using (4.21) for $v = \varphi(\xi,\alpha)$, $w = \overline{\varphi}(\xi,\alpha)$, one obtains

$$|a(\alpha)|^2 + |b(\alpha)|^2 = 1. \tag{4.26}$$

Finally, $a(\zeta)$ obeys a dispersion relation

$$a(\zeta) = e^{i\mu} \exp\left\{\frac{1}{2\pi i}\int_{-\infty}^{\infty} \frac{\log|a(\alpha)|^2}{\alpha - \zeta}\, d\alpha\right\} \prod_k \frac{\zeta - \zeta_k}{\zeta - \zeta_k^*}. \tag{4.27}$$

To solve the inverse scattering problem one introduces, following Reference 4, a function $\Phi(\zeta)$ having poles on the discrete part of the spectrum ($\zeta = \zeta_k$) and a cut on the continuous part (real ζ):

$$\Phi(\zeta) = \begin{cases} a^{-1}(\zeta)\varphi(\xi,\zeta)e^{i(\zeta-A')\xi} & \text{Im}\,\zeta > 0 \\ \begin{pmatrix} \psi_2^*(\xi,\zeta^*) \\ -\psi_1^*(\xi,\zeta^*) \end{pmatrix} e^{i(\zeta-A')\xi} & \text{Im}\,\zeta < 0 \end{cases} . \tag{4.28}$$

Its residue at the poles (assumed simple, and not real) is

$$R_k = \frac{e^{i(\zeta_k A')\xi}}{a'(\zeta_k)} \phi(\xi,\zeta_k)$$

and the discontinuity at the cut $\Delta(\alpha) \equiv \Phi(\alpha+i0) - \Phi(\alpha-i0)$ is

$$\Delta(\alpha) = \frac{b(\alpha)}{a(\alpha)} \psi(\xi,\alpha)e^{i(\alpha-A')\xi} .$$

This information gives, through Cauchy's theorem, the following representation for Φ:

$$\Phi(\zeta) = e^{-i\mu}\begin{pmatrix}1\\0\end{pmatrix} + \sum_k \frac{R_k c_k}{\zeta-\zeta_k}\psi(\xi,\xi_k) + \frac{1}{2\pi i}\int_{-\infty}^{\infty} \frac{\Delta(\alpha)}{\alpha-\zeta}\,d\alpha . \tag{4.29}$$

By letting in this equation $\zeta = \alpha - i0$ and $\zeta = \zeta_k$, one obtains equations for $\Delta(\alpha)$ and $\psi(\xi,\zeta_k)$ that give p and q when one compares the asymptotic behavior of ψ in the ζ plane obtained from (4.29) with that obtained in (4.17). Integral equations of the Marchenko type are obtained by writing the Jost solution ψ in the integral representation

$$\psi(\xi,\zeta) = \begin{pmatrix}0\\1\end{pmatrix} e^{i(\zeta-A')\xi+i\mu_+(\xi)} + \int_{\xi}^{\infty} K(\xi,s)e^{i(\zeta-A')s}ds. \tag{4.30}$$

The result is that the potentials are given in terms of the kernel $K(\xi,s)$ by

$$-q(\xi)e^{-i\mu_+(\xi)} = 2K_1(\xi,\xi) \tag{4.31}$$

and that the kernel itself is a solution of the linear integral èquation

$$\overline{K}(\xi,y) = -\begin{pmatrix}0\\1\end{pmatrix}F(\xi+y)e^{i\mu_+(\xi)} - \int_{x}^{\infty} K(\xi,s)F(s+y)ds. \tag{4.32}$$

where $F(z)$ is given in terms of the scattering data through

$$F(z) = -i\sum_k \frac{c_k}{a'(\zeta_k)} e^{i(\zeta_k - A')z} + \frac{1}{2\pi}\int_{\infty}^{\infty} \frac{b(\alpha)}{a(\alpha)} e^{i(\alpha - A')z}\, d\alpha. \tag{4.33}$$

For initial data such that the potential is reflectionless and has only one bound state at $\zeta_1 = \alpha_1 + i\beta_1$ with normalization coefficient $m_o = (c_1(0))/(a'(\zeta_1))$, one recovers the soliton (4.5)-(4.6) with parameters

$$v = \frac{1 - 4|\zeta_1|^2}{1 + 4|\zeta_1|^2} \tag{4.34a}$$

$$A = \frac{\alpha_1}{|\zeta_1|} \tag{4.34b}$$

$$x_o = \frac{1}{2\beta_1} \log \frac{2\beta_1}{|m_o|}. \tag{4.34c}$$

The solution corresponding to an arbitrary number of eigenvalues, which is an n-soliton solution, has also been obtained. See the last paper of Reference 8.

The equations of motion (4.1)-(4.2) admit an infinite set of conserved quantities. This is a consequence of the fact that $a(\zeta)$, the inverse of the transmission coefficient, is independent of time. To find them, one uses the procedure first employed by Zacharov and Faddeev [27] for the KdV equation: Introduce a function $\phi = \log \varphi_1 + i(\zeta - A')\xi$, where φ_1 is the upper component of the Jost solution with boundary conditions given at $\xi \to -\infty$ (4.15). This function has the following asymptotic behavior:

$$\phi \xrightarrow[\xi \to -\infty]{} 0 \qquad \phi \xrightarrow[\xi \to +\infty]{} \log a, \tag{4.35}$$

and as an immediate consequence one has

$$\log a(\zeta) = \int_{-\infty}^{\infty} \frac{\partial \phi}{\partial \xi}\, d\xi. \tag{4.36}$$

Using now the linear equations (4.11), it is simple to see that the derivative $\phi' \equiv \partial\phi/\partial\xi$ satisfies the Ricatti equation

$$\phi'' + (\phi')^2 - 2i(\zeta - A')\phi' = ip + \frac{1}{q}\frac{\partial q}{\partial \xi}(\phi' - i(p-A')) - |q|^2$$

$$+ 2\zeta(p-A') - (p^2 - (A')^2). \tag{4.37}$$

Expanding now ϕ' in a power series as $\zeta \to \infty$

$$\Phi' = \sum_{n=0}^{\infty} \frac{f_n}{\zeta^n} \tag{4.38}$$

and substituting into (4.37) the following recursion relation for the coefficients f_n is obtained

$$\frac{\partial f_n}{\partial \xi} + 2iA'f_n - 2if_{n+1} + \sum_{j+k=n} f_j f_k = \frac{1}{q}\frac{\partial q}{\partial \xi} f_n, \quad n \geqq 2 \tag{4.39}$$

with

$$f_o = i(p-A'), \quad f_1 = -\frac{i}{2}|q|^2. \tag{4.40}$$

Similarly expanding

$$\log a(\zeta) = \sum_{n=0}^{\infty} \frac{C_n}{n}, \tag{4.41}$$

the desired conserved quantities are obtained in the form

$$C_n = \int_{-\infty}^{\infty} f_n d\xi. \tag{4.42}$$

The first three are

$$C_n = i\int_{\infty}^{\infty}\left(\frac{\cos 2\theta}{2\sin^2\theta}\frac{\partial \lambda}{\partial \xi} - A'\right)d\xi$$

$$C_1 = -\frac{i}{2}\int_{-\infty}^{\infty}\left(\cot^2\theta\left(\frac{\partial \lambda}{\partial \xi}\right)^2 + \left(\frac{\partial \lambda}{\partial \xi}\right)^2\right)d\xi$$

$$C_2 = -\frac{i}{4}\int_{\infty}^{\infty}\left\{\cot\theta\frac{\partial^2\theta}{\partial \xi^2} - \frac{\partial \theta}{\partial \xi}\frac{\partial}{\partial \xi}(\cot\theta) - \right.$$

$$\left. - \frac{\cos 2\theta}{\sin^2\theta}\frac{\partial \lambda}{\partial \xi}\left(\cot^2\theta\left(\frac{\partial \lambda}{\partial \xi}\right)^2 + \left(\frac{\partial \theta}{\partial \xi}\right)^2\right)\right\} d\xi. \tag{4.43}$$

As usual, the physical significance of these quantities is completely obscure, except for that of the first two. When $\lambda = 0$ they reduce to the conserved quantities of the sine-Gordon equation, as they should. It is also curious to note that this same set forms a set of conserved quantities for the equations that are obtained from (4.1)-(4.2) by omitting the $\sin\theta$ term in the first of them [28].

As was mentioned in the beginning of this section, it is also possible to solve the Cauchy initial value problem for (4.1)-(4.2). What has been shown here in explicit form is the solution of the characteristic initial value problem for which the corresponding scattering operator is much simpler, as is also the case in sine-Gordon [10]. The canonical (i.e., Hamiltonian) structure of the system, however, is best understood in laboratory coordinates as in this case the temporal coordinate is well defined and the canonical momentum is an independent variable from the field. This is just a restatement of the fact that only the fields must be given initially when working in characteristic coordinates whereas both the field and its normal (time) derivative are the initial data needed for the Cauchy problem, the reason being that the normal to a characteristic surface lies along the surface itself. In the language of Dirac [29], there appears a second-class constraint among the canonical variables when working in light-cone coordinates and the Poisson brackets of the theory must be modified (to so-called Dirac brackets) to make them compatible with the constraint. The appearance of such a constraint signals a reduction in the number of degrees of freedom of the system, and in fact the characteristic initial value problem for Equations (4.1)-(4.2) has half as many solutions as the corresponding Cauchy problem. Another way of understanding this is to notice that either one of the light cone coordinates can be called "time" and that to one Cauchy problem there corresponds two characteristic problems. The solutions to both of the latter give the full set of solutions to the former.

In laboratory coordinates, then, it can be shown [8] that the transformation from fields to scattering data is canonical—that is, it preserves the Poisson brackets that are given naturally by the Lagrangian (2.41). These new variables are of the action-angle type, with the transmission coefficient and eigenvalues being of the action type and the reflection coefficient and normalization constants the corresponding angle variables. This denomination is justified since, in addition to having the correct Poisson brackets, they have the correct time evolution, namely, the action variables are time independent and the angle ones change linearly in time, as a consequence of the fact that the Hamiltonian of the system, again given by the Lagrangian (2.41) is expressible solely in terms of action-type variables. Finally, the conserved quantities are also expressible solely in terms of action variables showing that they are in involution—that is, their Poisson brackets vanish in pairs.

ACKNOWLEDGMENTS

Part of this work was prepared while staying at the Aspen Center for Physics. The hospitality of the Center is gratefully acknowledged.

REFERENCES

1. C. S. Gardner, J. Green, M. Kruskal and R. Miura, Phys. Rev. Lett. **19**, 1095 (1967).
2. D. J. Korteweg and G. de Vries, Phil. Mag. **39**, 422 (1895).
3. P. L. Lax, Commun. Pure and Appl. Math. **21**, 467 (1968).
4. V. E. Zacharov and A. B. Shabat, Zh. Eksp. Teor. Fiz. **61**, 118 (1971) [Sov. Physics JETP **34**, 62 (1972)].
5. See the article by R. Bullough in these Proceedings.
6. H. D. Wahlquist and F. B. Estabrook, J. Math. Phys. **16**, 1 (1975) and **17**, 1293 (1976). See also their "Prolongation Structures, Connection Theory and Bäcklund Transformation," to be published in the Proceedings of the International Symposium of Nonlinear Evolution Equations Solvable via the Inverse Scattering Transformation, Accademia dei Lincei, Rome, June 15-18, 1977, F. Calogero, Ed., and references therein. J. Corones and D. Levermore, Courant Institute Report (unpublished). J. Corones, "Using Pseudopotentials," Iowa State Report (unpublished).
7. G. E. Lamb, Jr., Phys. Rev. Lett. **37**, 235 (1976) and J. Math. Phys. **18**, 1654 (1977).
8. F. Lund and T. Regge, Phys. Rev. **D14**, 1524 (1976); F. Lund, Phys. Rev. **D15**, 1540 (1977); Phys. Rev. Lett. **38**, 1175 (1977); "Classically Solvable Field Theory Model," Princeton Preprint, 1977, Ann. Phys. (to be published).
9. L. Bianchi, Lezioni di Geometria Diferenziale (Spoerri, Pisa, 1922).
10. M. Ablowitz, D. J. Kaup, A. C. Newell and H. Segur, Phys. Rev. Lett. **30**, 1262 (1973) and Stud. Appl. Math. **53**,

249 (1974); L. A. Takhtadjan, Zh. Eksp. Teor. Fiz. 66, 476 (1974) [Sov. Phys. JETP 39, 228 (1974)]; V. E. Zakharov, L. A. Takhtadzhan and L. D. Faddeev, Dokl. Akad. Nauk. SSSR 219, 1334 (1974) [Sov. Phys. Dikl. 19, 824 (1975)]; L. A. Takhtadzhan and L. D. Faddeev, Teor. Mat. Fiz. 21, 160 (1974).

11. These results are standard. See, for example, J. J. Stoker, Differential Geometry, Wiley, New York-London-Sidney-Toronto, 1969, and M. Spivak, A Comprehensive Introduction to Differential Geometry, Publish or Perish, Boston, 1975.
12. A. Neveu and N. Papanicolaou, "Integrability of the Classical $(\bar{\psi}_i\psi_i)_2^2$ and $(\bar{\psi}_i\psi_i)_2^2 - (\bar{\psi}_i\gamma_5\psi_i)_2^2$ Interactions," Commun. Math. Phys. 58, 31 (1978).
13. Y. Nambu and G. Jona-Lasinio, Phys. Rev. 122, 345 (1961).
14. D. J. Gross and A. Neveu, Phys. Rev. D10, 3235 (1974).
15. R. F. Dashen, B. Hasslacher and A. Neveu, Phys. Rev. D12, 2443 (1975).
16. S. S. Shei, Phys. Rev. D14, 535 (1976).
17. This equation has also been considered by B. S. Getmanov, Pis'ma Zh. Eksp. Teor. Fiz. 25, 132 (1977) [JETP Lett. 25, 119 (1977)].
18. K. Pohlmeyer, Commun. Math. Phys. 46, 207 (1976).
19. M. Gell-Mann and M. Lévy, Nuovo Cim. 16, 705 (1960); F. Gürsey, Nuovo Cim. 16, 230 (1960); S. Weinberg, Phys. Rev. 166, 1568 (1968).
20. S. Weinberg, 1976 Erice Lecture Notes (unpublished) and references therein.
21. A. A. Belavin and A. M. Polyakov, Zh. Eksp. Teor. Fiz. Pis'ma Red. 22, 503 (1975), [JETP Lett. 22, 245 (1975)].
22. L. P. Eisenhart, Riemannian Geometry, Princeton University Press, Princeton, 1925, Chapter V.
23. P. L. Tchebychef, "Sur la Coupure des Vêtements," Oeuvres, Bd. II, S.708, 1878.
24. This problem has been independently considered by M. Jaulent as arising from a Schrödinger equation with a velocity-dependent potential. See Jaulent's contribution to these Proceedings.
25. F. Prats and J. Toll, Phys. Rev. 113, 363 (1959).
26. D. J. Kaup, Prog. Theor. Phys. 54, 396 (1975).
27. V. E. Zacharov and L. D. Faddeev, Func. Anal. and Its Applic. 5, 280 (1972).
28. B. Julia and F. Lund, in preparation.

29. P. A. M. Dirac, Lectures in Quantum Mechanics, Yeshiva University, New York, 1964. See also A. J. Hanson, T. Regge and C. Teitelboim, Constrained Hamiltonian Systems, Accademia dei Lincei, Rome, 1976.

HIROTA'S METHOD OF SOLVING SOLITON-TYPE EQUATIONS†

P. J. Caudrey
University of Manchester Institute of Science and Technology
Manchester M60 1QD, England

INTRODUCTION. Although it is a gross oversimplification to say that Hirota's method amounts to guesswork, this is basically true. It is extremely useful when more sophisticated methods have failed. Its principal drawback (apart from the guesswork element) is that it gives only soliton solutions and no background (or 'radiation').

1. THE KORTEWEG-DE VRIES EQUATION

The method is best illustrated by example. Let us consider the Korteweg-de Vries (KdV) equation.

$$u_{xxx} + 12uu_x + u_t = 0. \tag{1.1}$$

Using the Hopf-Cole transformation

$$u(x,t) = \frac{\partial^2}{\partial x^2} \ln f(x,t), \tag{1.2}$$

we find (after an integration with respect to x) that

$$ff_{xxxx} - 4f_x f_{xxx} + 3f_{xx}^2 + ff_{xt} - f_x f_t = 0. \tag{1.3}$$

† Lecture given at NATO Advanced Study Institute on Nonlinear Equations in Physics and Mathematics, Istanbul, August 1977.

A. O. Barut (ed.), Nonlinear Equations in Physics and Mathematics, 177–192.

This may be written in the form

$$\left(D_x^4 + D_x D_t\right)(f.f) = 0 \tag{1.4}$$

where the D-operators are defined by

$$D_x^m D_t^n (a\cdot b) =$$

$$= \left[\left(\frac{\partial}{\partial x} - \frac{\partial}{\partial x'}\right)^m \left(\frac{\partial}{\partial t} - \frac{\partial}{\partial x'}\right)^n a(x,t)\, b(x',t')\right]_{\substack{x'=x\\ t'=t}} . \tag{1.5}$$

A few properties of these operators are

i) $D_x^m D_t^n (a\cdot b) = (-1)^{m+n} D_x^m D_t^n (b\cdot a)$,

ii) $D_x^m D_t^n (a\cdot a) = 0$ if $m+n$ is odd,

iii) $D_x^m D_t^n (a\cdot 1) = (\partial^{m+n}/(\partial x^m \partial t^n))a$

iv) $D_x^m D_t^n (e^{kx+\omega t}\cdot e^{k'x+\omega' t}) = (k-k')^m(\omega-\omega')^n e^{(k+k')x+(\omega+\omega')t}$.

Hirota [1] lists many more.

It is obvious that $f=1$ is a solution of the bilinear equation (1.4) and so we 'perturb about $f=1$' by putting

$$f(x,t) = 1 + \epsilon f^{(1)}(x,t) + \epsilon^2 f^{(2)}(x,t) + \cdots \tag{1.6}$$

where ϵ is our 'perturbation parameter.' We substitute this series in (1.4) and equate the coefficients of powers of ϵ.

$$\epsilon^1 \Rightarrow \quad \left(D_x^4 + D_x D_t\right)\left(f^{(1)}\cdot 1 + 1.f^{(1)}\right) = 0.$$

I.e.,

$$f^{(1)}_{xxxx} + f^{(1)}_{xt} = 0. \tag{1.7}$$

We choose what can arguably be called the simplest nontrivial solution of this

$$f^{(1)} = e^{\theta} \tag{1.8}$$

where

$$\theta = kx - \omega t + \delta$$

with

$$\omega = k^3.$$

k, ω and δ are constants.

$$\epsilon^2 \Rightarrow \left(D_x^4 + D_x D_t\right)\left(f^{(2)}.1 + f^{(1)}.f^{(1)} + 1.f^{(2)}\right) = 0.$$

I.e.,

$$f_{xxxx}^{(2)} + f_{xt}^{(2)} = 0. \tag{1.9}$$

In this case we choose the trivial solution

$$f^{(2)} = 0. \tag{1.10}$$

Similarly, we can choose

$$f^{(3)} = f^{(4)} = \cdots = 0, \tag{1.11}$$

giving

$$f = 1 + e^{\theta} \tag{1.12}$$

where the ϵ has been absorbed into the arbitrary constant δ in θ. Then

$$u = D_x^2(f.f)/(2f^2)$$

$$= k^2 e^{\theta}/(1 + e^{\theta})^2$$

$$= \tfrac{1}{4}k^2 \operatorname{sech}^2(\theta/2), \tag{1.13}$$

which is the classic one-soliton solution.

To obtain the two-soliton solution, we choose $f^{(1)}$ a little more generally

$$f^{(1)} = \exp\theta_1 + \exp\theta_2 \tag{1.14}$$

where

$$\theta_i = k_i x - \omega_i t + \delta_i$$

with

$$\omega_i = k_i^3.$$

The equation for the coefficient of ϵ^2 now becomes

$$f^{(2)}_{xxxx} + f^{(2)}_{xt} = -(k_1-k_2)\left\{(k_1-k_2)^3 - (\omega_1-\omega_2)\right\}\exp(\theta_1+\theta_2). \tag{1.15}$$

The simplest choice is now

$$f^{(2)} = a(1,2)\exp(\theta_1+\theta_2) \tag{1.16}$$

where

$$a(1,2) = -\frac{(k_1-k_2)\{(k_1-k_2)^3-(\omega_1-\omega_2)\}}{(k_1+k_2)\{(k_1+k_2)^3-(\omega_1-\omega_2)\}}$$

$$= \left(\frac{k_1-k_2}{k_1+k_2}\right)^2 . \tag{1.17}$$

It still proves possible to choose

$$f^{(3)} = f^{(4)} = \cdots = 0 \tag{1.18}$$

so that

$$f = 1 + \exp\theta_1 + \exp\theta_2 + a(1,2)\exp(\theta_1+\theta_2) \tag{1.19}$$

(with ϵ absorbed into δ_1 and δ_2). Then

$$u = \left\{k_1^2\exp\theta_1 + k_2^2\exp\theta_2 + 2(k_1-k_2)^2\exp(\theta_1+\theta_2) + (k_2^2\exp\theta_1 + k_1^2\exp\theta_2)a(1,2)\exp(\theta_1+\theta_2)\right\} \Big/ \left\{1 + \exp\theta_1 + \exp\theta_2 + a(1,2)\exp(\theta_1+\theta_2)\right\}^2 . \tag{1.20}$$

We can without loss of generality assume that

$$0 < k_1 < k_2. \tag{1.21}$$

Then if $t \ll 0$,

$$u \approx \tfrac{1}{4}\, k_1^2 \operatorname{sech}^2\left(\frac{\theta_1+\beta}{2}\right) + \tfrac{1}{4}\, k_2^2 \operatorname{sech}^2\left(\frac{\theta_2}{2}\right), \tag{1.22}$$

and if $t \gg 0$,

$$u \approx \tfrac{1}{4}k_1^2 \operatorname{sech}^2\left(\frac{\theta_1}{2}\right) + \tfrac{1}{4}k_2^2 \operatorname{sech}^2\left(\frac{\theta_2+\beta}{2}\right), \tag{1.23}$$

where

$$\beta = \ln\left\{\left(\frac{k_1-k_2}{k_1+k_2}\right)^2\right\}. \tag{1.24}$$

This clearly illustrates the soliton collision property.

For the N-soliton solution we choose

$$f^{(1)} = \sum_{i=1}^{N} \exp\theta_i. \tag{1.25}$$

At each succeeding stage we choose $f^{(j)}$ to be a sum of exponentials with the least possible number of terms. Eventually, we find that $f^{(j)} = 0$ for $j > N$ and so

$$f = \sum_{m=0}^{N} \sum_{{}^N C_m} a(i_1, i_2, \ldots, i_m)\exp(\theta_{i_1} + \theta_{i_2} + \ldots + \theta_{i_m}) \tag{1.26}$$

where the summation over ${}^N C_m$ is over all combinations $(i_1, i_2, \ldots, i_m)$ chosen from $(1, 2, \ldots, N)$. We also find that

$$a(i_1, i_2, \ldots, i_m) = \prod_{j=1}^{N-1} \prod_{k=j+1}^{m} a(i_j, i_k). \tag{1.27}$$

2. THE SINE-GORDON EQUATION

As a second example, we consider the sine-Gordon (SG) equation,

$$\sigma_{xx} - \sigma_{tt} = \sin\sigma. \tag{2.1}$$

We put

$$\sigma = 4\tan^{-1}(g/f) \tag{2.2}$$

and get

$$\left(f^2-g^2\right)\left(D_x^2-D_t^2\right)(f.g) - fg\left(D_x^2-D_t^2\right)(f.f-g.g) = fg\left(f^2-g^2\right). \tag{2.3}$$

This can be reduced to two bilinear equations

$$\left(D_x^2 - D_t^2\right)(f.g) = \mu fg \tag{2.4}$$

$$\left(D_x^2 - D_t^2\right)(f.f - g.g) = (\mu - 1)\left(f^2 - g^2\right) \tag{2.5}$$

where μ is a constant (there is no loss of generality in this).

In this case the obvious solution is $f = 1$, $g = 0$ which requires $\mu = 1$.

$$\left(D_x^2 + D_t^2\right)(f.g) = fg \tag{2.6}$$

$$\left(D_x^2 - D_t^2\right)(f.f - g.g) = 0. \tag{2.7}$$

We now put

$$f = 1 + \epsilon^2 f^{(1)} + \epsilon^4 f^{(2)} + \cdots \tag{2.8}$$

and

$$g = \epsilon g^{(1)} + \epsilon^3 g^{(2)} + \epsilon^5 g^{(3)} + \cdots . \tag{2.9}$$

Again we equate coefficients of powers of ϵ.

$$\epsilon^1 \quad \Longrightarrow \quad g_{xx}^{(1)} - g_{tt}^{(1)} = g^{(1)}. \tag{2.10}$$

The simplest nontrivial solution to this is

$$g^{(1)} = e^{\theta} \tag{2.11}$$

where

$$\theta = kx - \omega t + \delta$$

with

$$k^2 - \omega^2 = 1.$$

All the other terms can be chosen to be zero so that

$$\sigma = 4 \tan^{-1} e^{\theta} \tag{2.12}$$

(with ϵ absorbed as before) which is the well-known one-kink solution.

For two kinks we put

$$g^{(1)} = \exp\theta_1 + \exp\theta_2 \tag{2.13}$$

where

$$\theta_i = k_i x - \omega_i t + \delta_i$$

with

$$k_i^2 - \omega_i^2 = 1.$$

Then the simplest choice for $f^{(1)}$ is

$$f^{(1)} = a(1,2)\exp(\theta_1 + \theta_2) \tag{2.14}$$

where

$$a(1,2) = \frac{1-k_1k_2+\omega_1\omega_2}{1+k_1k_2-\omega_1\omega_2} = -\left(\frac{k_1-k_2}{\omega_1+\omega_2}\right)^2 = -\left(\frac{\omega_1-\omega_2}{k_1+k_2}\right)^2. \tag{2.15}$$

Again all other terms can be chosen to be zero which gives

$$\sigma = -4\tan^{-1}\left(\left|\frac{k_1+k_2}{\omega_1-\omega_2}\right|\frac{\cosh\frac{1}{2}(\theta_1'-\theta_2')}{\sinh\frac{1}{2}(\theta_1'+\theta_2')}\right) \tag{2.16}$$

where

$$\theta_i' = \theta_i - \tfrac{1}{2}\ln\left|\frac{k_1+k_2}{\omega_1-\omega_2}\right|.$$

Solutions for higher numbers of kinks can be found in a similar way.

3. THE DOUBLE-SINE-GORDON EQUATION

Now let us turn to an equation for which no multisoliton solution is yet known. This is the double-sine-Gordon (DSG) equation,

$$\sigma_{xx} - \sigma_{tt} = \sin\sigma + 2\lambda\sin\tfrac{1}{2}\sigma. \tag{3.1}$$

Using the same substitution as before,

$$\sigma = 4\tan^{-1}(g/f), \tag{3.2}$$

we find

$$\left(D_x^2 - D_t^2\right)(f.g) = \mu fg \tag{3.3}$$

$$\left(D_x^2 - D_t^2\right)(f.f - g.g) = (\mu - 1)\left(f^2 - g^2\right) - \lambda\left(f^2 + g^2\right). \tag{3.4}$$

The obvious solution is again $f = 1$, $g = 0$, and it requires $\mu = \lambda + 1$.

$$\left(D_x^2 - D_t^2\right)(f.g) = (1+\lambda)fg \tag{3.5}$$

$$\left(D_x^2 - D_t^2\right)(f.f - g.g) = -2\lambda g^2. \tag{3.6}$$

Again we put

$$f = 1 + \epsilon^2 f^{(1)} + \epsilon^4 f^{(2)} + \cdots \tag{3.7}$$

$$g = \epsilon g^{(1)} + \epsilon^3 g^{(2)} + \epsilon^5 g^{(3)} + \cdots \tag{3.8}$$

$$\epsilon^1 \quad \Rightarrow \qquad g^{(1)}_{xx} - g^{(1)}_{tt} = (1+\lambda)g^{(1)}. \tag{3.9}$$

For one kink we put

$$g^{(1)} = e^{\theta} \tag{3.10}$$

where

$$\theta = kx - \omega t + \delta$$

with

$$k^2 - \omega^2 = 1 + \lambda.$$

$$\epsilon^2 \quad \Rightarrow \qquad f^{(1)}_{xx} - f^{(1)}_{tt} = -\lambda e^{2\theta}. \tag{3.11}$$

The simplest choice for $f^{(1)}$ is

$$f^{(1)} = -\frac{1}{4} \cdot \frac{\lambda}{1+\lambda}\, e^{2\theta} \tag{3.12}$$

and all the other terms are chosen to be zero. This gives (with ϵ absorbed).

$$\sigma = 4\tan^{-1}\left\{\frac{e^{\theta}}{1 - \frac{1}{4}\cdot(\lambda/(1+\lambda))e^{2\theta}}\right\}$$

$$= 4\tan^{-1} e^{\theta'+\Delta} + 4\tan^{-1} e^{\theta'-\Delta} \qquad (3.13)$$

where

$$\theta' = 0 + \tfrac{1}{2}\ln\left(\frac{1}{4}\cdot\frac{\lambda}{1+\lambda}\right)$$

and

$$\Delta = \sinh^{-1}\frac{1}{\sqrt{\lambda}} .$$

Thus we have one double kink in which the two halves are displaced by a distance $2\Delta/k$ relative to each other.

We would expect to obtain the two-kink solution by putting

$$g^{(1)} = \exp\theta_1 + \exp\theta_2 \qquad (3.14)$$

and that the series would terminate after the ϵ^4 terms. However, this is not what happens. The two series do not terminate and Hirota has found that they diverge. Thus in this example the method fails.

4. A HIERARCHY OF KdV EQUATIONS

Let us now return to the KdV equation. Lax [2] has used the inverse scattering method to obtain a hierarchy of soluble KdV-type equations of the form

$$K_x^{(n)} + u_t = 0 \qquad (4.1)$$

where the $K^{(n)}$ are given by the recurrence formula

$$K_x^{(n)} = K_{xxx}^{(n-1)} + 8uK_x^{(n-1)} + 4u_x K^{(n-1)} \qquad (4.2)$$

starting from

$$K^{(-1)} = \tfrac{1}{4} . \qquad (4.3)$$

The first few are

$$K^{(0)} = u$$

$$K^{(1)} = u_{xx} + 6u^2$$

$$K^{(2)} = u_{4x} + 20uu_{xx} + 10u_x^2 + 40u^3$$

$$K^{(3)} = u_{6x} + 28uu_{4x} + 56u_x u_{3x} + 42u_{xx}^2 + 280u^2 u_{xx} + 280uu_x^2 + 280u^4$$

where the notation is

$$u_{jx} = \frac{\partial^j}{\partial x^j} u.$$

$K^{(0)}$ gives the linear equation $u_x + u_t = 0$ while $K^{(1)}$ gives the KdV equation.

It will be noticed that there is a sort of 'homogeneity' in the terms of $K^{(n)}$.

Number of u's + $\frac{1}{2}$ (number of differentiations w.r.t.x.) = n+1. This can be expressed in terms of 'rank' as introduced by Kruskal, Miura, Gardner, and Zabusky in 1970. We can define a rank operator

$$\hat{R} = \sum_j (1 + \tfrac{1}{2}j) u_{jx} \frac{\partial}{\partial u_{jx}} \tag{4.4}$$

and then

$$\hat{R}K^{(n)} = (n+1)K^{(n)}. \tag{4.5}$$

A question which we asked ourselves in Manchester is: "Can we find another hierarchy of KdV-type equations of the form

$$A_x^{(n)} + u_t = 0 \tag{4.6}$$

which

i) has multisoliton solutions,
ii) is such that $A^{(n)}$ has rank n+1, and
iii) is distinct from Lax's hierarchy?"

For $n=1$ the most general form satisfying ii) is

$$A^{(1)} = \alpha u_{xx} + \beta u^2. \tag{4.7}$$

The scaling $t \to (1/\alpha)t$ and $u \to 6(\alpha/\beta)u$ gives the KdV equation again.

For $n=2$

$$A^{(2)} = u_{4x} + \alpha u u_{xx} + \beta u_x^2 + \gamma u^3 \tag{4.8}$$

where the coefficient in front of the first term has already been scaled to unity. We try the Hopf-Cole transformation

$$u = \frac{\partial^2}{\partial x^2} \ln f \tag{4.9}$$

and hope that Hirota's method will help. The resulting equation is homogeneous of degree six in f and its derivatives. However, we can still hope for a one-soliton solution of the form

$$f = 1 + e^{\theta} \tag{4.10}$$

where

$$\theta = kx - \omega t + \delta.$$

This gives

$$u = \tfrac{1}{4} k^2 \operatorname{sech}^2 \tfrac{1}{2}\theta \tag{4.11}$$

and requires

$$\beta = 30 - \alpha \tag{4.12}$$

$$\gamma = 2\alpha \tag{4.13}$$

and

$$\omega = k^5. \tag{4.14}$$

For two solitons the obvious suggestion is

$$f = 1 + \exp\theta_1 + \exp\theta_2 + a^{(2)}(1,2)\exp(\theta_1+\theta_2) \tag{4.15}$$

where

$$\theta_i = k_i x - \omega_i t + \delta_i$$

and

$$\omega_i = k_i^5.$$

This requires either

$$\alpha = 20, \tag{4.16}$$

$$a^{(2)}(1,2) = \left(\frac{k_1-k_2}{k_1+k_2}\right)^2 \tag{4.17}$$

which gives Lax's equation, or

$$\alpha = 30 \tag{4.18}$$

$$a^{(2)}(1,2) = -\left(\frac{k_1-k_2}{k_1+k_2}\right)\left(\frac{(k_1-k_2)^5-(\omega_1-\omega_2)}{(k_1+k_2)^5-(\omega_1+\omega_2)}\right) \tag{4.19}$$

which gives

$$A^{(2)} = u_{4x} + 30uu_{xx} + 60u^3. \tag{4.20}$$

The equation for f now reduces to

$$f^4\left(ff_{6x} - 6f_xf_{5x} + 15f_{xx}f_{4x} - 10f_{3x}^2 + ff_{xt} - f_xf_t\right) = 0.$$

I.e.,

$$\left(D_x^6 + D_xD_t\right)(f.f) = 0. \tag{4.21}$$

The obvious generalization of this and Equation (1.4) is

$$\left(D_x^{2n+2} + D_xD_t\right)(f.f) = 0 \tag{4.22}$$

which gives

$$A^{(n)} = \tfrac{1}{2}(2n+2)! \sum_{(m)} \prod_{j=1}^{m} \frac{1}{m_j!}\left(\frac{2u_{(2j-2)x}}{(2j)!}\right)^{m_j} \tag{4.23}$$

where the summation over (m) is over all sets of positive integers m_j satisfying

$$\sum_j jm_j = n+1. \tag{4.24}$$

The first few $A^{(n)}$ are

$$A^{(0)} = u$$

$$A^{(1)} = u_{xx} + 6u^2$$

$$A^{(2)} = u_{4x} + 30uu_{xx} + 60u^3$$

$$A^{(3)} = u_{6x} + 56uu_{4x} + 70u_{xx}^2 + 840u^2u_{xx} + 840u^4$$

$$A^{(4)} = u_{8x} + 90uu_{6x} + 420u_{xx}u_{4x} + 2520u^2u_{4x} + 6300uu_{xx}^2 + 25200u^3u_{xx} + 15120u.$$

$A^{(0)}$ and $A^{(1)}$ are the same as Lax's $K^{(0)}$ and $K^{(1)}$. The equation derived from $A^{(2)}$ was studied by Sawada and Kotera [3] in 1974.

The one- and two-soliton solutions for all of these are given by

$$u = \frac{\partial^2}{\partial x^2} \ln f \tag{4.25}$$

where for one soliton

$$f = 1 + \exp \theta_1 \tag{4.26}$$

and for two solitons

$$f = 1 + \exp \theta_1 + \exp \theta_2 + a^{(n)}(1,2)\exp(\theta_1+\theta_2) \tag{4.27}$$

where

$$\theta_i = k_i x - \omega_i t + \delta_i,$$

$$\omega_i = k_i^{2n+1}$$

and

$$a^{(n)}(1,2) = -\left(\frac{k_1-k_2}{k_1+k_2}\right)\left(\frac{(k_1-k_2)^{2n+1} - (\omega_1-\omega_2)}{(k_1+k_2)^{2n+1} - (\omega_1+\omega_2)}\right).$$

For the N-soliton solutions we would expect

$$f = \sum_{m=0}^{N} \sum_{{}^N C_m} a^{(n)}(i_1, i_2, \ldots, i_m) \exp\left(\theta_{i_1} + \theta_{i_2} + \cdots + \theta_{i_m}\right). \tag{4.28}$$

However, this turns out to be correct only for $n=1, 2$. I.e., for the KdV equation and the next one in the hierarchy. This does not necessarily mean that the higher equations have no N-soliton solutions but merely that this is not the correct form.

In the cases $n=1,2$

$$a^{(n)}\left(i_1, i_2, \ldots, i_m\right) = \prod_{j=1}^{m-1} \prod_{k=j+1}^{m} a^{(n)}(i_j, i_k). \tag{4.29}$$

5. POLYNOMIAL CONSERVED DENSITIES

If T and X are two polynomials of u and its derivatives such that

$$T_t + X_x = 0 \tag{5.1}$$

when u satisfies a given equation, then T is called a polynomial conserved density (PCD) of that equation. If there exists another polynomial F such that

$$T = F_x \tag{5.2}$$

and

$$X = -F_t , \tag{5.3}$$

then the conservation law is trivial. It is easy to see that for the equations in our new hierarchy terms of rank r in T require terms of rank $r+n$ in X and vice versa. Thus any PCD can be split into a sum of PCDs which are eigenfunctions of the rank operator.

For $n=1$, the inverse scattering method provides us with an infinity of nontrivial PCDs, one of each rank. At the time this work was done, no such method was available for $n=2$. For this reason we had to proceed the hard way by writing down all possible terms of appropriate rank for X and T, eliminating the trivial terms, and then differentiating and equating coefficients.

For $r=1$ the only nontrivial conservation law is the equation itself since u is the only possible term of rank 1.

$$T^{(1)} = u$$

$$X^{(1)} = u_{4x} + 30uu_{xx} + 60u^3.$$

For $r=2$,

$$T^{(2)} = \alpha u_{xx} + \beta u^2.$$

The u_{xx} term is trivial so we put $\alpha = 0$.

$$X^{(2)} = au_{6x} + buu_{4x} + cu_x u_{3x} + du_{xx}^2 + eu^2 u_{xx} + fuu_x^2 + gu^4.$$

Differentiating these gives

$$\begin{aligned} T_t^{(2)} &= 2\beta uu_t \\ &= -2\beta u\left(u_{5x} + 30uu_{3x} + 30u_x u_{xx} + 180u^2 u_x\right) \\ &= -X_x^{(2)} \\ &= -au_{7x} - buu_{5x} - (b+c)u_x u_{4x} - (c+2d)u_{xx}u_{3x} \\ &\quad - eu^2 u_{3x} - 2(e+f)uu_x u_{xx} - fu_x^3 - 4gu^3 u_x. \end{aligned} \tag{5.4}$$

Equating the coefficients gives

$$\beta = a = b = c = d = e = f = g = 0. \tag{5.5}$$

So there is no nontrivial PCD of rank 2.

$$T^{(3)} = u_x^2 - 2u^3$$

$$\begin{aligned} X^{(3)} &= 2u_x u_{5x} - 2u_{xx}u_{4x} + u_{3x}^2 + 72uu_x u_{3x} - 36uu_{xx}^2 \\ &\quad + 48u_x^2 u_{xx} - 6u^2 u_{4x} + 360u^2 u_x^2 - 180u^3 u_{xx} - 216u^5 \end{aligned}$$

$$T^{(4)} = u_{xx}^2 - 18uu_x^2 + 12u^4$$

$$\begin{aligned} X^{(4)} &= 2u_{xx}u_{6x} - 2u_{3x}u_{5x} + u_{4x}^2 - 36uu_x u_{5x} + 96uu_{xx}u_{4x} \\ &\quad - 48uu_{3x}^2 + 64u_{xx}^3 + 48u_x u_{xx} u_{3x} + 18u_x^2 u_{4x} \\ &\quad + 792u^2 u_{xx}^2 - 252uu_x^2 u_{xx} - 1224u^2 u_x u_{3x} + 48u^3 u_{4x} \\ &\quad + 108u_x^4 - 5400u^3 u_x^2 + 1440u^4 u_{xx} + 1440u^6. \end{aligned}$$

There is no nontrivial PCD of rank 5.

$$T^{(6)} = u_{4x}^2 - 42uu_{3x}^2 + 32u_{2x}^3 + 576u^2u_{xx}^2 - 240u_x^4 - 3600u^3u_x^2 + 576u^6$$

$$T^{(7)} = u_{5x}^2 - 54uu_{4x}^2 + 192u_{xx}u_{3x}^2 + 1044u^2u_{3x}^2 - 1956uu_{2x}^3 - 2592u_x^2u_{xx}^2 + 9082uu_x^4 - 9576u^3u_{xx}^2 - 45360u^4u_x^2 + 4320u^7 .$$

The fluxes $X^{(6)}$ and $X^{(7)}$ are too long to write down here as they have many terms and some of the numerical coefficients run into seven figures.

We speculate that there are nontrivial PCDs of ranks $3r$ and $(3r-2)$ but none of rank $(3r-1)$.

More recent work by Satsuma and Kaup [4] shows how conservation laws for all integer and half-integer ranks can be generated. However, up to rank 7 these are all either trivial or trivially equivalent to those given above. The speculation has still to be proved or disproved.

* * *

Most of this work was done by my colleagues, R. K. Dodd, J. D. Gibbon, and myself [5] in 1975 with a lot of helpful advice from R. K. Bullough. Credit for the rather tedious algebra in the last section goes to MIRA, a computer algebraic manipulation scheme due to Gawlik [6] which has been implemented on an I.C.L. 1906A computer in Manchester.

REFERENCES

1. R. Hirota, Prog. Theor. Phys. **52**, 1498 (1974).
2. P. D. Lax, Comm. Pure Appl. Math. **21**, 467 (1968); P. D. Lax, Comm. Pure Appl. Math. **28**, 141 (1975).
3. K. Sawada and T. Kotera, Prog. Theor. Phys. **51**, 1355 (1974).
4. J. Satsuma and D. J. Kaup, to be published.
5. P. J. Caudrey, R. K. Dodd, and J. D. Gibbon, Proc. Roy. Soc. (London) A351, 407 (1976).
6. J. H. Gawlik, Royal Armament Research and Development Establishment Technical Report, 10/77 (1977).

PROLONGATION STRUCTURE TECHNIQUES FOR THE NEW HIERARCHY OF KORTEWEG-DE VRIES EQUATIONS†

R. K. Dodd‡
Department of Mathematics
University of Manchester Institute of
Science and Technology
Manchester M60 1QD

In his lecture Dr. Caudrey produced a hierarchy of equations [1], [2], all the members of which have soliton solutions and which has the Korteweg-de Vries equation as the equation of lowest order. This hierarchy differs from the Lax hierarchy of equations [3] which can all be solved by the inverse method associated with the scalar Schrödinger equation as shown by Gardner, Greene, Kruskal, and Miura [4]. The next equation in this new hierarchy is the member of the one-parameter family of equations,

$$\left(q_{4x} + \alpha q q_{2x} + (30-\alpha) q_x^2 + 2\alpha q^3\right)_x + q_t = 0 \qquad (1)$$

for which $\alpha = 30$. I report work done in collaboration with J. D. Gibbon [5] which derives an inverse scattering problem and a Bäcklund transformation for this equation.

The method used is the prolongation technique due to Wahlquist and Estabrook [6], [7], [8]. The last reference in particular is an excellent review of this method and its interpretation in terms of differential geometry for the case of two independent variables. Consider the set of two forms [9],

---

† Lecture given at NATO Advanced Study Institute on Nonlinear Equations in Physics and Mathematics, Istanbul, August 1977.

‡ Present address: School of Mathematics, Trinity College, Dublin 2, Ireland.

A. O. Barut (ed.), Nonlinear Equations in Physics and Mathematics, 193–204.

$$\begin{aligned}
\Delta^1 &= dq \wedge dt - u\, dx \wedge dt \\
\Delta^2 &= du \wedge dt - r\, dx \wedge dt \\
\Delta^3 &= dr \wedge dt - v\, dx \wedge dt \\
\Delta^4 &= dv \wedge dt - s\, dx \wedge dt \\
\Delta^5 &= ds \wedge dt - dq \wedge dx + \left(\alpha qv + (60-\alpha)ur + 6\,\alpha q^2 u\right) dx \wedge dt,
\end{aligned} \tag{2}$$

which are defined on a manifold M with local coordinates $w^i = (x,t,q,u,r,v,s)$. If N is a submanifold of M with local coordinates x^i, dim N < dim M, then $w^j = w^j(x^i)$. ϕ, the corresponding map, $\phi: N \to M$ is injective, open in M and the Jacobian of its differential is nonsingular and of rank dim N. This map induces a map ϕ^* which "pulls back" forms on M into forms on N. In local coordinates it is obtained by the "substitution of coordinate functions." Thus if

$$\omega = a_i(w)dw^i$$

is a one-form on M, the induced one-form on N is

$$\phi^*(\omega) = \left(a_i(w(x)) \frac{\partial w^i}{\partial x^j}\right) dx^j$$

where we have assumed the usual summation convention. $\phi^*\omega$ is called the restriction of ω to N or we say that ω is sectioned into N. If $\phi^*(\omega) = 0$, then the submanifold N annuls ω and in this case N is called an integral manifold for ω.

Consider the two-dimensional submanifold of M, N(x,t) defined by

$$(x,t) \longrightarrow \Big(x,t,q(x,t), u = q_x(x,t),\; r = q_{2x}(x,t),\; v = q_{3x}(x,t),\; s = q_{4x}(x,t)\Big)$$

where $q_{nx}(x,t) = (\partial^n q/\partial_x^n)(x,t)$. Then the two forms (2) sectioned into N become

$$\begin{aligned}
\phi^*\Delta^1 &= (q_x - u)dx \wedge dt \\
\phi^*\Delta^2 &= (u_x - r)dx \wedge dt \\
\phi^*\Delta^3 &= (r_x - v)dx \wedge dt \\
\phi^*\Delta^4 &= (v_x - s)dx \wedge dt \\
\phi^*\Delta^5 &= \left(s_x + \alpha q_t + qv + (60-\alpha)ur + 6\alpha q^3 u\right)dx \wedge dt.
\end{aligned} \tag{3}$$

Thus comparing with (1), we see that N(x,t) is an integral manifold for the set of forms (2) on M when q is a solution of Equation (1).

Therefore, in investigating solutions to (1), we are led to the integral manifolds of the set of forms (2): the two problems are not equivalent, however. If we define $f \wedge g(x) = f(g(x))$, then the set of forms on M forms a ring. The condition that the exterior derivatives of the set of forms (2) are contained in the ideal of forms generated by (2) is the condition for equivalence, in terms of solutions, between the two problems as shown in the general case by Cartan [10]. The differential ideal I which has these properties is called an exterior system by Cartan.

Explicitly we require that

$$d\Delta^i = \omega^i_j \wedge \Delta^j \quad \text{where} \quad \omega^i_j \text{ are one-forms on M.} \tag{4}$$

It is easy to check that this condition is satisfied by all the Δ^i; for example,

$$d\Delta^1 = -du \wedge dx \wedge dt = dx \wedge \Delta^2;$$

thus the two-forms Δ^i, (2) generate an exterior system.

In general the equations that constitute an inverse method for (1), if it exists, will have the form,

$$\begin{aligned}
\underline{y}_x &= -R\left(q, q_x, \ldots, q_{\ell x}, q_t, \ldots, q_{mt}, q_{xt}, \ldots, q_{\ell xmt}\right)\underline{y} \\
\underline{y}_t &= -S\left(q, q_x, \ldots, q_{\ell x}, q_t, \ldots, q_{mt}, q_{xt}, \ldots, q_{\ell xmt}\right)\underline{y}
\end{aligned} \tag{5}$$

where $\underline{y}$ is an $n \times 1$ column vector and R, S are $n \times n$ matrix functionals of the potential q and its partial derivatives which do not explicitly depend on x and t. Equations (5) can be written as the vectorial one-form

$$d\underline{y} + R\underline{y}dx + S\underline{y}dt = 0 \tag{6}$$

on an integral manifold of (1).

Wahlquist and Estabrooks' method for determining R and Q is to prolong the exterior system generated by (2) by including the one-forms which when pulled back to N(x,t) have the form (6). The enlarged manifold $\overline{M}$ has the local coordinates

$$(x,t,q,u,r,v,s,y^1, \ldots, y^n),$$

and the variables $\{y^i\}$ are called the prolongation variables. The forms on $\overline{M}$ which pull back to the forms (6) are conveniently written as

$$\omega^i = dy^i + F^i(q,u,r,v,s,y)dx + G^i(q,u,r,v,s,y)dt. \tag{7}$$

For the set of forms (2) and (7) to generate an exterior system $\overline{I}$, requires that

$$d\omega^i = \eta_j^{\ i} \wedge \omega^j + f_j^{\ i}\Delta^j \tag{8}$$

where η^i, and f^i_j, are respectively one and zero (functions) forms on $\overline{M}$. Condition (8) requires F^i and G^i to satisfy the nonlinear partial differential equations

$$F^i = F^i(q,y) \qquad G^i = G^i(q,u,r,v,s,y)$$

$$G_s^i + F_s^i = 0$$

$$uG_q^i + rG_u^i + vG_r^i + sG_v^i + \left((60-\alpha)ur + \alpha qv + 6\alpha q^2u\right)F_q^i + [G,F]^i = 0$$

where

$$[G,F]^i = G^i \frac{\partial F^i}{\partial y^j} - F^j \frac{\partial G^i}{\partial y^i}. \tag{9}$$

Because of the first partial differential equation in (9) and the simple form of F^i, it proves possible to completely determine F^i and G^i.

$$F^i = \tfrac{1}{2}q^2Y_1^i + qY_2^i + Y_3^i$$

$$\left.\begin{aligned} G^i = &\left(-qs+uv-\frac{1}{2}r^2-\alpha q^2 r-\frac{3}{2}(20-\alpha)u^2 q-\frac{3}{2}\alpha q^4\right)Y_1^i \\ &-\left(\alpha qr-(\alpha-30)u^2+s+2\alpha q^3\right)Y_2^i+\left(v+\frac{3}{5}(40-\alpha)qu\right)Y_4^i \\ &+\left(qr-\frac{1}{2}u^2+\frac{1}{5}(40-\alpha)q^3\right)Y_5^i+\left(r+\frac{3}{10}(40-\alpha)q^2\right)Y_6^i \\ &+uY_7^i+Y_8^i+\frac{1}{2}q^2Y_9^i+qY_{10}^i \,. \end{aligned}\right\} \tag{10}$$

The functions $Y_j^i(y)$ depend only on the prolongation variables and arise as integration functions.

In fact in this case one finds that the integration procedure leads to the conclusion that

$$\begin{gathered} Y_4 = [Y_2,Y_2], \quad Y_5 = [Y_2,[Y_2,Y_3]] \\ [Y_2,[Y_2,[Y_2,Y_3]]] = 0, \quad Y_6 = [Y_3,[Y_2,Y_3]] \\ Y_7 = [Y_3,[Y_3,[Y_2,Y_3]]], \quad \frac{4}{5}(15-\alpha)[Y_2,Y_3] = [Y_2,[Y_3,[Y_2,Y_3]]] \\ Y_9 = [Y_2,[Y_3,[Y_3,[Y_2,Y_3]]]], \quad Y_{10} = [Y_3,[Y_3,[Y_3,[Y_2,Y_3]]]]. \end{gathered} \tag{11}$$

The Y_i's satisfy the following relationships:

$$\begin{aligned} &[Y_1,Y_m] = 0 \quad m = 1, \ldots, 10 \text{ excluding } m=8 \\ &[Y_2,Y_3] = Y_4,\ [Y_2,Y_4] = Y_5, [Y_2,Y_6] = [Y_3,Y_5] = \frac{4}{5}(15-\alpha)Y_4 \\ &[Y_2,Y_7] = Y_9,\ [Y_3,T_4] = Y_6, [Y_3,Y_6] = Y_7, [Y_3,Y_7] = Y_{10} \\ &[Y_3,Y_8] = 0,\ [Y_2,Y_8] = -[Y_3,Y_{10}], [Y_2,Y_9] = \lambda Y_4 \\ &[Y_3,Y_9] = -[Y_1,Y_8] - 2[Y_2,Y_{10}] + \frac{3}{5}(\alpha-40)Y_7 \end{aligned} \tag{12a}$$

and

$$Y_1 = 0 \quad \text{or} \quad \alpha - 20 = 0 \tag{12b}$$

where

$$\lambda = -\frac{4}{5}(\alpha-20)(\alpha-30) \tag{12c}$$

Further information is obtained by using the Jacobi identity,

$$\begin{aligned} &[Y_4,Y_5] = \frac{4}{5}(15-\alpha)Y_5, \quad [Y_4,Y_6] = Y_7 - \frac{4}{5}(15-\alpha)Y_6 \\ &[Y_5,Y_6] = \left(\lambda - \frac{16}{25}(\alpha-15)^2\right)[Y_5,Y_9] = 0, \\ &[Y_4,Y_7] = 3[Y_2,Y_{10}] + [Y_1,Y_8] - \frac{3}{5}(\alpha-40)Y_7 \\ &[Y_5,Y_7] = \lambda Y_6 + \frac{4}{5}(\alpha-15)Y_9. \end{aligned} \tag{12d}$$

In terms of the Y_i's it appears that this structure is open ended and one may go on increasing its size by defining new generators in terms of brackets of the existing generators. A structure like (11) emerges for all equations of the type $q_t = K(q)$ (in particular for the Korteweg-de Vries equation) where K is a local nonlinear operator involving q and its partial x-derivatives. We investigate (11) by trying to obtain a finite Lie algebra consistent with it by choosing the Lie algebra with the smallest number of generators, which is non-Abelian. In [11] this is called the minimal non-Abelian Lie algebra M. A search for M will, upon choosing a representation of the algebra as linear vector fields on a manifold with local coordinates $\{y^i\}$, lead to an inverse problem on an integral manifold of the form (5). Vectors which commute with the structure define conservation laws for the equation so we ignore these.

The brackets in (11) suggest that we start with $\dim M = 3$. The only nontrivial three-dimensional Lie algebra is $\underline{SL}(2,R)$ so we we use the ansatz,

$$Y_i = a_i X_1 + b_i X_2 + c_i X_3 \tag{13a}$$

where the $\underline{SL}(2,R)$ basis is defined by,

$$[X_1,X_2] = -2X_2, \quad [X_1,X_3] = 2X_3, \quad [X_2,X_3] = -X. \tag{13b}$$

Using (13) in (11) we find, up to a change in the basis of $\underline{SL}(2,R)$, the following two cases:

(i) $\alpha = 20, \; a \neq 0$

$$Y_2 = aX_2, \quad Y_3 = bX_1 - 2a^{-1}X_3, \quad Y_4 = 2abX_2 + 2X_1,$$

$$Y_5 = 4aX_2, \quad Y_6 = -4ab^2X_2 - 4bX_1 + 8a^{-1}X_3,$$

$$Y_7 = 8b^2X_1 + 8ab^3X_2, \quad Y_8 = -16b^4(bX_1 - 2a^{-1}X_3),$$

$$Y_9 = 16ab^2X, \quad Y_{10} = -16b^3X_1 - 16ab^4X_2 + 32b^2a^{-1}X_3.$$

Y_1 now commutes with the structure and so has been omitted. It corresponds to the conservation law,

$$(\tfrac{1}{2}q^2)_t + \left(-qq_{4x} + q_xq_{3x} - \tfrac{1}{2}q_{2x}^2 - 20q^2q_{2x} - 30q^4\right)_x = 0. \qquad (14)$$

(ii) $\alpha = 30$, $a \neq 0$

$$Y_2 = aX_2, \; Y_3 = -6a^{-1}X_3, \; Y_4 = 6X_1, \; Y_5 = 12aX_2, \; Y_6 = 72a^{-1}X_3,$$

$$Y_7 = Y_8 = Y_9 = Y_{10} = 0.$$

It is obvious that for $b=0$, cases (i) and (ii) are analagous; however, for nonzero b, α must be 20. The one-parameter family of Equations (1) are invariant under the scaling transformation,

$$x \mapsto ax, \; q \mapsto a^{-2}q, \; t \mapsto a^5t \qquad (15a)$$

and the corresponding automorphism of the ideal I is,

$$(x,t,q,u,r,v,s) \longmapsto (ax, a^5t, a^{-2}q, a^{-3}u, a^{-4}r, a^{-5}v, a^{-6}s). \qquad (15b)$$

Under this map the ω^i, (7) with (10), transform in the following manner:

$$\omega^i \mapsto dy^i + \left(\tfrac{1}{2}q^2a^{-3}Y_1^i + qa^{-1}Y_2^i + aY_3^i\right)dx + \left[\left(-qs + uv - \frac{1}{2}r^2\right.\right.$$

$$\left. - \alpha q^2r - \frac{3}{2}(20-\alpha)u^2q - \frac{3}{2}\alpha q^4\right)a^{-3}Y_1^i - \alpha qr - (\alpha-30)u^2 +$$

$$\left.\begin{aligned} & s+2\alpha q^3\Big)a^{-1}Y_2^i+\Big(v+\tfrac{3}{5}(40-\alpha)qu\Big)Y_4^i+\Big(qr-\tfrac{1}{2}u^2 \\ & +\tfrac{7}{5}(40-\alpha)q^3\Big)a^{-1}Y_5^i+\Big(r+\tfrac{3}{10}(40-\alpha)q^2\Big)aY_6^i \\ & +ua^2Y_7^i+\tfrac{1}{2}q^2aY_q^i+qa^3Y_{10}^i+a^5Y_8^i\Big]\,dt. \end{aligned}\right\} \quad (16)$$

It is clear that in both cases (i), after rescaling b, b $\longmapsto$ ab, and (ii) that the parameter a transforms away. The parameter b in case (i) will correspond to the eigenvalue in the associated scattering problem. In the case of the Korteweg-de Vries equation this degree of freedom is generated by a Galilean transformation,

$$q_t + 6qq_x + q_{xxx} = 0$$

$$q \longmapsto q + a, \quad t \longmapsto t, \quad x \longmapsto x + 6at. \quad (17)$$

Since an inverse method and Bäcklund transformations are known for the Lax hierarchy of equations [4], we omit Equation (1) with $\alpha = 20$ from further consideration. For the case (ii), Equation (1) with $\alpha = 30$ is

$$q_t + \left(q_{4x} + 30qq_{2x} + 60q^3\right)_x = 0. \quad (18)$$

The occurrence of the free parameter a in the representation of the Y^i's in terms of the SL(2,R) basis renders the effect of the scaling transformation trivial. However, in [5] we obtain a Lie algebra in which the effect of (15) is nontrivial. Explicitly, a three-dimensional representation of this Lie algebra as linear vector fields is given by,

$$\left.\begin{aligned} & Y_2 = -2592\,y^1b_3, \quad Y_3 = \tfrac{1}{432}y^3b_1 + 2\left(y^1b_2 + y^2b_3\right), \\ & Y_4 = -6\left(y^1b_1 - y^3b_3\right), \quad Y_5 = 12Y_2, \quad Y_6 = -\tfrac{1}{36}y^3b_1 \\ & + 12\left(y^1b_2 + y^2b_3\right), \quad Y_7 = \tfrac{1}{12}\left(y^3b_2 - y^2b_1\right), \end{aligned}\right\}$$

$$\left.\begin{aligned} Y_8 &= \frac{1}{2592}\left(y^2 b_1 + y^3 b_2\right) + \frac{1}{3} y^1 b_3, \\ Y_9 &= -216\left(y^1 b_2 + y^2 b_3\right), \quad Y_{10} = -\frac{1}{6}\left(y^3 b_3 - 2y^2 b_2 + y^1 b_1\right) \end{aligned}\right\}$$

where

$$b_k = \frac{\partial}{\partial y^k}, \quad k = 1, 2, 3. \tag{19}$$

It is left for the reader to check that this Lie algebra of vector fields agrees with the abstract prolongation structure defined by Equations (12). Because a free parameter does not enter in the definitions (19), the effect of scaling on ω^i, Equation (16) is now nontrivial and introduces an eigenvalue into the associated inverse problem. On an integral manifold Equations (16) yield the inverse scattering problem for Equation (18). The first of Equations (5) is obtained with

$$R = \begin{bmatrix} 0 & 0 & \frac{a}{432} \\ 2a & 0 & 0 \\ -2592aq & 2a & 0 \end{bmatrix}. \tag{20}$$

To obtain the Bäcklund transformation, it is convenient to deal with the scalar version of the scattering equations. These are obtained from the linear equations by eliminating the components y^1 and y^3 and obtaining equations involving only y^2 and its derivatives. By writing y for y^2 and introducing the new parameter $\zeta = (108a^3)^{-1}$, we find that the scalar inverse problem is defined by

$$y_{3x} + 6qy_x = -\zeta y, \tag{21a}$$

$$y_t - 6\left(q_{2x} - 6q^2\right)y_x + 36\,\zeta q y + 3\zeta\left(1 + \frac{3}{2}\zeta^{-1} q_x\right)y_{2x} = 0. \tag{21b}$$

Defining $\phi = \log y$ and eliminating q between Equations (21) leads to the equation

$$
\begin{aligned}
\phi_x^2\phi_t = & -10\,\phi_x^4\phi_{3x} - \phi_{3x}\phi_x^2 - \phi_x^7 + 5\zeta\,\phi_x\phi_{3x} + 5\zeta\,\phi_x\phi_{3x} \\
& + 5\zeta^2\phi_x - 5\zeta\phi_{2x}^2 - 5\zeta\phi_x^4 + 5\phi_{4x}\phi_x\phi_{2x} + 15\phi_x^4\phi_{3x} \\
& - 5\phi_{2x}^2\phi_{3x} - 5\phi_{2x}^2\phi_x^2,
\end{aligned} \tag{22}
$$

which is invariant under $(\phi,\zeta) \longrightarrow (-\phi,-\zeta)$. Following Chen [12], let $q \equiv w_x$ be the solution of (18) corresponding to (ϕ,ζ) and $\bar{q} \equiv \bar{w}_x$ be the solution corresponding to $(-\phi,-\zeta)$. Then (21a) for the two potentials q, $\bar{q}$ implies that

$$\phi_{2x} = (\bar{q} - q) = (\bar{w} - w)_x. \tag{23}$$

Consequently we find that the x-component of the Bäcklund transformation for (18) is

$$(\bar{w} - w)_{xx} + (\bar{w} - w)^3 + 3(\bar{w} - w)(\bar{w} + w)_x = -\zeta. \tag{24}$$

It is straightforward to obtain from the Bäcklund transformation by taking $w = \frac{1}{2}i\zeta^{\frac{1}{3}}$, the one soliton solution

$$\bar{q} = \tfrac{1}{4}b^2\operatorname{sech}^2(bx - b^5t + \eta). \tag{25}$$

This equation fits into the Zakharov and Shabat scheme [13] upon defining operators L_1, L_2 by

$$
\begin{aligned}
L_1 &\equiv \frac{\partial^3}{\partial x^3} + 6q\frac{\partial}{\partial x}, \\
L_2 &\equiv \frac{\partial}{\partial t} - 9\frac{\partial^5}{\partial x^5} - 90q\frac{\partial^3}{\partial x^3} - 90q_x\frac{\partial^2}{\partial x^2} - 60\left(3q^2 + q_{2x}\right)\frac{\partial}{\partial x}.
\end{aligned} \tag{26}
$$

With this form of the scattering equations, the Marchenko equation is

$$F(x,s,f) + K(x,s,t) + \int_x^\infty K(x,z,t)F(x,s,t)\,dz = 0 \tag{26a}$$

where F satisfies the equations

$$\frac{\partial^3}{\partial x^3} F(x,s,t) + \frac{\partial^3}{\partial s^3} F(x,s,t) = 0 \tag{26b}$$

$$\frac{\partial}{\partial t} F(x,s,t) + 9\left(\frac{\partial^5}{\partial x^5} F(x,s,t) + \frac{\partial^5}{\partial s^5} F(x,s,t)\right) = 0. \tag{26c}$$

In conclusion we note the scattering problem obtained from case (ii) by taking the representation of SL(2,R) as vector fields on the one-dimensional manifold with local coordinate y

$$X_1 \to 2y\frac{\partial}{\partial y}, \quad X_2 \to \frac{\partial}{\partial y}, \quad X_3 \to -y^2\frac{\partial}{\partial y}$$

means that we must take $\zeta = 0$ in Equations (21).

REFERENCES

1. P. J. Caudrey, R. K. Dodd, and J. D. Gibbon, "A New Hierarchy of Korteweg-de Vries Equations," Proc. Roy. Soc. (London) A 351, 407-422 (1976).
2. See article by P. J. Caudrey in this volume.
3. P. D. Lax, "Integrals of Nonlinear Equations of Evolution and Solitary Waves," Communs Pure Appl. Math. 31, 467-490 (1968).
4. C. S. Gardner, J. M. Greene, M. D. Kruskal, and R. M. Miura, "Korteweg-de Vries Equation and Generalisations VI. Methods for Exact Solution," Communs Pure Appl. Math. 27, 97-133 (1974).
5. R. K. Dodd and J. D. Gibbon, "The Prolongation Structure of a Higher Order Korteweg-de Vries Equation," Proc. Roy. Soc. (London) A 358, 287-296 (1977).
6. H. D. Wahlquist and F. B. Estabrook, "Prolongation Structures of Nonlinear Evolution Equations," J. Math. Phys. 16, 1-7 (1975).
7. F. B. Estabrook and H. D. Wahlquist, "Prolongation Structures of Nonlinear Evolution Equations II," J. Math. Phys. 17, 1293-1297 (1976).
8. F. B. Estabrook and H. D. Wahlquist, "Prolongation Structures, Connection Theory and Bäcklund Transformations," to appear in the proceedings of the international symposium on nonlinear evolution equations solvable via the inverse

scattering transform, Accademia Nazionale dei Lincei, Roma, F. Calogero (editor).

9. W. Slebodizinski, Formes Extérieures et Leurs Applications Pãnstwowe Wydawnictwo Naukowe, Warsaw, 1954 and 1963, Vols. I and II.

10. E. Cartan, Les Systèmes Différentials Extérieurs et Leurs Applications Géométriques, Hermann, Paris, 1955.

11. R. K. Dodd and B. J. D. Gibbon, "The Prolongation Structures of a Class of Nonlinear Evolution Equations," Proc. Roy. Soc. (London) A 359, 411-433 (1978).

12. H. Chen, General Derivation of Bäcklund Transformations from Inverse Scattering Problems," Phys. Rev. Lett. 33, 925-928 (1974).

13. V. E. Zakharov and A. B. Shabat, "A Scheme for Integrating the Nonlinear Equations of Mathematical Physics by the Method of the Inverse Scattering Problems, I," Funkt.-Analiz. Priloz. 8, 226 (1974). (Funct. Analysis Appl. 8, 226-235 (1975).)

PERTURBATION THEORY FOR THE DOUBLE SINE-GORDON EQUATION†

A. L. Mason

Department of Mathematics, UMIST,
Manchester M60 1QD U.K.

ABSTRACT. Perturbation techniques developed by D. J. Kaup and A. C. Newell are applied to an initial value problem (in laboratory coordinates) for the double sine-Gordon equation. A motion resembling the breather solution of the simple sine-Gordon equation is found.

Kaup and Newell, jointly and separately, have been developing perturbation techniques that should make the Inverse Scattering Transform a useful tool for the investigation of equations other than those that are exactly solvable. I shall illustrate some of their techniques by applying them to an initial value problem of particular interest to us at U.M.I.S.T.

The double sine-Gordon equation

$$u_{XX} - u_{TT} = \sin u + \tfrac{1}{2}\lambda \sin \tfrac{1}{2}u \tag{1}$$

arises naturally in nonlinear optics and also for spin waves in superfluid He^3. The energy density is

$$E = \tfrac{1}{2}u_X^2 + \tfrac{1}{2}u_T^2 + (1 - \cos u) + \lambda(1 - \cos \tfrac{1}{2}u), \tag{2}$$

and we can conveniently visualize the motion as that of a heavy

---

† Lecture given at NATO Advanced Study Institute on Nonlinear Equations in Physics and Mathematics, Istanbul, August 1977.

A. O. Barut (ed.), Nonlinear Equations in Physics and Mathematics, 205–218.

elastic string sliding on a corrugated-iron roof. For the sine-Gordon equation the corrugations are regular. For the double sine-Gordon equation, alternate ones are deeper—as in Figure 1. The string may lie at rest in any of the minima, but only the lower ones are true ground states. The others are "false vacua," in the words of Sidney Coleman.

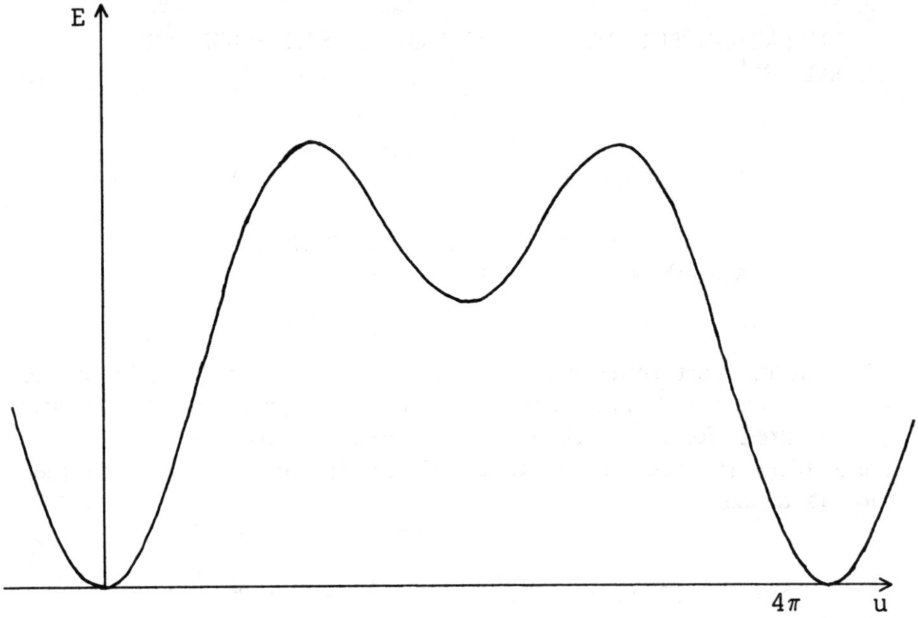

Figure 1. Potential energy for double sine-Gordon equation.

One fairly obvious stationary state has $u \to 0$ as $x \to -\infty$ and $u \to 4\pi$ as $x \to +\infty$ (see Figure 2). This is clearly stable, and small perturbations simply make it wobble. It is not clear whether or not the wobbling persists indefinitely. Another rather less obvious stationary state has the "false Vacuum" occupied for most of space, except for a small region hanging over into the true vacuum, as in Figure 3. This corresponds to a Coleman bubble of critical size and is obviously unstable. I shall investigate motions that start from rest in positions close to this critical bubble, which can be written explicitly as

$$u = 4 \tan^{-1}\left[\sqrt{\frac{\lambda}{4-\lambda}} \cosh \tfrac{1}{2}\sqrt{4-\lambda}\, X\right]. \tag{3}$$

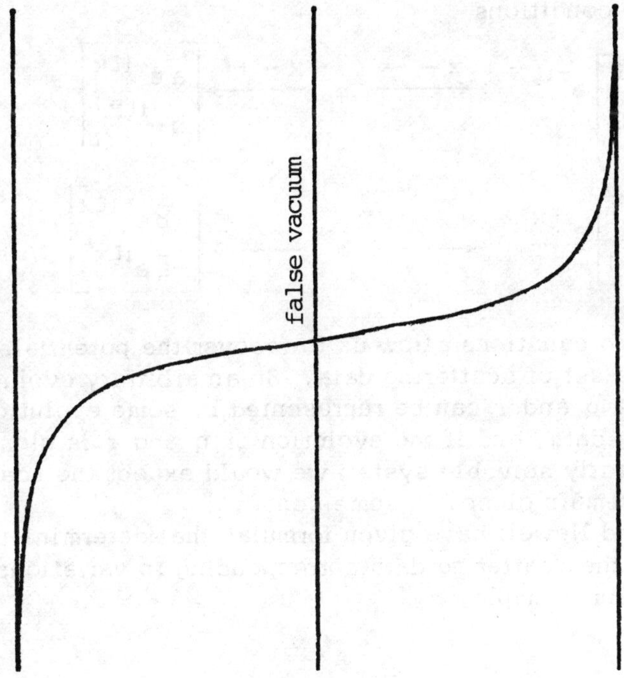

Figure 2. Stable stationary state.

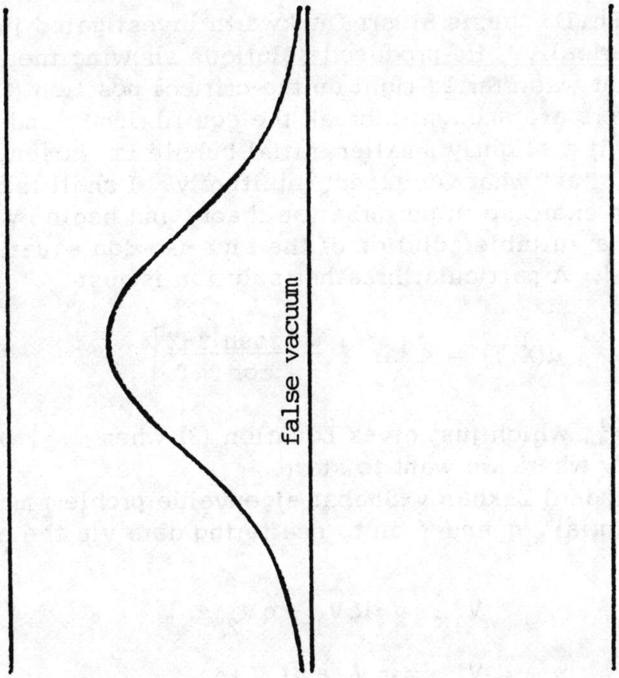

Figure 3. Unstable stationary state.

In his Ph.D. thesis Stuart Duckworth investigated just this problem numerically. He produced solutions showing the bubble "bursting" if it was started right on the critical position (the numerical errors are enough to break the equilibrium), and oscillating slowly if a slightly smaller initial bubble is chosen. This is, of course, just what we expect intuitively. I shall regard the problem as an exercise in perturbation theory and begin by searching for a suitable solution of the sine-Gordon equation to perturb around. A particular breather solution is just

$$u(X,T) = 4 \tan^{-1} \frac{\omega}{\eta} \left[\frac{\cosh 2\eta X}{\cos 2\omega T}\right] \tag{4}$$

with $\omega^2 + \eta^2 = \frac{1}{4}$, which just gives Equation (3) when $\omega = \frac{1}{4}\sqrt{\lambda}$, $T = 0$. This is clearly where we want to start.

The standard Zakharov-Shabat eigenvalue problem maps arbitrary potentials q and r on to scattering data via the equations

$$\begin{aligned} V_{1x} &= -i\zeta V_1 + q V_2 \\ V_{2x} &= r V_1 + i\zeta V_2 \end{aligned} \tag{5a}$$

and boundary conditions

$$\begin{aligned} &\begin{bmatrix} 1 \\ 0 \end{bmatrix} e^{-i\zeta x} \xleftarrow{x \to -\infty} \phi \xrightarrow{x \to +\infty} \begin{bmatrix} a\, e^{-i\zeta x} \\ b\, e^{i\zeta x} \end{bmatrix} \\ &\begin{bmatrix} 0 \\ -1 \end{bmatrix} e^{i\zeta x} \longleftarrow \bar{\phi} \longrightarrow \begin{bmatrix} \bar{b}\, e^{-i\zeta x} \\ -\bar{a}\, e^{i\zeta x} \end{bmatrix} \end{aligned} \tag{5b}$$

The Marchenko equations allow us to recover the potentials from some suitable set of scattering data. So an arbitrary evolution of the potentials q and r can be represented by some evolution of the scattering data, and if the evolution of q and r is close to that of an exactly solvable system we would expect the scattering data also to remain close, in some sense.

Kaup and Newell have given formulae that determine the variations in the scattering data corresponding to variations in the potentials. For example,

$$\delta\begin{bmatrix}b\\a\end{bmatrix} = \frac{-1}{a^2}\int_{-\infty}^{\infty}\begin{bmatrix}\delta q\\ \delta r\end{bmatrix}\cdot\begin{bmatrix}\phi_2^2\\ -\phi_1^2\end{bmatrix}dx, \tag{6}$$

and similarly for the variations of the eigenvalues and residues. If $q=-\frac{1}{2}u_{,x}$ $r=-q$, this reduces to

$$\frac{d}{dt}\begin{bmatrix}b\\a\end{bmatrix} = \frac{1}{a}\int_{-\infty}^{\infty} u_{xt}\left(\phi_1^2+\phi_2^2\right)dx. \tag{7}$$

If u_{xt} in the integral is replaced by $\sin u$, we can immediately write down the value of the integral since if u evolves according to the light cone sine-Gordon equation

$$u_{xt} = \sin u, \tag{8}$$

then we know that $\begin{bmatrix}b\\a\end{bmatrix}$ evolves according to

$$\frac{d}{dt}\begin{bmatrix}b\\a\end{bmatrix} = -\frac{i}{4\xi}\begin{bmatrix}b\\a\end{bmatrix}. \tag{9}$$

So if u evolves according to

$$u_{xt} = \sin u + F, \tag{10}$$

then

$$\frac{d}{dt}\begin{bmatrix}b\\a\end{bmatrix} = -\frac{i}{4\xi}\begin{bmatrix}b\\a\end{bmatrix} + \frac{1}{a^2}\int_{-\infty}^{\infty} F\left(\phi_1^2+\phi_2^2\right)dx. \tag{11}$$

We have found how to determine the perturbations of the scattering data about the light-cone sine-Gordon equation. However, our initial conditions are given in laboratory coordinates so we must consider a new scattering problem. The evolution equations

$$\begin{aligned} V_{1t} &= \frac{i}{4\zeta}\left(V_1\cos u + V_2\sin u\right)\\ V_{2t} &= \frac{i}{4\zeta}\left(V_1\sin u - V_2\cos u\right)\end{aligned} \tag{12}$$

determine the evolution of the wave functions if the potential u evolves according to the light-cone sine-Gordon equation. Light cone coordinates are related to laboratory coordinates by

$$x = \tfrac{1}{2}(X+T), \quad t = \tfrac{1}{2}(X-T). \tag{13}$$

We use Equations (5a) and (12) to obtain

$$\begin{aligned} V_{1X} &= \left[\frac{-i\zeta}{2} + \frac{i}{8\zeta}\cos u\right]V_1 + \left[\frac{i}{8\zeta}\sin u - \tfrac{1}{4}[u_X + u_T]\right]V_2 \\ V_{2X} &= \left[\frac{i}{8\zeta}\sin u + \tfrac{1}{4}[u_X + u_T]\right]V_1 + \left[\frac{i\zeta}{2} - \frac{i}{8\zeta}\cos u\right]V_2 \end{aligned} \tag{14a}$$

and discard the equations for V_{1T}, V_{2T}. Hence we have a new scattering problem if taken in conjunction with the boundary conditions;

$$\begin{aligned} &\begin{bmatrix} 1 \\ 0 \end{bmatrix} e^{-ikX} \xleftarrow{X \to -\infty} \phi \xrightarrow{X \to +\infty} \begin{bmatrix} a e^{-ikX} \\ b e^{ikX} \end{bmatrix} \\ &\begin{bmatrix} 0 \\ -1 \end{bmatrix} e^{ikX} \longleftarrow \phi \longrightarrow \begin{bmatrix} b e^{-ikX} \\ -\bar{a} e^{ikX} \end{bmatrix} \end{aligned} \tag{14b}$$

where

$$k = \frac{1}{2}\left[\zeta - \frac{1}{4\zeta}\right].$$

To solve it we take the solution of the light-cone sine-Gordon scattering problem (including the t dependence) and restrict this to fixed T. Kaup has shown that the analogues of the Marchenko equations exist, that the required scattering data are the same as before and that they determine u and u_T; that is, all the potentials in the problem.

Kaup and Newell have given the following equations of motion for the scattering data. Supposing u satisfies

$$u_{XX} - u_{TT} = \sin u + F. \tag{15}$$

We have

$$\zeta_{k,T} = -\frac{1}{4}\frac{1}{\gamma_k a'^2_k}\int_{-\infty}^{\infty} F.\left[\phi_1^2+\phi_2^2\right]_{\zeta_k} dX \tag{16}$$

$$\gamma_{k,T} = 2i\gamma_k\omega_k - \gamma_k\frac{a''_k}{a'_k}\zeta_{k,T}$$

$$-\frac{1}{4}\frac{1}{a'^2_k}\int_{-\infty}^{\infty} F.\frac{\partial}{\partial\zeta}\left[\phi_1^2+\phi_2^2\right]_{\zeta_k} dX \tag{17}$$

$$\frac{d}{dT}\left[\frac{b}{a}\right] = 2i\omega\left[\frac{b}{a}\right] - \frac{1}{4}\frac{1}{a^2}\int_{-\infty}^{\infty} F.\left[\phi_1^2+\phi_2^2\right]_{\zeta} dX \tag{18}$$

where

$$\omega_k = \frac{1}{2}\left[\zeta_k + \frac{1}{4\zeta_k}\right]$$

and

$$\gamma_k = b_k a'_k.$$

The scattering data for a breather consist of two eigenvalues ζ_1 and ζ_2, related by $\zeta_1 = -\zeta_2^*$ and the residues γ_1 and γ_2 similarly related. The reflection coefficients (continuum) are zero. I shall assume that the continuum remains negligible throughout the motion and investigate the evolution of ζ_k and γ_k under the perturbation $F = \frac{1}{2}\lambda\sin\frac{1}{2}u$. It is convenient to write

$$\begin{aligned}\zeta_1 &= |\zeta_1| e^{i\phi}\\ \gamma_1 &= |\gamma_1| e^{i\psi}\\ \theta &= \psi - \phi.\end{aligned} \tag{19}$$

It turns out that there are two constants of the motion that ensure that if the centre of the breather starts out stationary at the origin, it remains there.

It can be shown that

$$\begin{aligned}|\zeta_1| &= \tfrac{1}{2}\\ |\gamma_1| &= \tan\phi.\end{aligned} \tag{20}$$

I shall use these without proof to simplify the algebra. The most general expression for u(X), then, is

$$u(X) = 4\tan^{-1}\left[\cot\phi\,\frac{\cosh(2X\sin\phi)}{\cos(\psi-\phi)}\right]. \tag{21}$$

If there is no continuum, the eigenfunctions can be calculated algebraically in terms of the eigenvalues and residues, an elementary but exceedingly laborious exercise. However, the final results for the expressions that appear in the integrals of Equations (16)-(18) are fairly simple. If we set

$$\begin{aligned} N &= 2\cosh 2X\sin\phi \\ D &= 2\tan\phi\cos\theta, \end{aligned} \tag{22}$$

then we find that

$$\sin\tfrac{1}{2}u = \frac{2ND}{N^2+D^2} \tag{23}$$

$$\left[\phi_1^2+\phi_2^2\right]_{\zeta_1} = \left[\frac{e^{-i\theta}e^{-2X\sin\phi} - e^{i\theta}e^{2X\sin\phi}}{N^2+D^2}\right]e^{i\theta}, \tag{24}$$

etc. Eventually we arrive at the following equations of motion for ϕ and θ

$$\begin{aligned} \dot\phi &= -\tfrac{1}{4}\lambda\frac{\tan\theta}{\sin\phi}I_1(a^2) \\ \dot\theta &= \cos\phi\,\frac{\lambda}{4}\left[\frac{\frac{1}{2}\cos\phi}{\sin^2\phi}I_2(a^2) - \frac{I_1(a^2)}{\cos\phi\sin^2\phi}\right] \end{aligned} \tag{25}$$

where $a^2 = \cos^2\theta\tan^2\phi$ and

$$\begin{aligned} I_1(a^2) &= \int_{-\infty}^{\infty}\frac{a^2\cosh^2 X.\,dX}{[a^2+\cosh^2 X]^2} \\ I_2(a^2) &= \int\frac{2a^2\cosh X\sinh X.\,X\,dX}{[a^2+\cosh^2 X]^2}. \end{aligned} \tag{26}$$

When we began the calculation there was no reason to expect that the resulting integrals would be tractable. We expected to have to make the approximations that λ is small and the initial state close to the critical bubble, that is, $\cos\phi \sim \sqrt{\lambda}$. Then we could obtain a systematic expansion in powers of $\sqrt{\lambda}$. However, this procedure is not immediately necessary, since the integrals in (26) can be evaluated analytically quite easily. We find that

$$I_1 = \frac{a}{1+a} + \frac{J}{1+a}$$
$$I_2 = 2J \tag{27}$$

where

$$J = \sqrt{\frac{a}{1+a}}\tanh^{-1}\sqrt{\frac{a}{1+a}} . \tag{28}$$

We have now reduced the problem to an explicit pair of ordinary differential equations for ϕ and θ. Since they appear rather complicated I decided at this stage to solve them numerically. The equilibrium condition is

$$\theta_o = 0$$
$$\cos^2\phi_o = \frac{\lambda}{4} \tag{29}$$

and I chose to take initial conditions

$$\theta_{init} = 0$$
$$\phi_{init} \text{ close to } \phi_o . \tag{30}$$

When $\phi_{init} < \phi_o$, the motion turns out to be periodic. The parameters $\cos\theta$ and ϕ evolve as shown in Figures 4 and 5. In Figure 6 this is translated into the behaviour of the function u itself, and for comparison in Figure 7 I show a simple sine-Gordon breather evolving from the same initial conditions.

Since the initial conditions are chosen to be close to a stationary state, it takes some time to evolve away from that state. When it has done so, it evolves about as fast as the simple breather for a while until it approaches another nearly stationary state on the other side of the false vacuum. The variation in the eigenvalues has only a minor effect on the width of the pulse.

When $\phi_{init} > \phi_o$, our initial bubble is supercritical. Cos θ evolves much the same as before, and ϕ evolves as shown by the dotted line in Figure 5 until it reaches $\pi/2$. At this stage the eigenvalues have reached the imaginary axis and it is no longer appropriate to use the breather solutions. To study the continuing

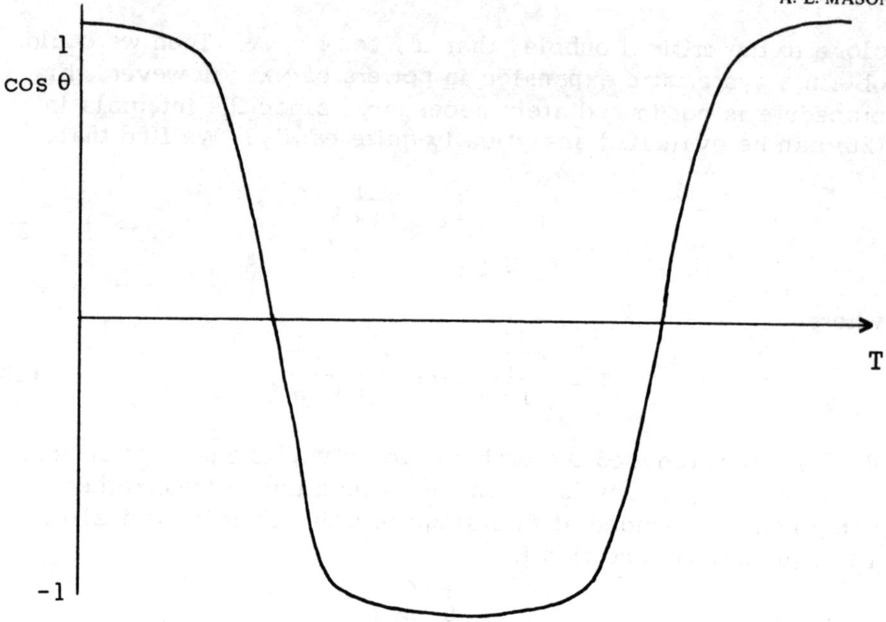

Figure 4. Time evolution of $\cos\theta$.

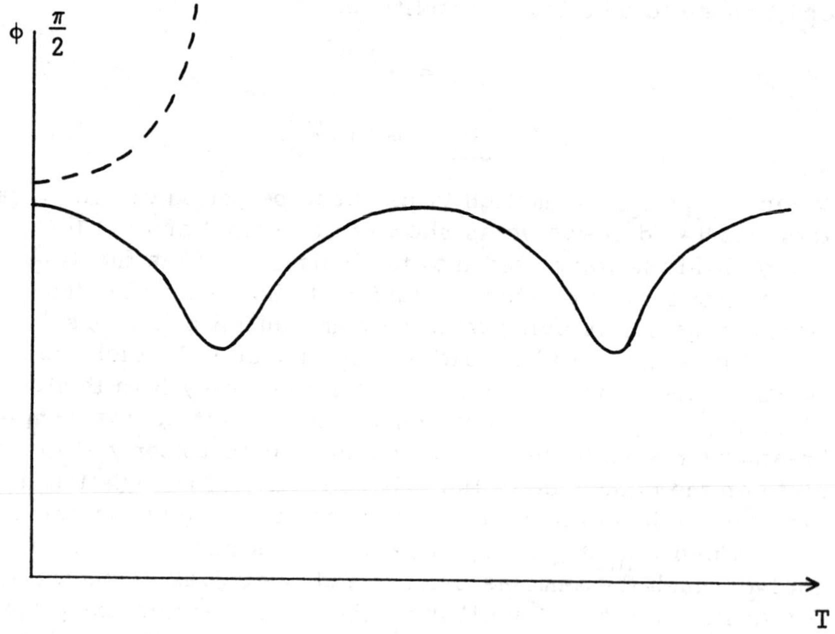

Figure 5. Time evolution of ϕ (dotted line shows case of break up).

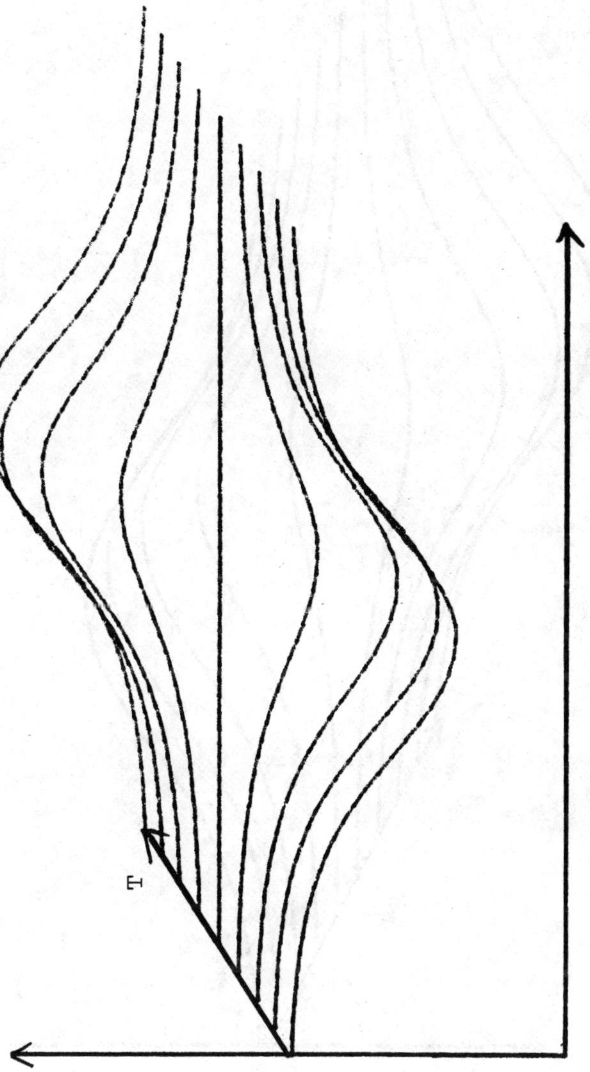

Figure 6. Breather-like solution of double sine-Gordon equation.

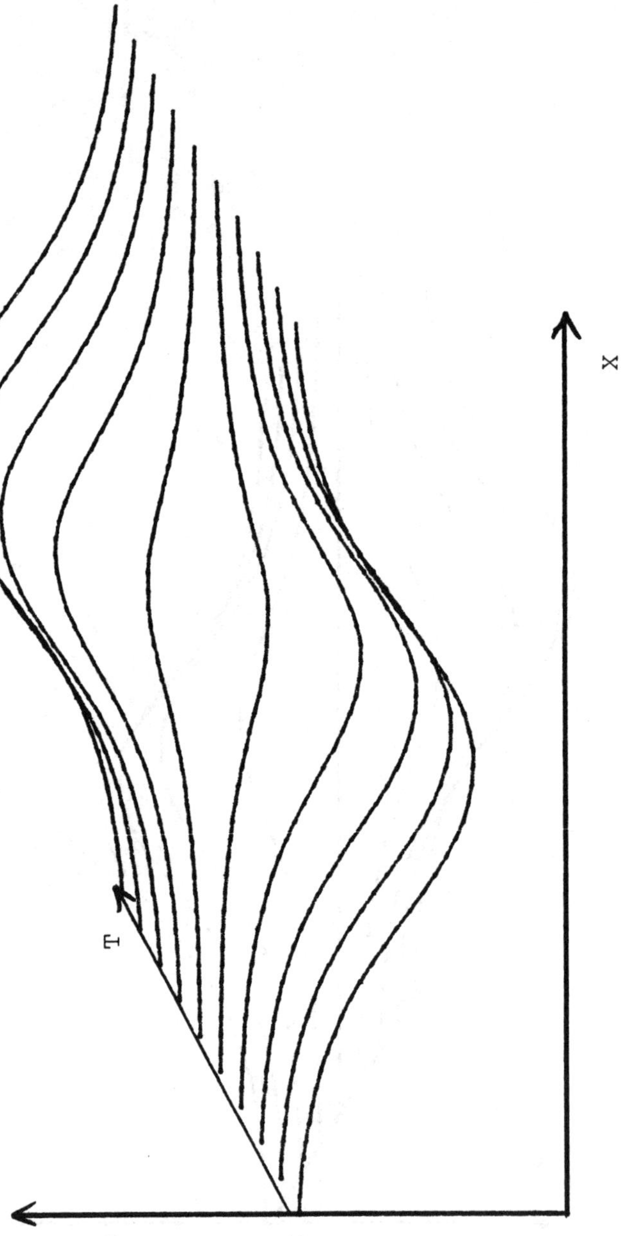

Figure 7. Breather solution of sine-Gordon equation.

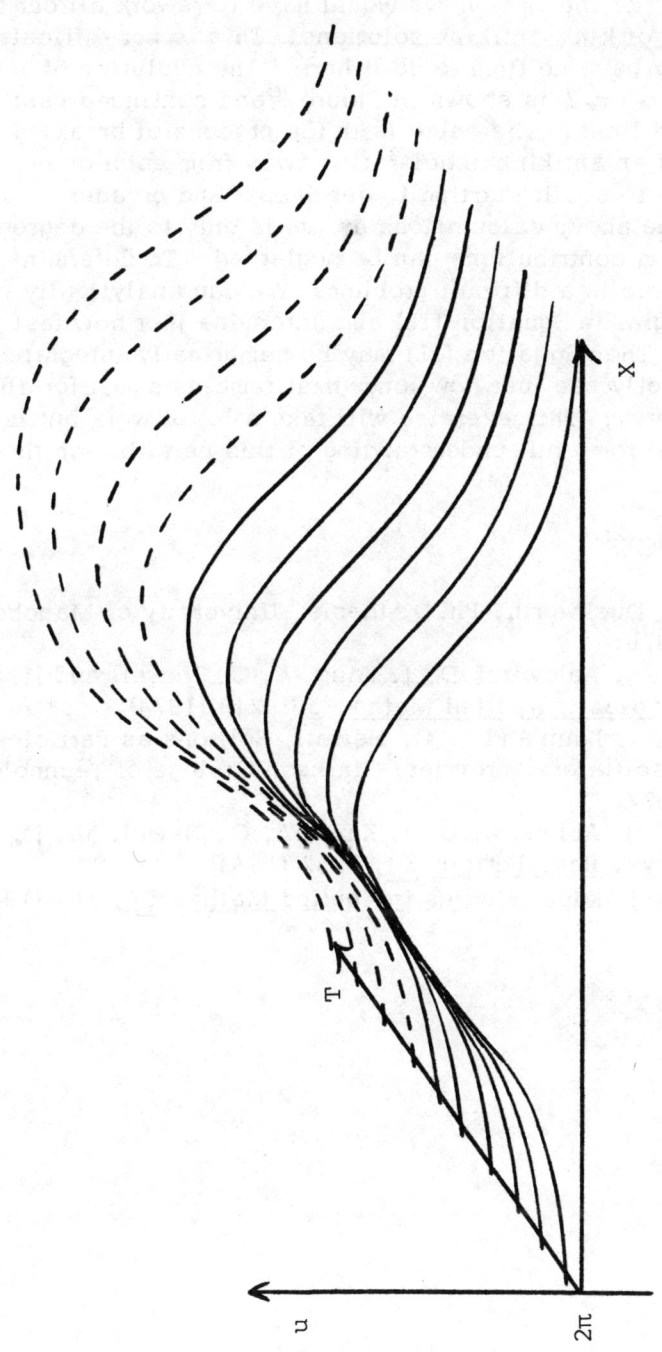

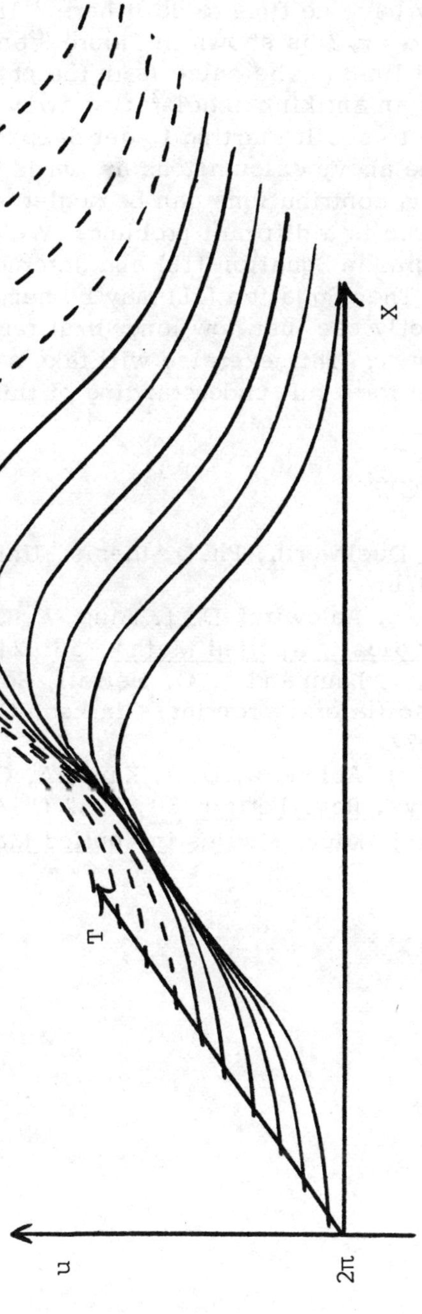

Figure 8. Pulse breaking up.

evolution of the motion we would have to rework all our formulae in terms of kink-antikink solutions. This is not difficult but I certainly have no time to do it here. The evolution of u up to the moment $\phi = \pi/2$ is shown in Figure 8 and continued past this time in dotted lines. The pulse is in the process of breaking up into a kink and an antikink accelerating away from each other. It is just possible to see it starting to get deeper and broader.

The above calculations are valid only to the degree that the continuum contributions can be neglected. To determine how far this is true is a difficult problem. We can analytically evaluate the integral in Equation (18) and determine just how fast (b/a) (ξ) varies. Then Equation (18) may be numerically integrated and we can directly see just how long (b/a) remains small for different values of λ. This exercise will take a lot of work but is really essential for a full understanding of this perturbation theory.

REFERENCES

1. S. Duckworth, Ph.D. thesis, University of Manchester, 1976.
2. M. J. Ablowitz, D. J. Kaup, A. C. Newell and H. Segur, Studies in Applied Maths. **53**, 249 (1974).
3. D. J. Kaup and A. C. Newell, Solitons as Particles and Oscillators, preprint, Clarkson College of Technology, 1977.
4. M. J. Ablowitz, D. J. Kaup, A. C. Newell and H. Segur, Phys. Rev. Letters **30**, 1262 (1973).
5. D. J. Kaup, Studies in Applied Maths. **54**, 165 (1975).

PART III

DISCRETE SYSTEMS AND CONTINUUM MECHANICS

PAINLEVÉ TRANSCENDENTS AND SCALING FUNCTIONS OF THE TWO-DIMENSIONAL ISING MODEL†‡

Craig A. Tracy
Institute for Theoretical Physics
State University of New York
Stony Brook, L.I., N.Y. 11794, U.S.A.

1. INTRODUCTION

In this lecture I should like to report on the work I have done in collaboration with Barry McCoy and Tai Tsun Wu (and on certain aspects with Eytan Barouch) on the correlation functions of the two-dimensional Ising model. In particular, I wish to demonstrate how a particular solution of the two-dimensional hyperbolic sine-Gordon equation (also known as the two-dimensional nonlinear Debye-Hückel equation),

$$\Delta\phi = \sinh \phi, \quad (r > 0), \tag{1.1}$$

plays a fundamental role in the scaled two-point function in both the one-phase and two-phase regions. Before I discuss these results, I would like first to review the definition of the 2-d. Ising model[1)-4)] and related quantities (Section 2); and then briefly recall the scaling theory hypothesis[5)-7)] for correlation functions (Section 3).

---

† Lectures given at NATO Advanced Study Institute on Nonlinear Equations in Physics and Mathematics, Istanbul, August 1977.

‡ Supported in part by National Science Foundation Grants Nos. PHY-76-15328 and PMR 73-07565 A01.

A. O. Barut (ed.), Nonlinear Equations in Physics and Mathematics, 221–237.

2. TWO-DIMENSIONAL ISING MODEL

Consider a two-dimensional square lattice of $\mathcal{M}$ rows and $\mathcal{N}$ columns and at each lattice site, which we label with the pair of indices (i,j), we define a variable $\sigma_{i,j}$ that can assume the values of ± 1. The two-dimensional Ising model with nearest neighbor interactions on this square lattice is specified by the energy of interaction

$$\mathcal{E} = -E_1 \sum_{j,k} \sigma_{j,k}\sigma_{j,k+1} - E_2 \sum_{j,k} \sigma_{j,k}\sigma_{j+1,k}. \tag{2.1}$$

The partition function $Z_{\mathcal{M}\mathcal{N}}(\beta)$ is

$$Z_{\mathcal{M}\mathcal{N}}(\beta) = \sum_{\{\sigma_{i,j}=\pm\}} \exp(-\beta\mathcal{E}) \tag{2.2}$$

and is related to the free energy per lattice site by

$$-\beta f(\beta) = \lim_{\mathcal{M},\mathcal{N}\to\infty} \frac{1}{\mathcal{M}\cdot\mathcal{N}} \log Z_{\mathcal{M}\mathcal{N}}(\beta) \tag{2.3}$$

where $\beta = (k_B T)^{-1}$, T = temperature and k_B is Boltzmann's constant.

This free energy, $f(\beta)$, was computed exactly by Onsager[2] who observed that there is a singularity at a temperature T_c determined from the equation

$$\sinh 2\beta E_1 \ \sinh 2\beta E_2 = 1. \tag{2.4}$$

Strictly speaking, to interpret this temperature T_c as the critical temperature, the spontaneous magnetization $\mathcal{M}_s(T)$ must be analyzed. This quantity was known to Onsager,[8] but the first published derivation of $\mathcal{M}_s(T)$ was given by Yang[3] who showed

$$\mathcal{M}_s(T) = \left[1 - (\sinh 2\beta E_1 \ \sinh 2\beta E_2)^{-2}\right]^{1/8}. \tag{2.5}$$

Actually, Yang considered the special case of the symmetric lattice $E_1 = E_2 = E$ (as we do from here on) and the general case $E_1 \neq E_2$ was derived by Chang.[9] The important point to note is $\mathcal{M}_s(T)$ goes continuously to zero as T approaches the temperature T_c determined by (2.4). This approach to zero is proportional to

$(1-T/T_c)^{1/8}$ and the 1/8 defines one of the critical exponents[5)-7)] of this model.

The correlation functions are defined as follows: Consider n points in the lattice

$$M_1, N_1; M_2, N_2; \cdots; M_n, N_n, \tag{2.6}$$

then the n-point function (in the thermodynamic limit) is

$$\langle \sigma_{M_1N_1} \sigma_{M_2N_2} \cdots \sigma_{M_nN_n} \rangle =$$

$$= \lim_{\substack{\mathcal{M} \to \infty \\ \mathcal{N} \to \infty}} \frac{\Sigma \sigma_{M_1N_1} \sigma_{M_2N_2} \cdots \sigma_{M_nN_n} e^{-\beta\mathcal{E}}}{\Sigma e^{-\beta\mathcal{E}}}$$

where the sums are over all configurations, i.e., $\{\sigma_{i,j} = \pm 1\}$.

3. SCALING LIMIT AND SCALING FUNCTIONS

The <u>scaling limit</u>[5)-7)] of the pair correlation function is the limit

$$R \to \infty, \quad \xi \to \infty \quad (T \to T_c), \quad \text{such that} \quad x = R/\xi \quad \text{is fixed} \tag{3.1}$$

where R is the radial distance $[=(M^2+N^2)^{\frac{1}{2}}$ for the symmetric case $E_1 - E_2$ of the previous section], $\xi = \xi(T)$ is the correlation length which goes to infinity as $T \to T_c^{\pm}$ [for the 2-d. Ising model[4)] $\xi(T)$ diverges as $(1-T/T_c)^{-1}$] and the variable x is called a scaling variable. The <u>scaling theory hypothesis</u>[5)-7)] is the assertion that the pair correlation function $\langle \sigma_{0,0}\sigma_{M,N}\rangle$ (which is in general a function of M, N, and T) assumes in the scaling limit the scaling form

$$\langle \sigma_{0,0}\sigma_{M,N}\rangle \sim \mathcal{M}^2 \hat{F}_{\pm}(x) \tag{3.2}$$

where $\mathcal{M} = |1-\sinh^{-4} 2\beta E|^{1/8}$ and the functions $\hat{F}_{\pm}(x)$ are called <u>scaling functions</u>. The + (-) sign denotes that $T \to T_c$ from above (below). The reader is referred to the articles by Fisher[5)] and by Kadanoff et al.[6),7)] for a complete discussion.

An alternative formulation of the scaling hypothesis can be given in momentum space: Let

$$\chi(\vec{k},T) \equiv \sum_{M=-\infty}^{\infty} \sum_{N=-\infty}^{\infty} \left[\langle\sigma_{0,0}\sigma_{M,N}\rangle - \mathcal{M}_s^2(T)\right] e^{i\vec{k}\cdot\vec{R}} \tag{3.3}$$

then the scaling limit is

$$k \to 0, \quad \xi \to \infty \quad \text{such that} \quad p = k\xi \quad \text{is fixed}$$

and the scaling theory hypothesis (again we restrict the statement of the hypothesis to the 2-d. Ising model) is the assumption that $\chi(\vec{k},T)$ in this limit becomes

$$\chi(\vec{k},T) \sim c\,\xi^{7/4} G_{\pm}(p^2) \tag{3.4}$$

where c is some lattice dependent constant.

The relation between $\hat{F}_{\pm}(x)$ and $G_{\pm}^{(2)}(p^2)$ is

$$G_{+}^{(2)}(p^2) = 2\pi \int_0^{\infty} dx\, x\, J_0(xp)\hat{F}_{+}(x)$$

and

$$G_{-}^{(2)}(p^2) = 2\pi \int_0^{\infty} dx\, x\, J_0(xp)[\hat{F}_{-}(x) - 1] \tag{3.5}$$

where $J_0(x)$ is the zeroth order Bessel function. For a discussion of the connection of $G_{\pm}^{(2)}(p^2)$ to critical scattering experiments, see Reference 10 and the references contained therein.

4. EXPLICIT FORMULAS FOR $\hat{F}_{\pm}(x)$

For the case of the two-dimensional Ising model on a square lattice, the scaling hypothesis (3.2) has been verified and explicit expressions have been obtained for $\hat{F}_{\pm}(x)$.[11)-15)] The generalization of Reference 12 to the triangular lattice has been given by Vaidya.[16)] We now present these results:

4.1 Result No. 1

(a) $$\hat{F}_{-}(x) = \exp(f_2) \tag{4.1a}$$

$$f_2(x) = \sum_{n=1}^{\infty} f_2^{(2n)}(x)\lambda^{2n}, \quad \lambda = \pi^{-1}, \tag{4.1b}$$

with

$$f_2^{(2n)}(x) = \frac{(-1)^{n+1}}{n}\int_1^{\infty} dy_1 \cdots \int_1^{\infty} dy_{2n} \prod_{j=1}^{2n}$$

$$\cdot \frac{e^{-xy_j}}{\sqrt{y_j^2-1}} \frac{1}{y_j+y_{j+1}} \prod_{j=1}^{n} (y_{2j}^2-1) \tag{4.1c}$$

where $y_{2n+1} \equiv y_1$.

(b) $$\hat{F}_+(x) = G(x)\hat{F}_-(x) \tag{4.2a}$$

$$G(x) = \sum_{n=0}^{\infty} g_{2n+1}(x)\lambda^{2n+1}, \quad \lambda = \pi^{-1} \tag{4.2b}$$

with

$$g_1(x) = \int_1^{\infty} dy \frac{e^{-xy}}{\sqrt{y^2-1}} = K_0(x) \tag{4.2c}$$

$$g_{2n+1}(x) = (-1)^n \int_1^{\infty} dy_1 \cdots \int_1^{\infty} dy_{2n+1}$$

$$\cdot \left[\prod_{j=1}^{2n+1} \frac{e^{-xy_j}}{(y^2-1)^{\frac{1}{2}}}\right]\left[\prod_{j=1}^{2n} (y_j+y_{j+1})^{-1}\right]$$

$$\cdot \left[\prod_{j=1}^{n} (y_{2j}^2-1)\right] . \tag{4.2d}$$

For large-x we have

$$f_2^{(2n)}(x) \sim c_n \frac{e^{-2nx}}{x^{2n}} \quad (x \to \infty) \tag{4.3a}$$

and

$$g_{2n+1}(x) \sim c_n \frac{e^{-(2n+1)x}}{x^{n+\frac{1}{2}}} \qquad (x \to \infty) \tag{4.3b}$$

and for small-x we have

$$f_2^{(2n)}(x) \sim c_{2n}(\ell nx)^{2n} + c_{2n-1}(\ell nx)^{2n-1}$$

$$+ \cdots + c_1(\ell nx) + c_0 + o(1) \quad (x \to 0) \tag{4.4a}$$

and

$$g_{2n+1}(x) \sim c_{2n+1}(\ell nx)^{2n+1} + \cdots + c_1(\ell nx) + c_0 + o(1) \quad (x \to 0), \tag{4.4b}$$

where we have used the same symbol 'c_n' to denote the various different constants.

The above representations for $\hat{F}_{\pm}(x)$ are most easily interpreted in momentum space [recall (3.5)]. If we look in the complex p^2-plane, then the propagator $G_+^{(2)}(p^2)$ has a single-particle pole at $p^2 = -1$ (in statistical mechanics this is referred to as the Ornstein-Zernike pole) and has continuum thresholds (which are square root type branch points) at $p^2 = -3^2, -5^2, -7^2, \ldots$. On the other hand, the propagator $G_-^{(2)}(p^2)$ in the two-phase region has only branch points which are located at $p^2 = -2^2, -4^2, -6^2, \cdots$. Thus, for example, the function $[f_4^{(2)} + \frac{1}{2}(f_2^{(2)})^2]$ when used in (3.5) gives the four-particle contribution to $G_-^{(2)}(p^2)$.

That these representations provide a rapidly convergent expansion for $G^{(2)}(p^2)$ for small p^2 can best be illustrated by comparing the exact value of $G_{\pm}^{(2)}(p^2)$ at $p^2 = 0$ with the contribution coming from the low-lying excitations. If we denote by $G_{+,2n+1}^{(2)}(p^2)$ the contribution to $G_+^{(2)}(p^2)$ coming from the $(2n+1)$-particle cut and by $G_{-,2n}^{(2)}(p^2)$ the analogous contribution to $G_-^{(2)}(p^2)$, we have

$$G_+^{(2)}(p^2) = \sum_{n=0}^{\infty} G_{+,2n+1}^{(2)}(p^2)$$

and (4.5)

$$G_-^{(2)}(p^2) = \sum_{n=1}^{\infty} G_{+,2n}^{(2)}(p^2).$$

Using (4.1) and (4.2) in (3.5), we can show[18)]

$$G^{(2)}_{+,1}(0) = 2,$$

$$G^{(2)}_{+,3}(0) = \frac{1}{\pi^2}\left\{\tfrac{1}{3}\pi^2 + 2 - 3\sqrt{3}\, C\ell_2(\pi/3)\right\}, \tag{4.6}$$

$$G^{(2)}_{-,2}(0) = \frac{1}{6\pi},$$

and

$$G^{(2)}_{-,4}(0) = \frac{1}{8\pi^3}\left\{\frac{4\pi^2}{9} - \frac{1}{6} - \frac{7}{2}\zeta(3)\right\}$$

where

$$\zeta(s) = \sum_{n=1}^{\infty} n^{-s} \quad \text{and} \quad C\ell_2(\theta) = \sum_{n=1}^{\infty} \frac{\sin n\theta}{n^2} .$$

In particular, $\zeta(3) = 1.2020569031 \cdots$ and $C\ell_2(\pi/3) = 1.0149417 \cdots$.[19] We now compare the LHS of (4.5) at $p^2 = 0$ to the first two terms of the RHS:

$$G^{(2)}_{+}(0) = 2.001\,630\,521$$

$$G_{+,1}(0) = 2.0$$

$$G^{(2)}_{+,1}(0) + G^{(2)}_{+,3}(0) = 2.001\,628\,925 \cdots \tag{4.7}$$

$$G^{(2)}_{-}(0) = .053\,102\,589 \cdots$$

$$G^{(2)}_{-,2}(0) = .053\,051\,648 \cdots$$

$$G^{(2)}_{-,2}(0) + G^{(2)}_{-,4}(0) = .053\,102\,545 \cdots . \tag{4.8}$$

The representations given above (Result No. 1) are not so useful if we wish to examine the short-distance behavior of $\hat{F}_{\pm}(x)$. That this is the case follows directly from (4.4a) and (4.4b), where we see the short distance behavior of the functions $f_2^{(2n)}(x)$ and $g_{2n+1}(x)$ have ever increasing powers of logarithms. We will see that $\hat{F}_{\pm}(x) \sim Cx^{-\frac{1}{4}}$ as $x \to 0$ so the logarithms must sum up to an algebraic power. This 'summing up' feature of logarithms is not unusual in quantum field theory. However, we point out that viewed in this language of summing logarithms, one must sum all logarithms not just leading logs.

To study the short distance behavior of $\hat{F}_{\pm}(x)$, we do not actually sum all these logarithms 'by hand.' Rather, we find that there is an underlying nonlinear differential equation whose solution, roughly speaking, sums the series representations (4.1) and (4.2). We now discuss this nonlinear differential equation. Since this workshop is on nonlinear differential equations, we present our results in their most general form and then specialize them in Section 6, when we apply these solutions to the Ising model.

5. PAINLEVÉ TRANSCENDENTS

The Painlevé equation of the third kind is

$$\frac{d^2w}{d\theta^2} = \frac{1}{w}\left(\frac{dw}{d\theta}\right)^2 - \frac{1}{\theta}\frac{dw}{d\theta} + \frac{1}{\theta}(\alpha w^2 + \beta) + \gamma w^3 + \frac{\delta}{w} \tag{5.1}$$

where α, β, γ, and δ are constants. The importance of (5.1) in in the theory of nonlinear equations is discussed in the papers by Painlevé[20] and Gambier[21] and in the book by Ince.[22] If we make the restriction

$$\alpha\sqrt{-\delta} + \beta\sqrt{\gamma} = 0 \tag{5.2}$$

on the constants appearing in (5.1), then (5.1) is easily reducible to

$$\frac{d^2w}{d\theta^2} = \frac{1}{w}\left(\frac{dw}{d\theta}\right)^2 - \frac{1}{\theta}\frac{dw}{d\theta} + \frac{2\nu}{\theta}(w^2 - 1) + w^3 - \frac{1}{w} \tag{5.3}$$

where ν is a constant. We call (5.3) the restricted Painlevé Equation of the third kind.

Let us denote by $\eta(\theta;\nu,\lambda)$ the one-parameter family of solutions of (5.3) that remains bounded as $\theta \to \infty$ along the positive real θ-axis. Then we have[14]

5.1 Result No. 2

For sufficiently large positive θ and $\mathrm{Re}\,\nu > -\frac{1}{2}$, the function $\eta(\theta;\nu,\lambda)$ defined above has the representation

$$\frac{1-\eta(\theta;\nu,\lambda)}{1+\eta(\theta;\nu,\lambda)} = G(x;\nu,\lambda) = \sum_{n=0}^{\infty} \lambda^{2n+1} g_{2n+1}(x;\nu) \tag{5.4}$$

where $2\theta = x$

$$g_1(x;\nu) = \int_1^{\infty} dy \frac{e^{-xy}}{(y^2-1)^{\frac{1}{2}}} \left(\frac{y-1}{y+1}\right)^{\nu}, \tag{5.5a}$$

and for $n \geqq 1$

$$g_{2n+1}(x;\nu) = (-1)^n \int_1^{\infty} dy_1 \cdots \int_1^{\infty} dy_{2n+1} \left[\prod_{j=1}^{2n+1} \frac{e^{-xy_j}}{(y_j^2-1)^{\frac{1}{2}}} \cdot \left(\frac{y_j-1}{y_j+1}\right)^{\nu}\right] \left[\prod_{j=1}^{2n} (y_j+y_{j+1})^{-1}\right] \left[\prod_{j=1}^{n} (y_{2j}^2-1)\right]. \tag{5.5b}$$

If we define $\psi(x;\nu,\lambda)$ by

(i) $\eta(\theta;\nu,\lambda) = e^{-\psi(x;\nu,\lambda)}, \quad x = 2\theta,$

(ii) $\psi(x;\nu,\lambda) \to 0 \quad \text{as} \quad x \to +\infty,$

then $\psi(x;\nu,\lambda)$ satisfies the differential equation

$$\psi'' + x^{-1}\psi' = \tfrac{1}{2}\sinh(2\psi) + 2\nu x^{-1}\sinh(\psi). \tag{5.6}$$

Furthermore, $\psi(x;\nu,\lambda)$ has the representation:

5.2 Result No. 3

$$\psi(x;\nu,\lambda) = \sum_{n=0}^{\infty} \lambda^{2n+1} \psi_{2n+1}(x;\nu) \tag{5.7}$$

with

$$\psi_1(x;\nu) = 2g_1(x;\nu) \tag{5.8a}$$

$$\psi_{2n+1}(x;\nu) = \frac{2}{2n+1}\int_1^\infty dy_1 \cdots \int_1^\infty dy_{2n+1}\left[\prod_{j=1}^{2n+1} \cdot \frac{e^{-xy_j}}{y_j+y_{j+1}}\right]\left[\prod_{j=1}^{2n+1}\left(\frac{y_j-1}{y_{j+1}}\right)^{\nu-\frac{1}{2}} + \prod_{j=1}^{2n+1}\left(\frac{y_j-1}{y_j+1}\right)^{\nu+\frac{1}{2}}\right] \quad (5.8b)$$

Notice that both $g_{2n+1}(x;\nu)$ and $\psi_{2n+1}(x;\nu)$ are in the form of iterated integrals: Define the <u>linear operator</u> $\mathcal{K}$

$$(\mathcal{K}f)(x) = \int_1^\infty d\sigma_\pm(y)\, e^{-\theta(x+y)}(x+y)^{-1} f(y) \quad (5.9a)$$

where the measure $d\sigma_\pm$ is

$$d\sigma_\pm = d\sigma_\pm(y) = \left(\frac{y-1}{y+1}\right)^{\nu\pm\frac{1}{2}} dy. \quad (5.9b)$$

Consider the eigenfunctions and eigenvalues ϕ_j and λ_j,

$$(\mathcal{K}\phi_j^\pm)(x) = \lambda_j^\pm(\theta,\nu)\,\phi_j^\pm(x;\theta,\nu). \quad (5.10)$$

Then we can rewrite (5.8b) as

$$\psi_{2n+1}(x;\nu) = \frac{2}{2n+1}\int_0^\infty d\zeta\left[(e,\mathcal{K}^{2n}e)_+ + (e,\mathcal{K}^{2n}e)_-\right] \quad (5.11)$$

where the vector $|e\rangle$ is

$$\langle y|e\rangle = e^{-(\zeta+\theta)y}$$

and the scalar product $(\ ,\ )_\pm$ is

$$(g,f)_\pm = \int_1^\infty d\sigma_\pm(y)\,\overline{g(y)}\,f(y).$$

Using (5.11) in (5.7) and recalling $\eta = e^{-\psi}$, we have

$$\eta(\theta;\nu,\lambda) = \prod_{j=1}^{\infty}\left(\frac{1-\lambda_j^{+}\lambda}{1+\lambda_j^{+}\lambda}\right)^{a_j^{+}} \prod_{j=1}^{\infty}\left(\frac{1-\lambda_j^{-}\lambda}{1+\lambda_j^{-}\lambda}\right)^{a_j^{-}} \tag{5.12}$$

where

$$a_j^{\pm} = a_j^{\pm}(\theta,\nu) = (\lambda_j^{\pm})^{-1}\int_{\theta}^{\infty} d\zeta \left|\int_1^{\infty} d\sigma_{\pm}(y)e^{-\zeta y}\phi_j^{\pm}(y;\theta,\nu)\right|^2 . \tag{5.13}$$

The representation (5.12) clearly displays the behavior of $\eta(\theta;\nu,\lambda)$ in the complex λ-plane (recall λ is the integration constant parameter). It is an open problem to find explicit formulas for $\lambda_j^{\pm}$ and $\phi_j^{\pm}(x)$.

If we examine the small distance behavior of the functions $\psi_{2n+1}(x;\nu)$ as given by (5.8), we find[14)]

$$\psi_{2n+1}(x;\nu) = \sigma_{2n+1}\ln\left(\frac{1}{x}\right) + B_{2n+1} + o(1) \quad (x\to 0^{+}) \tag{5.14}$$

That is to say, the logarithms do not increase in order when we go to the ψ-representation. Thus

$$\psi(x;\nu,\lambda) = \sum_{n=0}^{\infty}\lambda^{2n+1}\psi_{2n+1}(x;\nu)$$

$$= \sigma\ln\left(\frac{1}{x}\right) + B + o(1) \quad (x\to 0^{+}) \tag{5.15}$$

with

$$\sigma = \sum_{n=0}^{\infty}\lambda^{2n+1}\sigma_{2n+1}$$

and

$$\ln B = -\sum_{n=0}^{\infty}\lambda^{2n+1}B_{2n+1} . \tag{5.16}$$

We expect (5.15) to be valid whenever the series expansions for σ and B converge. There is a very simple physical argument for the case $\nu=0$ that gives (5.15). For $\nu=0$ the nonlinear differential equation (5.6) is essentially the spherically symmetric version of (1.1), the nonlinear Debye-Hückel equation. In Debye-Hückel theory $\phi(x)$ is the electrostatic potential of the field of the ion cloud surrounding the test charge at the origin. As $x \to 0$, we must see the bare test charge that is located at the origin. Hence the factor $\sigma \ln(1/x)$. The quantity B represents the potential due to all other ions of the cloud at the position of the test charge.

The mathematical problem is to extract from (5.8) the coefficients σ_{2n+1} and B_{2n+1} and then sum the resulting series (5.16). We have[14)]

5.3 Result No. 4

$$\sigma = \sigma(\lambda) = \frac{2}{\pi} \arcsin(\pi\lambda)$$

$$\begin{aligned} B &= B(\sigma,\nu) \\ &= 2^{-3\sigma} \frac{\Gamma^2((1-\sigma)/2)}{\Gamma^2((1+\sigma)/2)} \frac{\Gamma(((1+\sigma)/2)+\nu)}{\Gamma(((1-\sigma)/2)+\nu)}. \end{aligned} \tag{5.17}$$

We notice that σ does not depend upon ν. This is easy to understand since the logarithm in (5.14) comes from the region of integration variables y_j large in (5.8). In this region $((y_j-1)/(y_j+1))^\nu \sim 1$. Also we see that $\lambda = 1/\pi$ $(\sigma = 1)$ plays a distinguished role in these formulas. It is precisely $\lambda = 1/\pi$ that is needed in the Ising model.

Using Result No. 4, we can determine the behavior of $\eta(x/2;\nu,\lambda)$ at $x=0$:

5.3.1 $0 < \lambda < 1/\pi$

$$\begin{aligned} \eta(x/2;\nu,\lambda) = Bx^\sigma \Big\{ & 1 - \nu B^{-1}(1-\sigma)^{-2} x^{1-\sigma} + B\nu(1+\sigma)^{-2} x^{1+\sigma} \\ & + \left[\frac{1}{4}\nu^2 B^{-2}(1-\sigma)^{-4} - \frac{1}{16} B^{-2}(1-\sigma)^{-2}\right] \\ & \cdot x^{2-2\sigma} + O(x^2)\Big\}. \end{aligned} \tag{5.18}$$

5.3.2 $\lambda = 1/\pi$

$$\eta(x/2;\nu,\pi^{-1}) \sim \frac{1}{2} x\left[\nu \ln^2 x - C(\nu) \ln x + \frac{1}{4\nu}(C^2(\nu)-1)\right]$$

where

$$C(\nu) = 1 + 2\nu\left[3 \ln 2 - 2\gamma - \psi(\nu+1)\right] \tag{5.19}$$

and $\psi(x) = (d/dx) \ln \Gamma(x)$, $\Gamma(x)$ being the gamma function.

5.3.3 $\lambda > 1/\pi$ (for simplicity we set $\nu = 0$)

$$\eta(x/2;0,\lambda) \sim -\frac{1}{4\mu} x \sin[2\mu \ln(x/8) + 2\phi(\mu)]. \tag{5.20}$$

with $\sigma = 1 + 2i\mu$ and

$$\Gamma(iy) = |\Gamma(iy)| e^{i\phi(y)}$$
$$= \left(\frac{\pi}{y \sinh y}\right)^{\frac{1}{2}} e^{i\phi(y)} .$$

5.3.4 For $\lambda < 0$ use

$$\eta(x/2;\nu,-\lambda) = \frac{1}{\eta(x\ 2;\nu,\lambda)}. \tag{5.21}$$

In Case 5.3.3, the origin $x = 0$ is the limit point of a set of zeros and in Case 5.3.4 for $\lambda < -(1/\pi)$, the origin is the limit point of a set of poles on the positive real axis. The asymptotic spacing of these zeros and poles follows from the above formulas. These zeros and poles are intimately connected with the eigenvalues $\lambda_j^{\pm}(\theta;\nu)$ as can be seen by recalling (5.12).

6. $\hat{F}_{\pm}(x)$ IN TERMS OF $\psi(x;0,\pi^{-1})$

Comparing Result No. 1 with Result No. 2, we see

$$\frac{\hat{F}_+(x)}{\hat{F}_1(x)} = \frac{1-\eta(x/2;0,\pi^{-1})}{1+\eta(x/2;0,\pi^{-1})} = \tanh[\tfrac{1}{2}\psi(x;0,\pi^{-1})]. \tag{6.1}$$

This relates the ratio of $\hat{F}_+(x)$ to $\hat{F}_-(x)$ to the Painlevé function $\eta(x/2;0,\pi^{-1})$. We still need a relation of $\hat{F}_\pm(x)$ to $\eta(x/2;0,\pi^{-1})$ [or $\psi(x;0,\pi^{-1})$]. This we have in the next result.[14]

6.1 Result No. 5

$$\hat{F}_-(x) = \exp\left(\sum_{n=1}^{\infty} \lambda^{2n} f_2^{(2n)}(x)\right)$$

$$= \cosh \tfrac{1}{2}\psi(x,0,\lambda) \exp\left[-\frac{1}{4}\int_x^{\infty} dr\, r\, \mathcal{L}(r)\right] \tag{6.2}$$

where

$$\mathcal{L}(r) = \left(\frac{d\psi}{dr}\right)^2 - \sinh^2\psi \tag{6.3}$$

the ψ being $\psi(r;0,\lambda)$.

From this representation and the short-distance behavior of $\psi(x;0,\lambda)$ of the previous section, we have[12] for $\lambda = 1/\pi$

$$\hat{F}_\pm(x) = Cx^{-\frac{1}{4}}\left\{1 \pm x\Omega + \frac{1}{16}x^2 \pm \frac{1}{32}x^3\Omega\right.$$

$$\left. + \frac{1}{256}x^4(-\Omega^2 + \Omega + \frac{1}{8}) + 0(x^5\Omega^4)\right\} \tag{6.4}$$

with $\Omega = \ln(x/8) + \gamma$, γ = Euler's constant.

Some comments:

(1) The RHS of (6.2) looks almost like an action. If one varies the $\mathcal{L}(r)$ in (6.3), one gets

$$\frac{d^2\psi}{dr^2} + \frac{1}{r}\frac{d\psi}{dr} = -\tfrac{1}{2}\sinh(2\psi) \tag{6.5}$$

which disagrees with (5.6) by a minus sign. What is the correct interpretation of the RHS of (6.2)?

(2) As Result No. 5 is stated the identity holds for all λ though its application to the Ising model is only for $\lambda = 1/\pi$. Does the variable λ have a physical interpretation? If we keep λ as a variable and compute the short distance behavior of $\hat{F}_\pm(x;\lambda)$, we find that the anamolous dimension is now a function of λ. This suggests the Baxter model.[23]

(3) In Reference 14, (6.2) is generalized to $\nu \neq 0$ in which case $\mathcal{L}(r)$ of (6.3) becomes

$$\mathcal{L}(r) = \left(\frac{d\psi}{dr}\right)^2 - \sinh^2\psi - \frac{4\nu}{r}\sinh^2\tfrac{1}{2}\psi. \tag{6.6}$$

Recently Ablowitz and Segur[24] have shown a deep connection of the Painlevé transcendent of second kind

$$\frac{d^2W}{dz^2} = zW + 2W^3 \tag{6.7}$$

and the long-time behavior of the modified KdV equation. Equation (6.7) is a special case[25] of Result No. 2 (or Result No. 3) and thus the solution that is needed in the modified KdV analysis is known.[25] Ablowitz and Segur[26] have also obtained this same solution to (6.7) using the inverse spectral transform method. Does the generalized $\nu \neq 0$ identity (6.2) when analyzed in the KdV limit have any applications?

7. n-POINT FUNCTIONS

The generalization of Result No. 1 to the n-point functions has been recently given by a number of authors.[27)-29)] We will not write down the formulas here but merely note the form of the answer in the scaling limit: First for $T \to T_c^-$

$$\lim \mathcal{M}^{-n} \langle \sigma_{M_1N_1} \cdots \sigma_{M_nN_m} \rangle = \exp(f_n)$$

$$f_n = \sum_{k=2}^{\infty} f_n^{(k)}$$

where $f_n^{(k)}$ is a 2k-dimensional integral. For $T \to T_c^+$

$$\lim \mathcal{M}^{-n} \langle \sigma_{M_1N_1} \cdots \sigma_{M_nN_n} \rangle = g_n \exp(f_n)$$

$$g_n = |\det g_{(n)ij}|^{\frac{1}{2}}$$

$$g_{(n)ij} = \sum_{k=1}^{\infty} g_{(n)ij}^{(k)},$$

where $g^{(k)}_{(n)ij}$ are 2k-dimensional integrals.

We point out that the integrals $f^{(k)}_n$ and $g^{(k)}_{(n)ij}$ require special care in treating singularities of the integrand. This point is discussed in References 27 and 28, but the integrals appearing in Reference 29 are ambiguous since no prescription for interpreting the singular integrals is given.

REFERENCES AND FOOTNOTES

1. E. Ising, Z. Phys. 31, 253 (1925).
2. L. Onsager, Phys. Rev. 65, 117 (1944).
3. C. N. Yang, Phys. Rev. 85, 808 (1952).
4. See also, B. M. McCoy and T. T. Wu, The Two-Dimensional Ising Model, Harvard University, Cambridge, Massachusetts, 1973.
5. L. P. Kadanoff, Physics (N.Y.) 2, 263 (1966).
6. M. E. Fisher, Rep. Prog. Phys. 30, 615 (1967).
7. L. P. Kadanoff, W. Gotze, D. Hamblen, R. Hecht, E. A. S. Lewis, V. V. Palciauskas, M. Rayl, J. Swift, D. Aspens, and J. Kane, Rev. Mod. Phys. 39, 395 (1967).
8. L. Onsager, discussion, Nuovo Cimento 6, Suppl., 261 (1949).
9. C. H. Chang, Phys. Rev. 88, 1422 (1952).
10. C. A. Tracy and B. M. McCoy, Phys. Rev. B12, 368 (1975).
11. E. Barouch, B. M. McCoy, and T. T. Wu, Phys. Rev. Letters 31, 1409 (1973);
    C. A. Tracy and B. M. McCoy, Phys. Rev. Letters 31, 1500 (1973).
12. T. T. Wu, B. M. McCoy, C. A. Tracy, and E. Barouch, Phys. Rev. B13, 316 (1976).
13. B. M. McCoy, C. A. Tracy, and T. T. Wu in Statistical Mechanics and Statistical Methods in Theory and Application, U. Landman (ed.), Plenum Press, New York, 1977.
14. B. M. McCoy, C. A. Tracy, and T. T. Wu, J. Math. Phys. 18, 1058 (1977).
15. The closely related scaling functions in the one-dimensional XY-Model can be found in H. G. Vaidya and C. A. Tracy, Stony Brook preprint ITP-SB-77-48.
16. H. Vaidya, Phys. Lett. 57A, 1 (1976).
17. The numbers $G^{(2)}_{\pm}(0)$ when multiplied by a lattice dependent number are equal to the susceptibility coefficients $C_{0\pm}$

where $\chi(T) \sim C_{0\pm}|1-T/T_c|^{-7/4}$. For the square lattice with symmetric interactions this constant is $2^{3/8}[2\ell n(1+\sqrt{2})]^{-7/4}$. For the triangular lattice this lattice dependent constant can be found in Reference 16.

18. B. M. McCoy, C. A. Tracy and T. T. Wu, unpublished notes.
19. The direct numerical evaluation of $G_\pm^{(2)}(0)$ is done using the Painlevé transcendent introduced below.
20. P. Painlevé, Acta Math. 25, 1 (1902).
21. B. Gambier, Acta Math. 33, 1 (1910).
22. E. L. Ince, Ordinary Differential Equations, Dover, New York, 1945, Chapter 14.
23. R. J. Baxter, Phys. Rev. Letters 26, 832 (1971) and Annals of Phys. (New York) 70, 193 (1972).
24. M. J. Ablowitz and H. Segur, Stud. App. Math. 57, 13 (1977).
25. B. M. McCoy, C. A. Tracy, and T. T. Wu, Phys. Lett. 61A, 283 (1977).
26. M. J. Ablowitz and H. Segur, Phys. Rev. Letters 38, 1103 (1977).
27. B. M. McCoy, C. A. Tracy, and T. T. Wu, Phys. Rev. Letters 38, 793 (1977).
28. M. Sato, T. Miwa, and M. Jimbo, Proc. Japan Acad. 53A, 6 (1977).
29. D. B. Abraham, Phys. Lett. 61A, 271 (1977).

Note Added in Proof

For some new results, see B. M. McCoy and T. T. Wu, Phys. Lett. 72B, 219 (1977); R. Z. Bariev, 64A, 169 (1977); D. Wilkinson, to appear in Phys. Rev. D; and R. Haberman, Stud. Appl. Math. 57, 247 (1977).

STATISTICAL MECHANICS OF NONLINEAR LATTICE DYNAMIC MODELS EXHIBITING PHASE TRANSITIONS†

T. Schneider
IBM Zurich Research Laboratory
8803 Rüschlikon ZH,
Switzerland

1. INTRODUCTION

To imitate real systems exhibiting distortive phase transitions, it becomes customary to consider Hamiltonians of the form [1]

$$\mathcal{H} = \sum_{\ell,\alpha} \frac{P_{\ell\alpha}^2}{2M} + \frac{A}{2} \sum_{\ell,\alpha} X_{\ell\alpha}^2 + \frac{B}{4n} \sum_{\ell} \left(\sum_{\alpha} X_{\ell\alpha}^2\right)^2$$

$$+ \frac{B_1}{4} \sum_{\ell,\alpha} X_{\ell\alpha}^4 + \sum_{\ell,m,\alpha} V_{\ell,\ell+m,\alpha\alpha} X_{\ell\alpha} X_{\ell+m\alpha} . \qquad (1)$$

ℓ denotes the lattice sites, $P_{\ell\alpha}$, $X_{\ell\alpha}$ are the α-th component ($\alpha = 1, \ldots, n$) of momentum and displacement vector of the ℓ-th particle with mass M, with respect to a rigid reference lattice. M, A, B, B_1 and $V_{\ell,\ell+m,\alpha\alpha}$ are the model parameters. They are chosen in such a way that one of the mean displacement components

$$\langle X_\alpha \rangle = \frac{1}{N} \sum_{\ell} \langle X_{\ell\alpha} \rangle \qquad (2)$$

does not vanish at zero temperature. Thus, the system is

† Lectures given at NATO Advanced Study Institute on Nonlinear Equations in Physics and Mathematics, Istanbul, August 1977.

A. O. Barut (ed.), Nonlinear Equations in Physics and Mathematics, 239–270.

expected to undergo a distortive phase transition at $T = T_C > 0$, where the order parameter $\langle X_\alpha \rangle$ [Equation (2)] vanishes.

Such model Hamiltonians may be understood if one considers the situation of a crystal consisting of two sublattices. The particles of one sublattice give rise to the single-particle potential

$$\frac{A}{2} \sum_{\ell,\alpha} X_{\ell\alpha}^2 + \frac{B}{4n} \sum_{\ell} \left(\sum_{\alpha} X_{\ell\alpha} \right)^2 + \frac{B_1}{4} \sum_{\ell,\alpha} X_{\ell\alpha}^4 \tag{3}$$

in which the particles of the other sublattice, which are considered explicitly, move. These latter particles are then coupled by means of a nearest-neighbor interaction. Higher-order terms in the single-particle potential [Equation (3)] may give rise to multicritical points.

In recent years, the static and dynamic properties of these systems have been studied by means of different techniques. They include the renormalization-group approach [1], the Monte Carlo [2] and the molecular-dynamics techniques [3]-[8]. Even exact results have been established for quartic long-range interactions [9], in the limit $n \to \infty$ [10], for the existence of a phase transition [11] and for tricritical points [12]. Moreover, quantum effects have been considered [13].

It should be emphasized, however, that the exactly soluble versions represent limits, where Hartree-like approximations become exact. Consequently, the equations of motion become linear. The nonlinearities are taken into account in the very powerful renormalization-group approach [1]. The results are limited, however, to a very narrow region around the critical temperature T_C.

Until recently, the molecular-dynamics technique, representing a brute-force numerical solution of the equations of motion, has not been applied to such systems. Recently, we have undertaken such studies on one-, two-, and three-dimensional, one- and two-component models [3]-[8]. One of the interesting results was the demonstration of solitary-wave and soliton-like features in the excitation spectrum [3]-[8] and in the heat-pulse propagation [8].

In these Lectures, we want to do three things: (i) We establish the link between two particular models and important nonlinear wave equations; (ii) we demonstrate the relevance of solitary wave solutions in a statistical description of these models by means of molecular-dynamics results; (iii) we demonstrate envelope soliton-like heat-pulse propagation in a three-dimensional one-component model.

It should be emphasized that the models considered here differ from the Toda lattice or the Fermi-Pasta-Ulam problem, in that a phase transition occurs.

2. THE MODELS AND THEIR CONTINUUM LIMIT, THE EQUATIONS OF MOTION AND PARTICULAR SOLUTIONS

In this section we define the models, the dynamic variables of interest and study the equations of motion in the continuum limit. Moreover, we establish the link with the corresponding nonlinear wave equations.

2.1 The two-dimensional X-Y model with quartic anisotropy

The Hamiltonian of this model belongs to the family defined by Equation (1) and reads

$$\mathcal{H} = \frac{M}{2}\sum_{\ell}\left(\dot{X}_\ell^2 + \dot{Y}_\ell^2\right) + \frac{A}{2}\sum_{\ell}\left(X_\ell^2 + Y_\ell^2\right) + \frac{B}{8}\sum_{\ell}\left(X_\ell^2 + Y_\ell^2\right)^2$$
$$+ \frac{B_1}{4}\sum_{\ell}\left(X_\ell^4 + Y_\ell^4\right) - C\sum_{\ell,m}\left(X_\ell X_{\ell+m} + Y_\ell Y_{\ell+m}\right). \tag{4}$$

The momentum and displacement vector $\vec{X}_\ell$ has two components, $\dot{\vec{X}}_\ell = (\dot{X}_\ell, \dot{Y}_\ell)$ and $\vec{X}_\ell = (X_\ell, Y_\ell)$, respectively. We consider nearest-neighbor interactions only. For $B_1 = 0$, the model would be rotationally invariant. The rigid reference lattice is assumed to be a simple square lattice with lattice constant a. The equations of motion read

$$-M\ddot{X}_\ell = (A - 8C)X_\ell + \frac{B}{2}\left(X_\ell^2 + Y_\ell^2\right)X_\ell + B_1 X_\ell^3 + 2C\sum_m (X_\ell - X_{\ell+m});$$
$$-M\ddot{Y}_\ell = (A - 8C)Y_\ell + \frac{B}{2}\left(X_\ell^2 + Y_\ell^2\right)Y_\ell + B_1 Y_\ell^3 + 2C\sum_m (Y_\ell - Y_{\ell+m}). \tag{5}$$

At zero temperature, where $\ddot{X}_\ell = \ddot{Y}_\ell = 0$ within the framework of classical mechanics, the mean displacements are easily calculated, yielding

$$X_\ell^2 = Y_\ell^2 = \langle X_\ell \rangle^2_{T=0} = \langle Y_\ell \rangle^2_{T=0} = \frac{8C - A}{B + B_1}, \quad B_1 > 0;$$
$$Y_\ell = \langle Y_\ell \rangle_{T=0} = 0, \; X_\ell^2 = \langle X_\ell \rangle^2_{T=0} = \frac{8C - A}{B/2 - B_1}, \quad B_1 < 0. \tag{6}$$

Introducing the displacement field

$$\vec{Q} = \rho(\vec{R}) \left[\cos\varphi\,(\vec{R}),\; \sin\varphi\,(\vec{R}) \right], \tag{7}$$

the equation of motion (5) becomes in the continuum limit

$$-M\ddot{\rho} = (A - 8C)\rho + \frac{B}{2}\rho^3 + 2C\,a^2\rho\,(\nabla\varphi)^2$$
$$+ B_1\rho^3(1 - \tfrac{1}{2}\sin^2 2\varphi) - 2C\,a^2\nabla^2\rho; \tag{8}$$

$$\nabla^2\varphi - \frac{M}{2C\,a^2}\ddot{\varphi} - \frac{M\dot{\rho}\dot{\varphi}}{\rho C a^2} = -\frac{B_1\rho^2}{8C\,a^2}\sin 4\varphi + \frac{M\dot{\rho}\dot{\varphi}}{\rho C a^2}. \tag{9}$$

For $\rho = \text{const.}$, Equation (9) is just the sine-Gordon equation. To simplify the model, we assume that the potential barrier in the radial field is so high that the relevant motions occur only in the angular field. For this purpose, the model parameters have been chosen as follows [7]:

$$A = -1, \quad B = 4/5, \quad B_1 = -1/15, \quad C = 1/4. \tag{10}$$

The resulting shape of the single-particle potential

$$V(X_\ell, Y_\ell) = \frac{A - 8C}{2}\left(X_\ell^2 + Y_\ell^2\right) + \frac{B}{8}\left(X_\ell^2 + Y_\ell^2\right)^2$$
$$+ \frac{B_1}{4}\left(X_\ell^4 + Y_\ell^4\right) \tag{11}$$

is illustrated in Figure 1. The four minima on the brim of the Mexican hat arise from the quartic anisotropy term. The height of the hat indicates that for not too high temperatures, the motions should be determined by the angular field only. Accordingly, we keep the radial component of the field constant.

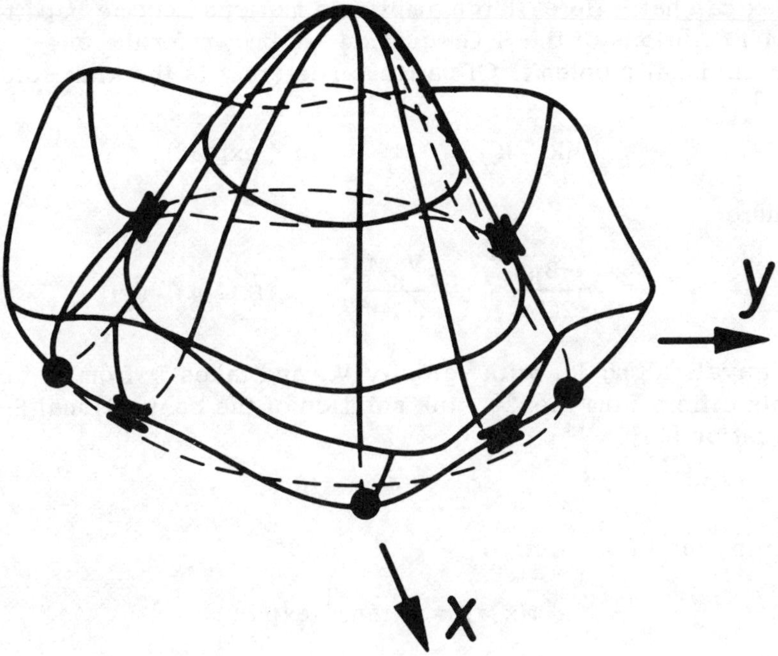

Figure 1. Shape of the single-particle potential $V(X_\ell, Y_\ell)$ as given by Equation (11). • denotes the minima and ≍ passes due to the quartic anisotropy term.

At very low temperature, most particles will be located in the minimum [Equation (6)]

$$Y_\ell = 0, \quad X_\ell = \left(\frac{8C - A}{B/2 - B_1}\right)^{\frac{1}{2}}, \quad B_1 < 0 \tag{12}$$

and perform small amplitude oscillations. Here, we assume that the preparation of the system at $T = 0$ was such that all particles are in the minimum $X_\ell > 0$. For the small amplitude oscillations, the sine-Gordon (S-G) equation (9) can be linearized. The resulting frequency is given by

$$M\omega_T^2 = -B_1\rho^2 + 2C\,a^2q^2, \tag{13}$$

where $\vec{q}$ is the wave vector. At higher temperatures, the particles will more frequently overcome the passes along the brim of the Mexican hat. Here, large amplitude motions become important. Such solutions of the S-G equation are known for the one-dimensional problem. Of particular interest is the kink solution [14]

$$\varphi(R_x, R_y = 0, t) = \tan^{-1} \exp(\alpha\xi), \tag{14}$$

where

$$\alpha^2 = \frac{-B_1\rho^2}{2C\,a^2}\left(1 - \frac{V_x^2 M}{2C\,a^2}\right)^{-1}, \quad \xi = R_x - V_x t. \tag{15}$$

It travels along R_x with velocity V_x and takes φ from 0 to $\pi/2$. This differs from the 2π kink solution of the conventional S-G equation [14],

$$\Phi_{xx} - \ddot{\Phi} = \sin\Phi, \tag{16}$$

having the kink solution

$$\Phi(x,\tau) = 4\tan^{-1}(\exp\alpha\xi), \tag{17}$$

where

$$\xi = x - vt, \quad \alpha^2 = (1 - v^2)^{-1}. \tag{18}$$

By the change of variables

$$X = \left(\frac{|B_1|\rho^2}{2C\,a^2}\right)^{\frac{1}{2}} R_x; \quad \tau = \left(\frac{|B_1|\rho^2}{M}\right)^{\frac{1}{2}} t; \quad \Phi = 4\varphi, \tag{19}$$

it becomes evident, however, that Equations (9) and (16) are equivalent. Hence, because the kink solution (17) is known to be a soliton [14], the particular solution of our problem will be a soliton as well. Moreover,

$$-\Phi_{sol} = 4\tan^{-1}(\exp - \alpha\xi) = \Phi_{antisol} \tag{20}$$

will also be a soliton, which may be identified as an antisoliton. The solutions (17) and (20) are defined modulo 2Π but

$$\Phi_{sol}(\infty,t) - \Phi_{sol}(-\infty,t) = 2\Pi$$

$$\Phi_{\text{antisol}}(\infty,t) - \Phi_{\text{antisol}}(-\infty,t) = -2\Pi . \tag{21}$$

There is also the soliton-antisoliton solution [15]

$$\Phi_{SH}(x,t) = 4\tan^{-1}\left[\frac{\sinh(vt/\sqrt{1-u^2})}{v\cosh(x\sqrt{1-u^2})}\right] \tag{22}$$

and the soliton-soliton scattering solution

$$\Phi_{ss}(x,t) = 4\tan^{-1}\left[\frac{v\sinh(x/\sqrt{1-u^2})}{\cosh(vt/\sqrt{1-u^2})}\right] \tag{23}$$

It is also possible to construct solutions involving an arbitrary number of solitons [14].

On physical grounds, however, we expect that the soliton (14) and the corresponding antisoliton will be of particular relevance in our two-dimensional system. In fact, if the temperature is not too low, the particles may overcome the passes and any successful attempt corresponds to a $\Pi/2$ kink or antikink solution.

To summarize this short discussion on the X-Y model with quartic anisotropy, the excitation spectrum is expected to be influenced by small amplitude oscillations, large amplitude motions, where the kink solitons may play an important role. Clearly, the original system is two-dimensional and a discrete lattice model. Moreover, the radial component of the field is only approximately constant. On these grounds, it is not yet clear whether or not the soliton features will survive.

Additional effects are expected from energy conservation. This problem will be discussed in the context of the next model.

2.2 The three-dimensional one-component model

The Hamiltonian of this model is again a special case of the family defined by Equation (1). It reads

$$\mathcal{H} = \frac{M}{2}\sum_{\ell}\dot{X}_{\ell}^{2} + \frac{A}{2}\sum_{\ell}X_{\ell}^{2} + \frac{B}{4}\sum_{\ell}X_{\ell}^{4} - C\sum_{\ell,m}X_{\ell}X_{\ell+m} . \tag{24}$$

Obviously, the displacement vector has one component only. The rigid reference lattice is assumed to be a primitive cubic one with lattice constant a. The equation of motion reads

$$-M\ddot{X}_\ell = (A - 12C)X_\ell + BX_\ell^3 + 2C \sum_m (X_\ell - X_{\ell+m}) . \quad (25)$$

At zero temperature, the mean displacement is easily calculated, yielding

$$X_\ell^2 = \frac{12C - A}{B} . \quad (26)$$

Introducing the scalar displacement field $f(\vec{R},t)$, the equations of motion (25) become in the continuum limit

$$-M\ddot{f} = -(12C - A)f + Bf^3 - 2C\, a^2\nabla^2 f. \quad (27)$$

Linearization of Equation (27) is adequate at very low temperatures where almost all particles perform small amplitude oscillations in one of the double wells defined by

$$-\frac{12C - A}{2} X^2 + \frac{B}{4} X^4 . \quad (28)$$

Which double well is preferred depends on the preparation of the system at $T=0$. The frequencies of the small amplitude oscillations are given by

$$M\omega^2 = -(12C - A) + 3B\langle f^2\rangle f + 2C\, a^2 q^2 , \quad (29)$$

where $\vec{q}$ is the wave vector.

With increasing temperature, the particles can easier overcome the potential barrier. Hence, the large amplitude motions become important. To obtain a qualitative picture, we have to restrict the analysis again to one dimension. As pointed out in this context by Krumhansl and Schrieffer [16], the kink-like solution

$$f\left(R_x, R_y = 0,\ R_z = 0;\ t\right) = U_0 \tanh\left(\frac{R_x - V_x t}{\xi_x\sqrt{2}}\right), \quad (30)$$

where

$$U_0 = \sqrt{\frac{12C - A}{B}},\quad \xi_x^2 = \frac{2C\, a^2 - MV_x^2}{12C - A} \quad (31)$$

is of particular interest. This kink takes f from U_0 to $-U_0$; an antiking (changes the sign of f) takes -f to f. Between +f and

-f there is a wall which we call a cluster wall of approximate thickness $2\sqrt{2}\,\xi_X$. The cluster wall moves with velocity V_X.

Note that the kink solution is not a soliton, although it resembles the kink solution (17) of the S-G equation in shape. It is, however, a solitary wave. In fact, numerical calculations indicate that on collision they slow down and radiate ripples [17].

3. DYNAMIC VARIABLES, CONSERVATION LAWS AND SPECTRAL DENSITIES

For simplicity, we consider only the one-component model explicitly. To describe the static and dynamic properties of the system, we might consider the following variables:

$$X(\vec{q}) = \frac{1}{\sqrt{N}} \sum_{\ell} (X_\ell - \langle X_\ell \rangle) e^{i\vec{q}\cdot\vec{R}_\ell}, \tag{32}$$

$$\mathcal{H}(\vec{q}) = \frac{1}{\sqrt{N}} \sum_{\ell} (\mathcal{H}_\ell - \langle \mathcal{H}_\ell \rangle) e^{i\vec{q}\cdot\vec{R}_\ell}, \tag{33}$$

where, according to Equation (24),

$$\mathcal{H}_\ell = \frac{M}{2}\dot{X}_\ell^2 + \frac{A}{2}X_\ell^2 + \frac{B}{4}X_\ell^4 - C\sum_m X_\ell X_{\ell+m}. \tag{34}$$

These variables describe displacement and energy flucuations, respectively. $\vec{R}_\ell$ is a lattice vector of the rigid reference lattice and $\vec{q}$ the wave vector. From Equations (32)-(34), it follows

$$\dot{X}(q) = \frac{1}{\sqrt{N}} \sum_{\ell} \dot{X}_\ell e^{i\vec{q}\cdot\vec{R}_\ell}, \tag{35}$$

$$\ddot{X}(q) = \frac{1}{\sqrt{N}} \sum_{\ell} \ddot{X}_\ell e^{i\vec{q}\cdot\vec{R}_\ell} \tag{36}$$

$$\dot{\mathcal{H}}(q) = 2C\frac{1}{\sqrt{N}} \sum_{\vec{q}} \left[X(\vec{q}-\vec{q}\,')\dot{X}(\vec{q}\,') - \dot{X}(\vec{q}-\vec{q}\,')X(\vec{q}\,')\right] \cdot F(\vec{q}\,')$$
$$+ 2C\langle X_\ell\rangle[F(\vec{0}) - F(\vec{q})]X(\vec{q}), \tag{37}$$

where

$$F(\vec{q}) = \cos a q_x + \cos a q_y + \cos a q_z . \tag{38}$$

It is seen that $\dot{\mathcal{H}}(\vec{0},t)$ vanishes,

$$\dot{\mathcal{H}}(\vec{0},t) = \frac{1}{\sqrt{N}} \sum_\ell \dot{\mathcal{H}}_\ell = \frac{1}{\sqrt{N}} \frac{d\mathcal{H}}{dt} = 0, \tag{39}$$

so that energy is conserved, as it should be for a Hamiltonian system. Displacement and momentum are not conserved, however, because

$$\dot{X}(\vec{0},t) = \frac{1}{\sqrt{N}} \sum_\ell \dot{X}_\ell ,$$

$$\ddot{X}(\vec{0},t) = \frac{1}{\sqrt{N}} \sum_\ell \ddot{X}_\ell = - \frac{1}{\sqrt{N}} \frac{1}{M} \sum_\ell \left[(A - 12C) X_\ell + B X_\ell^3 \right] . \tag{40}$$

In the last step, we used Equation (25). In these models, momentum is not conserved because the Hamiltonian (24) is not invariant with respect to continuous translations due to the rigid reference lattice. As a consequence, the system will not have acoustic modes. Hydrodynamic effects can, therefore, arise only from the conserved energy in terms of diffusive heat conduction or second sound. These phenomena are not restricted to the energy fluctuations below T_c, because the coupling between order parameter and energy fluctuations does not vanish. In fact,

$$\langle X(-\vec{q}) \mathcal{H}(\vec{q}) \rangle \neq 0 \ : \ T < T_c , \tag{41}$$

$$\langle \dot{X}(-q) \mathcal{H}(\vec{q}) \rangle = 2C \langle X_\ell \rangle \frac{k_B T}{M} [F(\vec{0}) - F(\vec{q})] \neq 0 \ : \ T < T_c . \tag{42}$$

To investigate the excitation spectrum, we consider the spectral densities

$$\hat{S}_{XX}(q,\omega) = \frac{\int_{-\infty}^{+\infty} dt\, e^{-i\omega t} \langle X(-\vec{q},t) X(\vec{q},0) \rangle}{\langle X(-\vec{q},0) X(\vec{q},0) \rangle} \tag{43}$$

and

$$\hat{S}_{\mathcal{H}\mathcal{H}}(\vec{q},\omega) = \frac{\int_{-\infty}^{+\infty} dt\, e^{-i\omega t} \langle \mathcal{H}(-\vec{q},t)\, \mathcal{H}(\vec{q},0) \rangle}{\langle \mathcal{H}(-\vec{q},0)\, \mathcal{H}(\vec{q},0) \rangle} . \tag{44}$$

The excitation spectrum, determined by the peak structure of these spectral densities at fixed wave vector is expected to be rather rich and complicated. From the analysis in Section 2, we expect structure arising from the small amplitude and large amplitude motions. Moreover, we have seen that energy conservation will give rise to additional structure at small wave vectors.

4. MOLECULAR-DYNAMICS TECHNIQUE

To simulate a canonical ensemble, we assume that the particles suffer collisions with much lighter ones which represent the heat bath. The collisions are described by a friction $-\Gamma P_\ell$ and a random force $\eta_\ell(t)$. The associated equations of motion are then coupled Langevin equations,

$$M\ddot{X}_\ell = -\frac{\partial \mathcal{H}}{\partial X_\ell} - \Gamma M\dot{X}_\ell + \eta_\ell(t), \tag{45}$$

where

$$\langle \eta_\ell(t)\eta_{\ell'}(t')\rangle = 2M\Gamma k_B T\,\delta_{\ell\ell'}\delta(t-t'). \tag{46}$$

T denotes the temperature of the bath. The stationary solutions of the associated Fokker-Planck equation is the canonical distribution

$$P_{eq}\left(\dot{X}_1, \ldots, \dot{X}_N; X_1 \ldots X_N\right) \sim e^{-\beta\mathcal{H}}, \quad \beta=1/k_B T. \tag{47}$$

Starting from initial values for positions and velocities, the particles are then allowed to move under the influence of the computer-generated random force. The temporal evolution of the variables are then calculated with a set of difference equations approximating the Langevin equations (45). On this basis, one obtains

$$X_\ell(t),\ \dot{X}_\ell(t),\ \ddot{X}_\ell(t),\ \ldots,\ \text{etc.}$$

A more detailed description of the algorithm and the random-force generation is given in Reference 8.

The system is then allowed to age or, in other words, to reach equilibrium. After this interval, the subsequent 10^5 steps are used to perform time averages, representing estimates for canonical-ensemble averages. An example is

$$\langle X\rangle = \langle\frac{1}{N}\sum_{\ell} X_{\ell}\rangle \approx \frac{1}{\tau}\int_0^{\tau}\left[\frac{1}{N}\sum_{\ell} X_{\ell}(t)\right]dt. \tag{48}$$

It is obvious that the dynamic properties will be modified, in particular, due to the damping term in Equation (45). To reduce these modifications, Γ must be chosen in such a way that

$$\frac{1}{\Gamma} \gg \tau_c, \tag{49}$$

where τ_c denotes the characteristic times of the dynamics. This implies, for example, that the excitations do not become overdamped due to the friction term. Another important constraint on Γ evolves from the energy conservation of a Hamiltonian system. Because our system evolves according to the Langevin equations, we obtain from Equation (45)

$$\frac{d\mathcal{H}}{dt} = \sum\left(\frac{\partial\mathcal{H}}{\partial p_{\ell}}\frac{\partial p_{\ell}}{\partial t} + \frac{\partial\mathcal{H}}{\partial x_{\ell}}\frac{\partial x_{\ell}}{\partial t}\right) = -\sum_{\ell}\left[\Gamma M\dot{X}_{\ell}^{2} - \dot{X}_{\ell}\eta_{\ell}(t)\right]. \tag{50}$$

Consequently, energy is not conserved because the Hamiltonian system is in contact with the heat bath. To avoid artificial features due to the random noise pulses, the mean time between two pulses must be small compared to τ_c. In view of this, we may average Equation (50) over some pulses. This leads to the expression

$$\frac{d\mathcal{H}}{dt} = -\Gamma\sum_{\ell}\left(M\dot{X}_{\ell}^{2}(t) - k_BT\right) = -\Gamma\left[2E_k(t) - Nk_BT\right] \tag{51}$$

implying that energy is nearly conserved in the time interval τ,

$$\frac{1}{\Gamma} \gg \tau_c, \tag{52}$$

which is equivalent to Equation (49). Because the system evolves according to the Langevin equations, the time interval, over which its evolution is followed, must clearly be larger than $1/\Gamma$ so that.

$$\frac{1}{\Gamma} \ll \tau_{ch}. \tag{53}$$

τ_{ch} denotes the equilibrium chain length. Combining

Equations (49) and (53), we finally obtain

$$\tau_{ch} \gg \frac{1}{\Gamma} \gg \tau_c . \tag{54}$$

From this relation, it becomes evident that energy can be nearly conserved provided Γ and the chain length τ_{ch} are appropriately chosen. An exception is the region very close to T_c, where the characteristic times become very long. In this region, however, numerical methods become difficult in any case because the linear dimension of the system must exceed the correlation length to avoid finite size effects.

To summarize this section, the above molecular-dynamics technique allows the simulation of the canonical ensemble where the global temperature is fixed. Energy conservation can also be nearly realized by choosing the damping constant Γ appropriately. The numerical solution of the coupled Langevin equations allows then estimates of all properties which can be derived from the variables entering the Hamiltonian. This includes static and dynamic properties. For a detailed discussion of the algorithm and the calculation of time-dependent correlation functions, we refer to References 7 and 8.

5. MOLECULAR-DYNAMICS RESULTS; STATIC AND DYNAMIC EQUILIBRIUM PROPERTIES

5.1 The X-Y model with quartic anisotropy

To clarify the relevance of the kink and antikink solitons of the one-dimensional continuum version [Equation (14)], we analyzed the temporal evolution of the angular displacements. Systems of 1600 and 3600 particles subjected to periodic boundary conditions have been studied. We found that in a temperature window around the critical temperature T_c, clusters are formed, built-up of particles connected by a nearest-neighbor bond with positions lying in the same quadrant. Figure 2 illustrates the four cluster patterns separated by 90-degree walls which may be distinguished. Figure 3 shows snapshots of cluster configurations in the model with parameters given by Equation (10). The system was prepared so that

$$\langle Y_\ell \rangle^2_{T=0} = 0, \ \langle X_\ell \rangle^2_{T=0} = \frac{8C - A}{B/2 - B_1} .$$

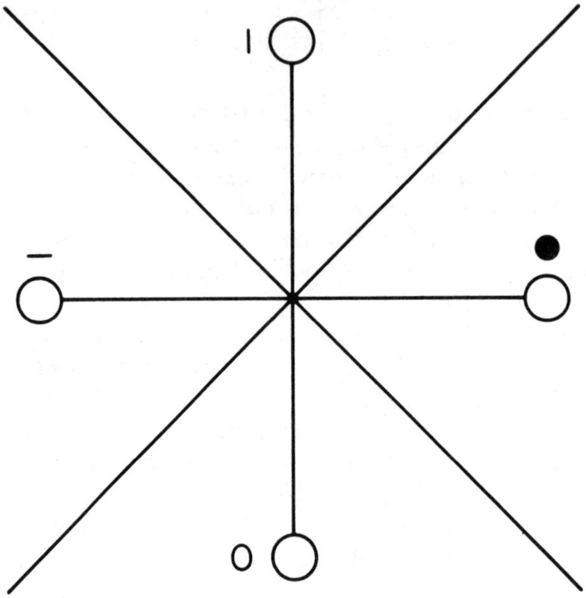

Figure 2. The four quadrants of the local angular displacement denoted by —, ○, ●, and |. The instantaneous position of a particle lies in one of the quadrants.

It is seen that with increasing temperature, clusters denoted by open circles and vertical bars are formed at first, while closer to $T_C \approx 2.7$, clusters marked by full circles also appear. Moreover, the cluster size increases when T_C is approached from above or below. Above T_C, the number of large clusters decreases with increasing temperature. Finally, clusters are almost exclusively separated by 90-degree walls. This feature reveals that the kink and antikink solitons of the one-dimensional continuum analogue are also relevant in the two-dimensional discrete lattice model, at least in a temperature window around T_C. For more details, we

---

Figure 3 (opposite page). Snapshots of cluster patterns at various temperatures: $k_B T$ equals (a) 2, (b) 2.5, (c) 3, and (d) 6. According to Figure 2, we distinguish four cluster patterns. Clusters are particles connected by a nearest-neighbor bond with positions lying in the same quadrant.

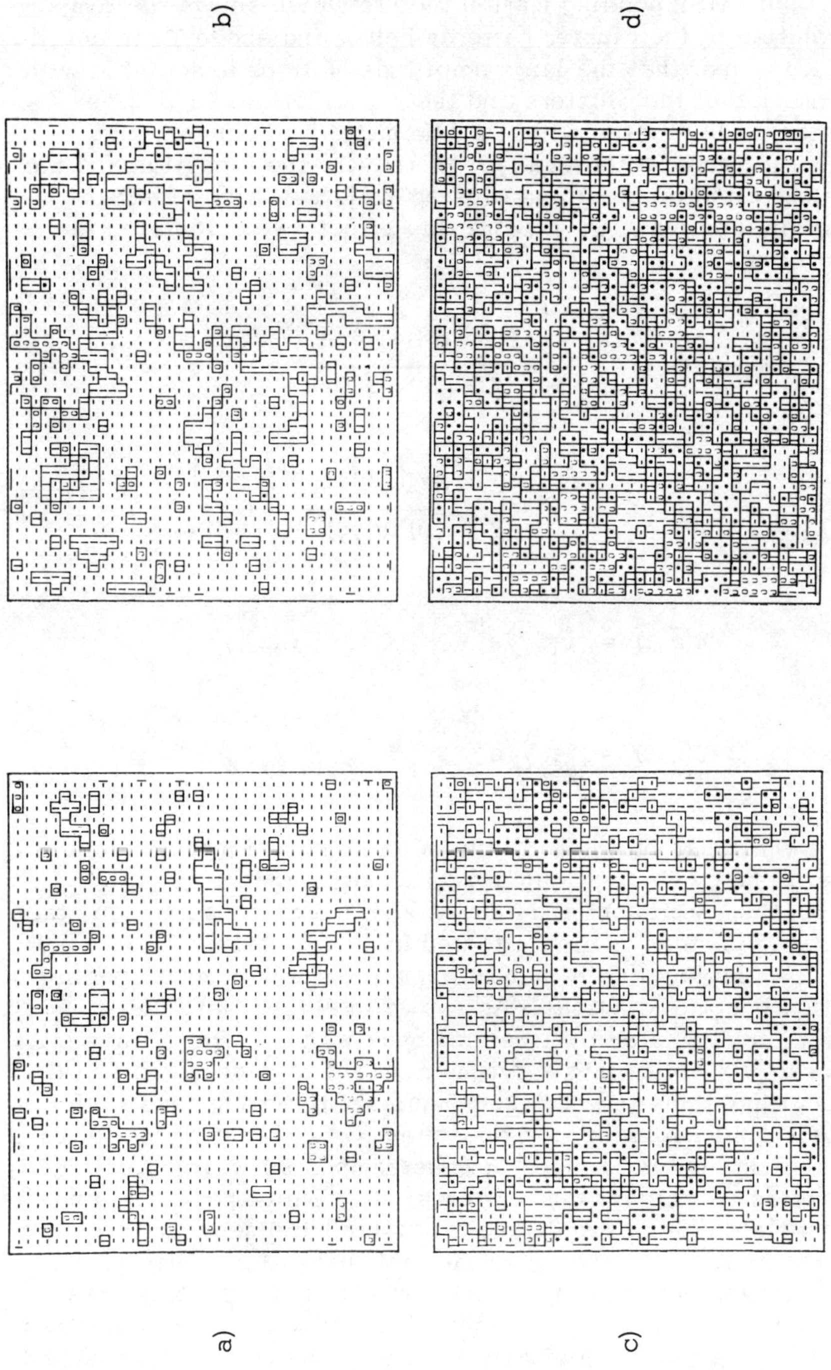
b)
d)
a)
c)

refer to our corresponding motion picture which shows the temporal evolution of the cluster patterns below and above T_c in detail.

It is clear that the large amplitude motions associated with the formation of the clusters and the motion of the 90-degree walls cannot be treated within the framework of conventional anharmonic perturbation theory. Moreover, the formation and the dynamics of the clusters and cluster walls will also affect the excitation spectrum. To explore this effect, we calculated the spectral densities

$$S_L(\vec{k},\omega) = \frac{\int_{-\infty}^{+\infty} \langle X(-\vec{q},t)X(\vec{q},0)\rangle e^{-i\omega t}dt}{\langle X(-\vec{q},0)X(\vec{q},0)\rangle}, \tag{55}$$

$$S_T(\vec{k},\omega) = \frac{\int_{-\infty}^{+\infty} \langle Y(-\vec{q},t)Y(\vec{q},0)\rangle e^{-i\omega t}dt}{\langle Y(-\vec{q},0)Y(\vec{q},0)\rangle}, \tag{56}$$

where

$$X(\vec{q},t) = \frac{1}{\sqrt{N}} \sum_{\ell} e^{i\vec{q}\cdot R_\ell}\left[X_\ell(t) - \langle X_\ell\rangle\right],$$

$$Y(\vec{q},t) = \frac{1}{\sqrt{N}} \sum_{\ell} e^{i\vec{q}\cdot R_\ell}\left[Y_\ell(t) - \langle Y_\ell\rangle\right]. \tag{57}$$

Here, we discuss only the spectrum for $T > T_c$, where displacement and energy fluctuations are no longer coupled. Below T_c, this coupling is present and, due to the energy conservation, the excitation spectrum is also affected by heat diffusion. Moreover, above T_c, the longitudinal and transverse displacement spectral densities become identical. As shown in Figure 4(a), the $\vec{q} = \vec{0}$ spectrum is dominated by a central peak which at finite wave vectors $(q_x, 0)$ splits into a symmetric $(\pm\omega)$ four-peak structure. The resulting higher branch [see Figure 4(b)] corresponds to the transverse phonon branch. Consequently, we identify the lower branch as due to overdamped or underdamped 90-degree cluster-wall motion. At higher temperatures, the central peak (CP) and the associated cluster-wall excitation branch disappear. This feature may be understood from the fact that at high temperatures, the interaction between the oscillators or, in other words, the dispersion, is no longer important. With increasing temperatures for $T > T_c$, the system gradually adopts the properties of independent oscillators.

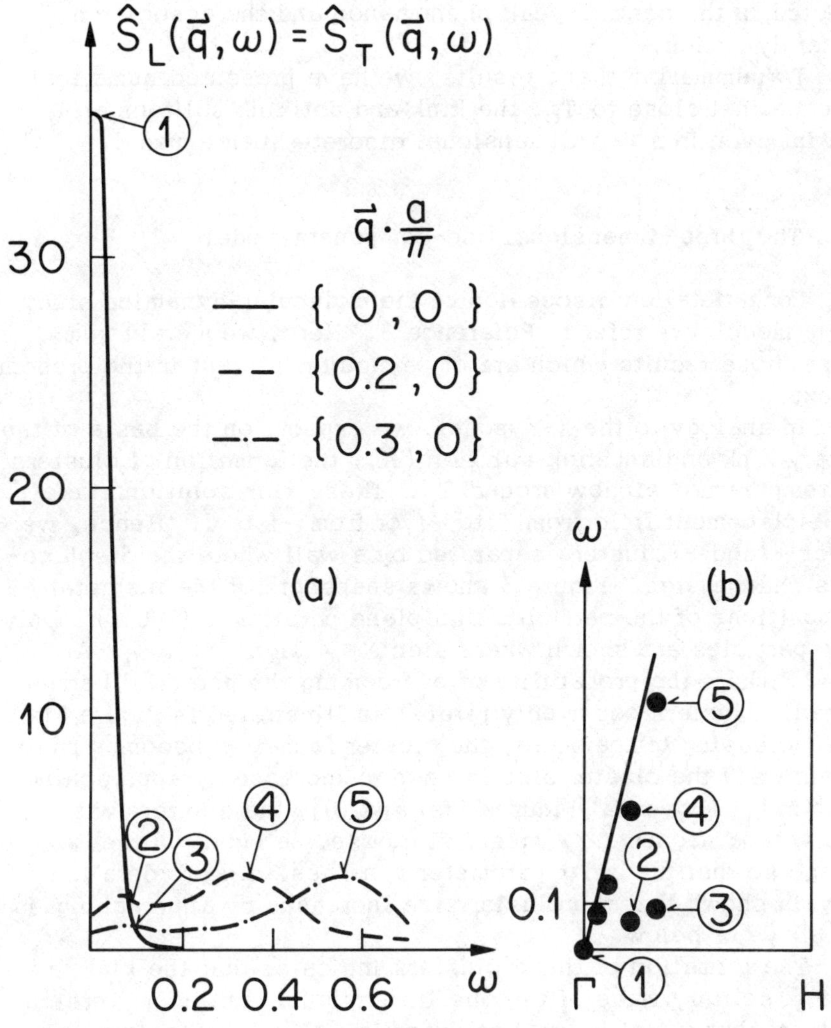

Figure 4. (a) Frequency dependence of $S_L(\vec{q},\omega) = S_T(\vec{q},\omega)$ for various $(q_x, 0)$ at $k_BT = 3 > k_BT_C \approx 2.7$. Peaks 4 and 5 are due to phonons, and peaks 2 and 3 arise from underdamped 90-degree cluster-wall motion.

(b) Dispersion curves determined by the peak maxima; the numbers label corresponding peak maxima.

We also note that the central-peak half width is found to decrease when T_c is approached from above. Hence, the critical slowing down of the order-parameter fluctuations is intimately reflected in the central-peak phenomenon and the associated cluster dynamics.

To summarize these results, we have presented numerical evidence that close to T_c, the kink and antikink solitons are relevant even in a two-dimensional discrete lattice model.

5.2 The three-dimensional one-component model

For a detailed discussion of the molecular-dynamics study of this model, we refer to Reference 8. Here, we merely summarize those results which are of particular interest in the present context.

In analogy to the X-Y model, we expect, on the basis of the solitary kink and antikink solution (30), the formation of clusters in a temperature window around T_c. These kink solutions take the displacement field from f to -f or from -f to f. Hence, we expect + and - clusters separated by a wall where the displacements change sign. Figure 5 shows snapshots of the instantaneous positions of the particles in a plane parallel to (1,0,0). Only those particles are shown where sign $X_\ell \neq$ sign $\langle X_\ell \rangle_{T=0}$. At $k_BT = 2$, where the probability of overcoming the potential barrier is small, clusters occur only rarely, as illustrated in Figure 5(a). With increasing temperature, the cluster formation becomes more probable and the cluster size is seen to increase by approaching $k_BT_c \approx 7.1$. Above T_c [Figures 5(e) and (f)], the clusters with positive and negative X_ℓ must, of course, be equal on the average so that the order parameter vanishes. A more detailed analysis shows that the cluster size increases by approaching T_c from above or below.

The formation of these clusters indicates that the kink and antikink solitary wave of the one-dimensional continuum version of the present model is even relevant in a three-dimensional discrete lattice model. Again, associated large amplitude motions cannot be described by conventional anharmonic perturbation theory. We expect, therefore, a rather rich excitation spectrum illustrating the effects of small and large amplitude motions and energy conservation.

Before turning to examine the explicit results, it is helpful to summarize those features of the excitation spectrum that might be expected on general grounds. At low temperatures, the

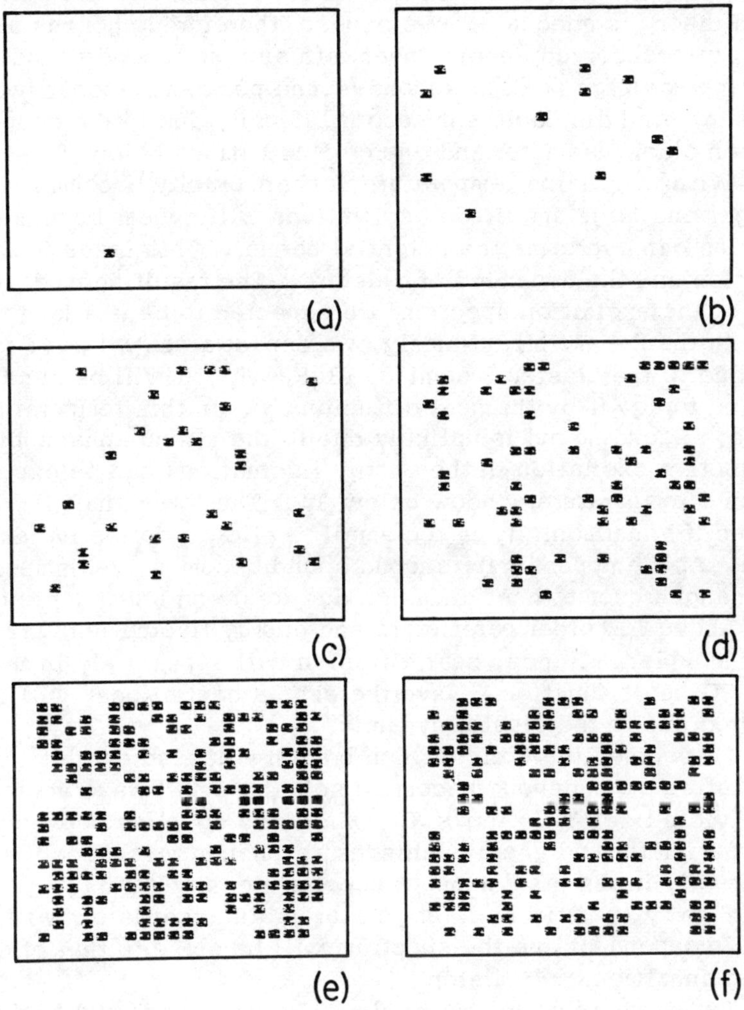

Figure 5. Snapshots of instantaneous cluster configuration in a plane perpendicular to (1,0,0). Clusters are particles connected by a nearest-neighbor bond whose local displacement X_ℓ has a sign opposite to $\langle X_\ell \rangle_{T=0}$. (a) $k_BT=2$; (b) $k_BT=4$; (c) $k_BT=5$; (d) $k_BT=6$; (e) $k_BT=8$; (f) $k_BY=15$. Only those particles are shown where sign $X_\ell \neq$ sign$\langle X_\ell \rangle_{T=0}$.

particles oscillate about a mean position determined by the order parameter. The displacements are small and anharmonic perturbation theory is adequate. We expect, therefore, phonons and, due to the conserved energy, heat diffusion or second sound. The resulting Rayleigh peak or second-sound peaks may occur in both the energy and displacement spectral density due to the coupling between order parameter and energy fluctuations below T_C.

With increasing temperature, anharmonicity becomes stronger and large amplitude oscillations will appear because the particles can overcome the potential barrier. This leads to the formation and the dynamics of clusters. The resulting modifications of the excitation spectrum are expected to be similar to those in the X-Y model. Namely, we expect a central peak at $\vec{q}=\vec{0}$ due to the cluster dynamics. Below T_C, it will be superimposed for $\vec{q}\neq\vec{0}$ by the heat diffusion CP. In this temperature regime, second sound is unlikely due to the strong anharmonicity. A collective excitation in the energy fluctuations can be expected only in a temperature window below T_C. The lower limit is reached if anharmonicity is too small to allow a collective excitation. At the upper limit, second sound becomes overdamped and changes over to heat diffusion due to strong anharmonicity.

Above T_C, order parameters and energy fluctuations are no longer coupled. Hence, heat diffusion will appear only in the energy Green's function. Nevertheless, a central peak will be expected due to the cluster dynamics.

Of course, the optic phonon branch will also exhibit strong temperature dependence, becoming soft (for small wave vectors) in the vicinity of T_C. Above T_C, where the coupling between order parameter and energy vanishes, a phonon resonance will appear only in the displacement spectral density.

Finally, at very high temperatures, the oscillators will be nearly independent and the spectrum will be close to that of a quartic anharmonic oscillator.

Let us now turn to the results. In view of the rich and complicated excitation spectrum, we present here an overview only. For more detailed results, we refer to Reference 8. In Figure 6, we sketched the main peaks appearing in $\hat{S}_{XX}(\vec{q},\omega)$ and $\hat{S}_{\mathcal{H}\mathcal{H}}(\vec{q},\omega)$ for various temperatures at wave vectors $\vec{q}=(0,0,0)$ and $\vec{q}=((\pi/10a),0,0)$, respectively. At $k_BT=15$ and 30, there is a broad phonon peak in $\hat{S}_{XX}$ but no central peak (CP) occurs. $\hat{S}_{\mathcal{H}\mathcal{H}}$ only exhibits a central peak with half width $\Delta\omega\sim q^2$ due to heat diffusion. This phenomenon does not occur in $\hat{S}_{XX}$ as expected because the displacement-energy coupling vanishes above T_C. Moreover, the broad phonon resonance shifts to lower frequencies

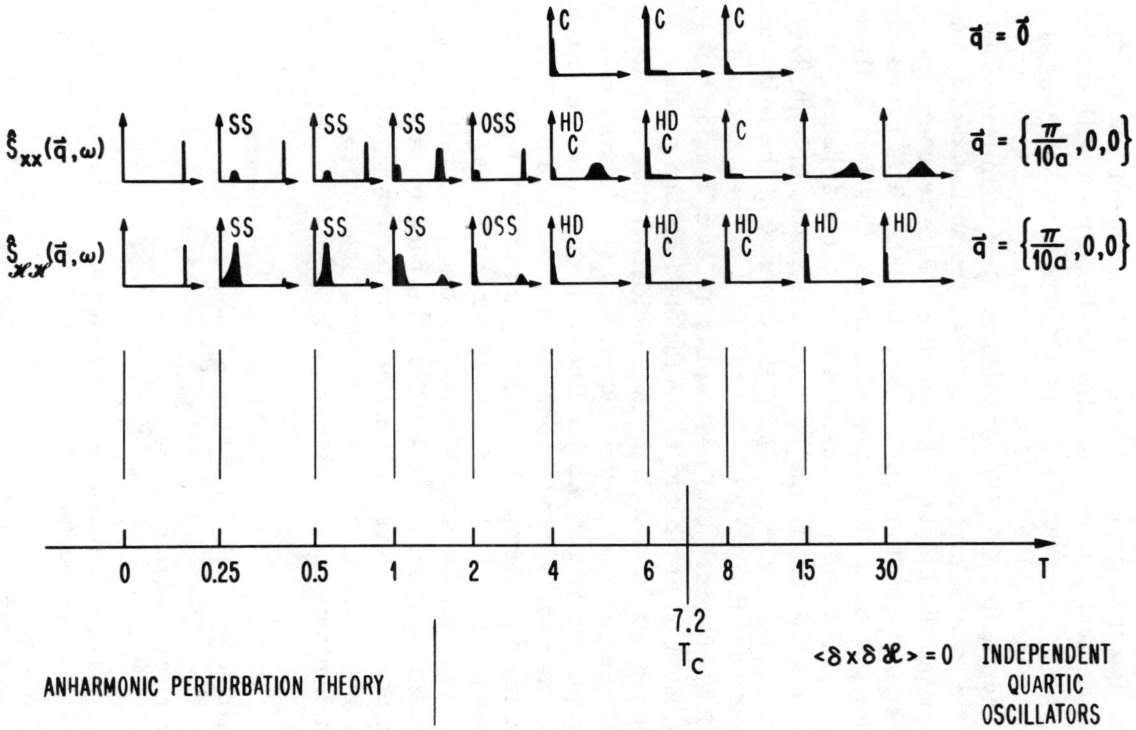

Figure 6. Sketch of the essential features of the excitation spectrum in $\hat{S}_{XX}(\vec{q},\omega)$ and $\hat{S}_{\mathcal{H}\mathcal{H}}(\vec{q},\omega)$ at various temperatures for wave vectors $\vec{q}=\vec{0}$ and $\vec{q}=((\pi/10a),0,0)$. HD denotes the central peak due to heat diffusion and C the central peak arising from the cluster dynamics. SS denotes the second sound and OSS overdamped and second sound.

with decreasing temperature. At $k_BT = 8$, which is close to $k_BT_C \approx 7.1$, new features appear: a CP at $\vec{q} = (0,0,0)$ in $\hat{S}_{XX}$ and the soft mode becomes overdamped. At $\vec{q} = \vec{0}$, heat diffusion does not exist. Consequently, the $\vec{q} = \vec{0}$-CP must be attributed to the cluster dynamics. In $\hat{S}_{\mathcal{H}\mathcal{H}}$ and $\vec{q} \neq \vec{0}$, it is superimposed by the Rayleigh peak.

Below T_C, the coupling between displacement and energy fluctuations no longer vanishes. The Rayleigh peak will appear, therefore, in both $\hat{S}_{XX}$ and $\hat{S}_{\mathcal{H}\mathcal{H}}$. Accordingly, at $k_BT = 6$ and $\vec{q} \neq 0$, we have the Rayleigh and cluster central peaks superimposed. The cluster CP also appears as it is seen from $\vec{q} = \vec{0}$. The soft mode is overdamped. At $k_BT = 4$, the soft mode becomes underdamped and the height of the cluster CP is reduced, as expected. In fact, for the formation of clusters, it is necessary that particles overcome the potential barrier. This fact is illustrated at $k_BT = 2$, where the cluster CP no longer appears. At $\vec{q} \neq \vec{0}$, there is a CP due to the energy fluctuations in both $\hat{S}_{XX}$ and $\hat{S}_{\mathcal{H}\mathcal{H}}$ but weakly in S_{XX}. The q-dependence of the half width reveals, however, that this CP cannot be explained in terms of heat diffusion. Because this peak splits at lower temperatures ($k_BT = 1$), it must be attributed to overdamped second sound. The phonon resonance is now well defined indicating that we enter the regime where anharmonic perturbation theory should work. At $k_BT = 1$, the phonon is very well defined. It dominates $\hat{S}_{XX}$, but at small ω, the weak second-sound resonance appears. This peak dominates $\hat{S}_{\mathcal{H}\mathcal{H}}$ where the phonon peak is very weak. At the lower temperature $k_BT = 0.5$, the features are similar to those at $k_BT = 1$. The only difference is a zero-frequency tail of the second-sound resonance in $\hat{S}_{\mathcal{H}\mathcal{H}}$. This tail must be attributed to two phonon processes arising from the first term in Equation (37). Finally, at $T = 0$, where the spectrum can be calculated exactly, only the phonon resonance survives.

6. NONLINEAR HEAT-PULSE PROPAGATION

6.1 Theoretical analysis

Heat-pulse techniques have been used extensively in the past to study second sound [18]. Experimentally, the systems are usually studied at very low temperatures where the available power input is insufficient to drive the nonlinearities. In this regime, the propagation of heat can be described by a diffusion or wave equation, or, in other words, within the framework of linear response theory.

In this section, we study the possibility of nonlinear heat propagation. In doing so, we consider the effect of the nonlinearities on the phonon-like solutions and rewrite the equation of motion (27) in the form

$$-M\ddot{g}(\vec{R},t) = \langle X\rangle\left(A - 12C + B\langle X\rangle^2\right) + \left(A - 12C + 3B\langle X\rangle^2\right)g$$
$$+ 3B\langle X\rangle g^2 + Bg^3 - 2C\, a^2\nabla^2 g, \tag{58}$$

where $g(\vec{R},t)$ denotes the fluctuating part of the displacement field $(g = f - \langle f\rangle)$.

We wish to consider solutions which reduce to phonons. For this purpose, we follow Varma [19] and introduce a small parameter ϵ so that for $\epsilon = 0$, we have a linear problem. The equation of motion then reads

$$-M\ddot{g} = \langle X\rangle\left(A - 12C + B\langle X\rangle^2\right) + \left(A - 12C + 3B\langle X\rangle^2\right)g$$
$$+ 3\epsilon\, B\langle X\rangle g^2 + B\epsilon^2 g^3 - 2C\,\nabla^2 g. \tag{59}$$

Following Varma, we look for solutions which reduce to phonons in the linear limit. Restricting the analysis to one-space dimensions, we set

$$g(X,t) \equiv \epsilon\left[\varphi^{(0)}(\bar{X},T) + \overset{*}{\varphi}{}^{(0)}(\bar{X},T)\right]$$
$$+ \sum_{\substack{n=-\infty \\ n\neq 0}}^{n=+\infty} \varphi^{(n)}(X,T)\epsilon^{|n|-1} e^{in(qx-\omega_0 t)}. \tag{60}$$

It is assumed that the variations in $\varphi^{(1)}(\bar{X},T)$ are slow compared to the phonon part. Accordingly, we introduce

$$T = \epsilon t, \quad \bar{X} = \epsilon X. \tag{61}$$

Equation (59) then becomes

$$g_{tt} - 2Ca^2 g_{XX} - A_2 g + 3\epsilon B\langle X\rangle g^2 + \epsilon^2 B g^3 + \epsilon^2 g_{TT}$$

$$- 2Ca^2\epsilon^2 g_{\overline{XX}} + 2\epsilon g_{tT} - 4\epsilon C a^2 g_{X\overline{X}} = 0, \tag{62}$$

where

$$A_2 = \left(12C - A - 3B\langle X\rangle^2\right). \tag{63}$$

Moreover, we assumed

$$12C - A - B\langle X\rangle^2 \approx 0 \tag{64}$$

which holds for very low temperatures [see Equation (26)].

Inserting (60) into (62), equating d.c. first and second harmonic terms, we get

$$\varphi^{(0)} = 3\,\frac{B\langle X\rangle}{A_2}\,|\varphi^{(1)}|^2, \quad \varphi^{(2)} = \frac{B\langle X\rangle}{A_2}\,(\varphi^{(1)})^2 \tag{65}$$

and

$$\left(-2i\omega_0\varphi_T^{(1)} - 4iqC\,a^2\varphi_{\overline{X}}^{(1)}\right) + \epsilon\left(\frac{42\ B\langle X\rangle}{A_2} + 3B\right)|\varphi^{(1)}|^2\varphi^{(1)}$$

$$+ \left(\varphi_{TT}^{(1)} - 2C\,a^2\varphi_{\overline{XX}}^{(1)}\right) = 0. \tag{66}$$

Introducing the group velocity

$$V_G = \frac{d\omega_0}{dq} = \frac{2C\,a^2 q}{\omega_0}, \quad \omega_0^2 = \left(A - 12C + 3B\langle X\rangle^2\right) + 2C\,a^2 q^2 \tag{67}$$

and the new scales

$$z = \overline{X} - V_G T, \quad s = \epsilon T, \tag{68}$$

Equation (66) reduces in lowest order in ϵ to the nonlinear Schrödinger equation (NLS) [14]

$$i\varphi_s^{(1)} + P\varphi_{zz}^{(1)} + R|\varphi^{(1)}|^2\varphi^{(1)} = 0, \tag{69}$$

where

$$P = \frac{2C\,a^2 - V_G^2}{2\omega_0}, \quad R = -\frac{B}{2\omega_0}\left(3 + \frac{42\langle X\rangle^2}{A_2}\right). \tag{70}$$

The solution of the NLS equation depends, in a crucial way, on the sign of PR. In our case, PR is negative. In this case, one class of stable solutions is plane waves with frequency

$$\omega = Pq^2 - R. \tag{71}$$

Other stable solutions can be obtained by noting that for $PR < 0$, the NLS can be reduced to the Korteweg-de-Vries equation (KdV). This can be achieved by following the procedure of Taniuti and Yajima [20]. One introduces real functions ρ and σ by

$$\varphi^{(1)} = \rho^{\frac{1}{2}} \exp i \int \frac{\sigma(z', s)}{2P} dz'. \tag{72}$$

Substituting (72) into (69) and separating real and imaginary parts, one finds

$$\begin{aligned} &\rho_s + (\rho\sigma)_z = 0 \\ &\sigma_s + \sigma\sigma_z = 2PR\rho_z + P^2\left[\rho^{-\frac{1}{2}}(\rho^{-\frac{1}{2}}\rho_z)_z\right]_z . \end{aligned} \tag{73}$$

Now, we introduce

$$\xi = z - vs, \tag{74}$$

and choose the boundary conditions

$$\rho(\xi = \infty) = \rho_0, \tag{75}$$

$$\sigma(\xi = \infty) = \sigma_\infty . \tag{76}$$

Since we are interested in a solitary wave, we also require that the first and second derivatives of ρ and σ vanish as $\xi \to \pm\infty$. Putting

$$\rho = \rho_0 F(\xi), \tag{77}$$

we find from Equation (73)

$$\sigma(\xi) = v - \frac{v - \sigma_\infty}{F(\xi)}, \tag{78}$$

and

$$F^{-\frac{1}{2}}\left(F^{-\frac{1}{2}}F_\xi\right)_\xi = \tfrac{1}{2}(1-\alpha)\gamma^2(1/F^2 - 1) + \gamma^2(F - 1), \tag{79}$$

where

$$\gamma = \frac{2|PR|\rho_0}{P^2},$$
$$\alpha = 1 - \frac{(v-\sigma_\infty)^2}{2|PR|\rho_0}. \tag{80}$$

Setting

$$F = 1 - G, \tag{81}$$

it is easily verified that Equation (79) reduces to the KdV equation

$$G_{\xi\xi\xi} + G_\xi(3\gamma^2 G - \alpha\gamma^2) = 0, \tag{82}$$

yielding the soliton solution [14]

$$G = 1 - F = \alpha \operatorname{sech}^2\left(\frac{\alpha^{\frac{1}{2}}\gamma}{2}\xi\right). \tag{83}$$

From our point of view, the important result is that the effect of the nonlinearities on the phonon-like solutions leads to a modulation of the amplitude. In lowest order of asymptotic expansions, the original equation was reduced to the KdV equation, which exhibits solitons, determining the amplitude modulation. One expects, therefore, that the effect of the nonlinearities on the phonon-like solutions will lead to an envelope or modulation of plane waves in terms of envelope solitons [21]. Accordingly, an initial perturbation, such as a heat pulse, might break up into envelope solitons. In our case, the heat pulse can be described by the kinetic energy describing the temperature field

$$k_B T(x,t) = M\dot{g}(x,t)^2 = M\dot{f}(x,t)^2. \tag{84}$$

From Equations (60), (65), (72), (77), (78), and (83), we find that the height H of the envelope soliton is proportional to the third power of its velocity

$$H \sim \alpha^3, \tag{85}$$

and its half width Δ obeys

$$\Delta \sim \alpha^{-\frac{1}{2}} \tag{86}$$

so that

$$\Delta \sim H^{-1/6} . \tag{87}$$

This analysis indicates that nonlinear heat propagation might be associated with envelope solitons provided that the effects of nonlinearity are sufficiently small. Moreover, the approximate solution (60) with $\varphi^{(0)}$ and $\varphi^{(2)}$ given by Equation (65), and $\varphi^{(1)}$ given by Equation (72), is expected to describe the motion of the particles at low temperatures where the kink solution (30) is no longer important because clusters appear only rarely. It should be emphasized, however, that we considered only the one-dimensional case in the continuum approximation.

6.2 Molecular-dynamics results

To study the propagation of heat pulses, we adopted the following procedure. The system, subjected to periodic boundary conditions, is brought into contact with a gas of particles having temperature T_p and mass M. After switching-off this contact, we follow the propagation of the resulting temperature pulse characterized by the distribution of the kinetic energy of the particles. The heat source is simulated by assuming collisions between the gas and layer particles. They occur at random time intervals t_i, distributed according to $1/\tau \exp - t_i/\tau$, $0 < t_i < \infty$. The mean time between two collisions is τ. The resulting new momentum $M\dot{X}_i(t_i)$ of the i-th layer particle is then chosen according to a Gaussian distributed random number with variance Mk_BT_p. This procedure applies to any particle in the first layers of the two opposite planes of the unit cube. We chose $\tau = 1/3$ time units and the "heat source" was switched on over five time units. Consequently, we have fifteen collisions per layer atom on the average. The remaining part of the system is assumed to evolve according to the Langevin equations (45). As a consequence, this part of the system is in contact with a heat bath, and its temperature is fixed.

Figure 7 shows the propagation of the temperature pulse of initial height $k_BT_p = 0.6$ in terms of a hypsometric plot. The intensity of the blackening measures temperature and is discretized into four levels. The ambient temperature is $k_BT = 0.125$ belonging to the second-sound regime. According to Equation (79), the profile is represented by the kinetic energy in the ℓ-plane

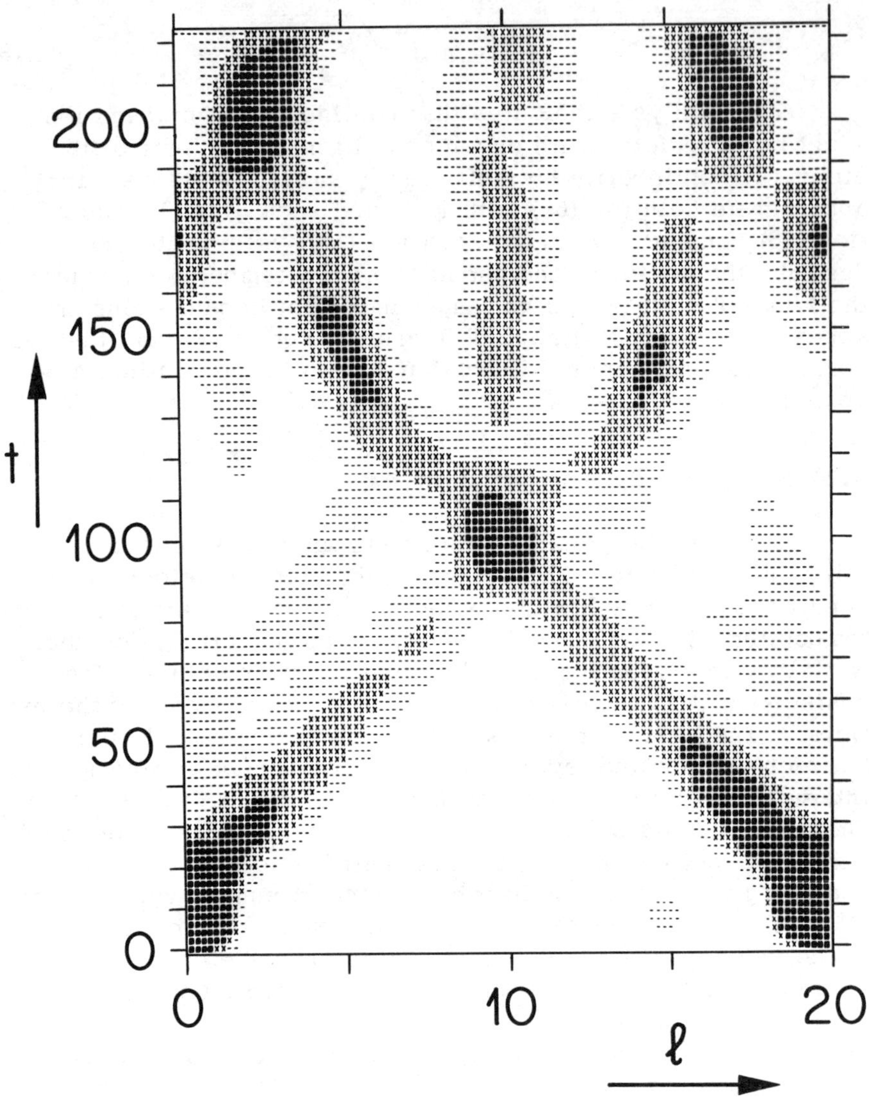

Figure 7. Hypsometric plot of the temporal evolution of the temperature profile with initial height $k_B T_p = 0.6$ at ambient temperature $k_B T = 0.125$. The intensity of the blackening measures temperature and is discretized into four levels. Temperature is measured in terms of the kinetic energy in the ℓ-th plane perpendicular to (1,0,0).

perpendicular to (1,0,0) in the cube. The profile is seen to propagate with velocity u according to $p = p(t - x/u)$. It then collides with a pulse coming from the opposite side. The collision has some features reminiscent of solitons; namely, they go through one another without considerable change of shape or velocity but with a small phase shift defined by

$$p(x,t) = p(t - x/u + \delta).$$

An important difference with respect to the collision of solitons is the appearance of a localized pulse ($u \approx 0$) after the collision. It should be kept in mind, however, that the system is a three-dimensional discrete lattice model where infinitely long-lived solitons are unlikely.

To substantiate the soliton-like properties, we also estimated the relation between height and half width of the profile before the collision occurred. The results are shown in Figure 8. For small pulse heights, the prediction $\Delta \sim H^{-1/6}$ [Equation (87)] fits remarkably well. For higher pulses, however, systematic

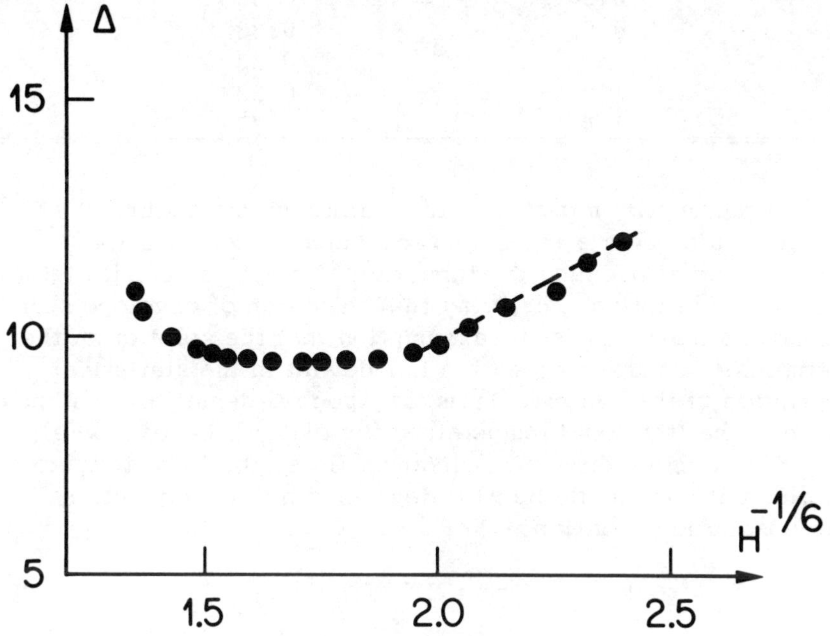

Figure 8. Profile half width Δ versus height H to the power $-1/6$ for $k_BT = 0.125$.

deviations occur. This indicates that the low order expansion, outlined in Section 2.2, is reasonable only for small amplitudes. Nevertheless, for small amplitudes, the pulses propagate like the envelope solitons derived from the continuum version of the equation of motion [Equation (78)].

From Table I, it is seen that the profile velocity u, evaluated before the collision, approaches the second-sound velocity if the initial pulse height T_p approaches the ambient temperature $k_BT = 0.125$.

Table I. Profile velocities for various initial pulse heights k_BT_p at ambient temperature $k_BT = 0.125$. These values should be compared with the second-sound velocity $C_{SS} \cdot \Pi/a \approx 0.28$, as determined by the second-sound peaks in the spectral densities.

k_BT_p	$u \cdot \frac{\Pi}{a}$
0.2	0.30
0.4	0.31
0.6	0.32
0.8	0.33
1	0.33
1.2	0.33
1.4	0.35

At the upper limit of the temperature window, where second should become overdamped, inherent nonlinearity becomes crucial because the formation of clusters sets in. Here, the basic assumption underlying the derivation of envelope solitons breaks down. It is the assumption that the solution of the deterministic equation of motion is relevant in the statistical description of the system. From the spectral densities, we know, however, that the local temperature fluctuations become purely relaxational above the upper window. Here, the local temperature fluctuations should be well described by the conventional heat-conduction equation

$$\frac{\partial T}{\partial t} = D\nabla^2 T. \tag{88}$$

This is illustrated in Figure 9 showing the propagation of a temperature pulse with initial height $k_BT_p = 3$ and ambient temperature $k_BT = 2$ where second sound is overdamped. There is no

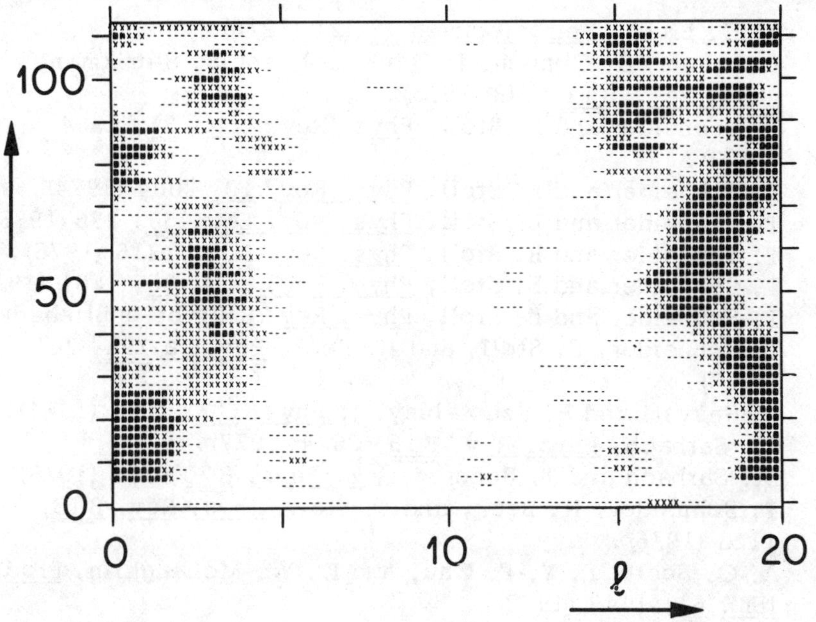

Figure 9. Hypsometric plot of the temporal evolution of the temperature profile with initial height $k_BT_p = 3$ at ambient temperature $k_BT = 2$.

longer a soliton-like propagation. The pattern is in qualitative agreement, however, with the relevant solution of Equation (88), namely:

$$T(x,t) = \left(T_p(4D\Pi t)^{-\frac{1}{2}} \exp - \frac{x^2}{4Dt}\right) + T. \tag{89}$$

ACKNOWLEDGMENT

I am grateful to E. Stoll for many comments on the subject matter of these lectures.

REFERENCES

1. A. D. Bruce, Ferroelectrics 12, 21 (1976).
2. I. Aviram, S. Goshen, D. Mukamel, and S. Shtrikman, Phys. Rev. B 12, 438 (1975).
3. T. Schneider and E. Stoll, Phys. Rev. Lett. 31, 1254 (1973).
4. T. Schneider and E. Stoll, Phys. Rev. 10, 2004 (1974).
5. T. Schneider and E. Stoll, Phys. Rev. Lett. 35, 296 (1975).
6. T. Schneider and E. Stoll, Phys. Rev. B 13, 1216 (1976).
7. T. Schneider and E. Stoll, Phys. Rev. Lett. 36, 1501 (1976).
8. T. Schneider and E. Stoll, Phys. Rev. B (to be published).
9. T. Schneider, E. Stoll, and H. Beck, Physica 79A, 201 (1975).
10. L. Sasvari and P. Szepfalusy, J. Phys. C 7, 1061 (1974).
11. St. Sarbach, Phys. Rev. B 15, 2694 (1977).
12. St. Sarbach and T. Schneider, Z. Phys. B 20, 399 (1975).
13. T. Schneider, H. Beck, and E. Stoll, Phys. Rev. B 13, 1123 (1976).
14. A. C. Scott, F. Y. F. Chu, and D. W. McLaughlin, Proc. IEEE 61, 1443 (1973).
15. A. Seeger, H. Donth, and A. Kochendorfer, Z. Physik 34, 173 (1953).
16. J. A. Krumhansl and J. R. Schrieffer, Phys. Rev. B 11, 3535 (1975).
17. A. E. Kudryavtsev, JETP Lett. 22, 82 (1975).
18. V. Narayanamurti, R. C. Dynes, and K. Andres, Phys. Rev. 11, 2500 (1975).
19. C. M. Varma, Phys. Rev. B 14, 244 (1976).
20. T. Taniuti and Y. Yajima, J. Math. Phys. 10, 1369 (1969).
21. F. D. Tappert and R. H. Hardin, SIAM Rev. 13, 282 (1971).

NONLOCAL CONTINUUM MECHANICS AND SOME APPLICATIONS†

A. Cemal Eringen
Princeton University
Princeton, New Jersey

ABSTRACT. The recent nonlocal continuum field theories are summarized and applied to the problems of wave propagations, fracture mechanics, dislocations and the secondary flow. The power and potential of these theories are shown in their capability for predicting various physical phenomena in the atomic scale.

1. INTRODUCTION

In all branches of physical sciences classical field theories have played prominent roles not only in understanding and prediction of physical phenomena but also in their versatility and suggestive power for opening new vistas to cast light on grounds beyond the domain of applicability of these theories. When these theories appeared to fall short in verifying the experimental results, usually a clever patchwork was made restoring the confidence in their validity. Such is the situation, for example, in the case of fracture of solids containing sharp-edged cracks, notches and line and surface discontinuities (e.g., dislocations). In fluid mechanics the problems of secondary flow and structure of turbulence and in electromagnetic field theories the dispersion of high frequency waves, spin waves and other long-range phenomena are but few examples falling into the same category.

---

† Lectures given at NATO Advanced Study Institute on Nonlinear Equations in Physics and Mathematics, Istanbul, August 1977.

A. O. Barut (ed.), Nonlinear Equations in Physics and Mathematics, 271–318.

To demonstrate the situation more clearly consider, for example, the fracture of a brittle solid. If a solid happens to have no internal crack or external sharp notches (an impossibility, since in the microscopic level large numbers of microcracks and dislocations exist), it fails under a uniform tension when the tensile stress exceeds the ultimate stress. In an elastic plate containing cracks or notches with sharp edges, under a uniform tensile load, the maximum stress predicted by the classical elasticity is infinite no matter how small the applied load may be. Yet such a plate under tension has considerable endurance and does not fail immediately. Noticing this discrepancy between the theory and the experiments on glass plates, Griffith [1] proposed a fracture criterion based on the energy considerations rather than maximum stress hypothesis. In essence, Griffith theory constitutes the basis of contemporary fracture mechanics. Thus over the past half century the fracture of uniform solids and solids with cracks and notches has had to be treated on two different bases. A similar dichotomy exists for the dislocations in solids, especially at the dislocation core. In the case of fluids, the outstanding problems are those involving secondary flow, and turbulence. These problems defied any rational approach on the basis of the otherwise successful Navier-Stokes theory of fluid dynamics.

The failures of classical field theories mentioned are not accidental, but are deeply ingrained in certain fundamental concepts underlying all physical theories. The applicability of a physical theory can be traced to the comparison of a characteristic length scale, a, involving the inner structure of the bodies (e.g., granular distance, lattice parameter) and the external characteristic length, ℓ (e.g., crack length, crack tip radius, wave length).* When $\ell/a \gg 1$ local continuum field theories predict sufficiently accurate outcome. On the other hand, when $\ell/a \sim 1$, "local" theories fail. In this range one may either resort to lattice dynamics or construct nonlocal theories which can account for the long-range interatomic interactions. Lattice dynamics, as useful as it has been, often leads to very complicated systems of equations for the treatment of real world problems. Moreover, such detailed and extensive calculations are not necessary since the interest is often in some average measurable quantity rather than in the individual motions of atoms. Also, the precise forms of the interatomic forces are not known. Moreover, the existence

* A different criterion, in place of this characteristic lengths ratio, is the ratio of an internal and external characteristic times (e.g., ratio of relaxation time to period of waves).

of a large number of dislocations, impurity atoms, vacancies, etc., makes the calculations prohibitive. Hence is the raison d'être for nonlocal continuum physics.

Nonlocal continuum theories, while speculated in some form or another previously, have begun only recently to be established on a rational basis. An account of the recent work in this field up to 1975 is to be found in Eringen [2] together with references to literature. The raison d'être of the present article is to present a brief formulation of the nonlocal continuum mechanics and to exhibit the power and potential of the theory by means of solutions of some crucial problems. The formulation presented here is simpler and more unified in nature as compared to our previous work. Although the problems solved are not all new, they have been cast in a different frame.

In Section 2 we summarize the basic balance laws and the second law of thermodynamics. These laws are valid for all types of bodies (whether fluid, solid or gas) having different geometries. In Section 3, constitutive theory is developed for nonlocal elastic-solids. Thermodynamic restrictions and the development of nonlocal, nonlinear elasticity is given in Section 4. Section 5 is devoted to the linear theory. The integro-partial differential field equations are obtained for anisotropic and isotropic nonlocal elastic solids. By comparing the dispersion relations of plane waves in nonlocal elasticity with those of the lattice dynamics, in Section 6, we determine the nonlocal elastic moduli of the theory. Sections 7 to 9 are devoted to the solution of various crucial problems that defied solutions by means of classical field theories. These include surface waves with small wave lengths (Section 7), the core problem of screw dislocation (Section 8), and crack tip problems (Section 9). In Section 10, we turn our attention to the theory of nonlocal fluid mechanics and its application to the problem of secondary flow and turbulence.

These solutions show that stress and energy singularities disappear in the nonlocal theory. The physical length scale involved is comparable to the lattice parameter. Fracture criterion based on the maximum stress hypothesis is restored, unifying the theory for bodies with and without cracks. Moreover, the results enable us even to predict the theoretical strength of solids. As far as the secondary flow of fluids in channels is concerned, the prediction is in accordance with experiments. Remembering that Navier-Stokes theory does not lead to any secondary flow at all, these unexpected results are certainly very gratifying.

2. BALANCE LAWS

Balance laws of the nonlocal continuum mechanics are identical to those of the classical global continuum mechanics. In their localized forms they are given by [3], [2] (Part IV)

$$\dot{\rho} + \rho v_{k,k} = \hat{\rho} \tag{2.1}$$

$$t_{k\ell,k} + \rho(f_\ell - \dot{v}_\ell) = \hat{\rho} v_\ell - \rho \hat{f}_\ell \tag{2.2}$$

$$\epsilon_{\ell mn}(t_{mn} - \rho x_m \hat{f}_n) = 0 \tag{2.3}$$

$$-\rho\dot{\epsilon} + t_{k\ell} v_{\ell,k} + q_{k,k} + \rho h - \hat{\rho}(\epsilon - \tfrac{1}{2}\mathbf{v}\cdot\mathbf{v}) - \rho \hat{f}_k v_k + \rho\hat{h} = 0 \tag{2.4}$$

$$\rho\dot{\eta} - (q_k/\theta)_{,k} - (\rho h/\theta) - \rho\hat{b} + \hat{\rho}\eta \geqq 0 \tag{2.5}$$

valid in $\mathcal{V}-\sigma$, where

ρ = mass density
v_k = velocity vector
$t_{k\ell}$ = stress tensor
f_ℓ = body force density
ϵ = internal energy density
q_k = heat vector
h = heat source density
θ = absolute temperature
η = entropy density
$\hat{\rho}$ = nonlocal mass production
$\hat{f}_k$ = nonlocal body force density
$\hat{h}$ = nonlocal heat source density
$\hat{b}$ = nonlocal entropy source density

Equations (2.1) to (2.4) are, respectively, the equations of balance of mass, momentum, moment of momentum and energy. The inequality (2.5) is the expression of the second law of thermodynamics. These expressions differ from their classical counterparts in the terms carrying a hat "^" which are sometimes named as the <u>nonlocal residuals</u>. The nonlocal residuals are the contributions of distant points $\mathbf{x}'$ in the body to the reference point $\mathbf{x}$ at which the balance laws are written. They are subject to the conditions that

$$\int_{\mathcal{V}-\sigma} (\hat{\rho}, \rho\hat{\underset{\sim}{f}}, \rho\hat{h}, \rho\hat{b})\, dv = 0 \tag{2.6}$$

where the integral is over the volume $\mathcal{V}$ occupied by the body excluding a discontinuity surface σ (e.g., a shock wave) which may be sweeping the body with its own velocity. There exist balance laws corresponding to (2.1) to (2.5) across the discontinuity surface σ. We refrain reproducing these expressions here, for the sake of brevity. For these, see Reference 2, page 213.

We introduce Helmholtz free energy ψ by

$$\psi = \epsilon - \theta\eta \tag{2.7}$$

and eliminate h between (2.4) and (2.5) to obtain the generalized Clausius-Duhem (C-D) inequality

$$-\rho(\dot{\psi} + \dot{\theta}\eta) + t_{k\ell}v_{\ell,k} + (1/\theta)q_k\theta_{,k} - \hat{\rho}(\psi - \tfrac{1}{2}v^2)$$

$$- \rho\hat{f}_k v_k + \rho\hat{h} - \rho\theta\hat{b} \geqq 0. \tag{2.8}$$

This inequality places restriction on the possible state of the body and the changes that are allowed to take place in a physical situation. Convenient for our purpose is the material form of this inequality which is obtained by introducing Piola-Kirchhoff stress tensor T_{KL}, Q_K and recalling the expression of the rate of the Green deformation tensor C_{KL} (cf. [4], pages 109, 71), i.e.,

$$T_{KL} = J t_{k\ell} X_{K,k} X_{L,\ell}, \quad Q_K = J q_k X_{K,k}$$

$$\dot{C}_{KL} = 2 v_{k,\ell} x_{k,K} x_{\ell,L} \tag{2.9}$$

where $J = \det(x_{k,K}) > 0$ is the Jacobian, $x_k = x_k(\underset{\sim}{X}, t)$ is the motion that takes a material point $\underset{\sim}{X}$ in the reference state to a spatial point $\underset{\sim}{x}$ at time t. The inverse motion $X_K = X_K(\underset{\sim}{x}, t)$ is assumed to exist and uniquely determined throughout the body, at all times.

Employing (2.9) in (2.8) we get

$$-\rho J(\dot{\psi} + \dot{\theta}\eta) + \tfrac{1}{2}T_{KL}\dot{C}_{KL} + \frac{1}{\theta}Q_K\theta_{,K} - \hat{\rho}J(\psi - \tfrac{1}{2}v^2)$$

$$- \rho J\hat{\underset{\sim}{f}}\cdot\underset{\sim}{v} + \rho J(\hat{h} - \theta\hat{b}) \geqq 0. \tag{2.10}$$

3. CONSTITUTIVE EQUATIONS

According to the axiom of Causality [5], the state of a body at a material point $\underset{\sim}{X}$ is fully determined by the motions and temperatures of all points $\underset{\sim}{X}'$ of the body at all past times, $t' \leqq t$. For solids that are not memory dependent this implies the dependence on the motions and temperatures at time t. To specify the state of the body, we must therefore express the constitutive dependent variables and nonlocal residuals

$$\begin{gathered} \psi, \eta, T_{KL}, Q_K, \\ \hat{\rho}, \hat{\underset{\sim}{f}}, \hat{h}, \hat{b} \end{gathered} \tag{3.1}$$

as functionals of the independent variables

$$\underset{\sim}{x}' \equiv \underset{\sim}{x}(\underset{\sim}{X}', t), \qquad \theta' \equiv \theta(\underset{\sim}{X}', t). \tag{3.2}$$

Thus, for example, for the free energy we write a constitutive equation of the form:

$$\psi(\underset{\sim}{X}, t) = \Psi\left[\underset{\sim}{x}(\underset{\sim}{X}', t), \theta(\underset{\sim}{X}', t)\right], \qquad \underset{\sim}{X}' \in V \tag{3.3}$$

where $\underset{\sim}{X}'$ covers all points of the body in the reference state V. Similar expression containing the same arguments, (3.2), are written for all the other members of (3.1). The response functionals (such as Ψ) must obey the Galilean invariance requirement which expresses the fact that they are form-invariant under rigid motions of the spatial frame of reference. Thus, if

$$\bar{\underset{\sim}{x}}' = \underset{\sim}{Q}\underset{\sim}{x}' + \underset{\sim}{V}_o t + \underset{\sim}{b}_o \tag{3.4}$$

where $\underset{\sim}{V}_o$ and $\underset{\sim}{b}_o$ are constant vectors and $\underset{\sim}{Q}$ is a constant orthogonal transformation

$$\underset{\sim}{Q}\underset{\sim}{Q}^T = \underset{\sim}{Q}^T\underset{\sim}{Q} = \underset{\sim}{I}, \qquad \det \underset{\sim}{Q} = \pm 1 \tag{3.5}$$

then

$$\Psi\left[\underset{\sim}{Q}\underset{\sim}{x}' + \underset{\sim}{V}_o t + \underset{\sim}{b}_o, \theta'\right] = \Psi[\underset{\sim}{x}', \theta'] \tag{3.6}$$

for all members of the full group $\{\underset{\sim}{Q}\}$ and for all vectors $\{\underset{\sim}{V}_o\}$ and $\{\underset{\sim}{b}_o\}$.

By applying the Galilean invariance in succession

a) $\quad \mathbf{Q} = \mathbf{I}, \quad \mathbf{V}_o = \mathbf{0}, \quad \mathbf{b}_o = -\mathbf{x}$

b) $\quad \mathbf{Q} = \mathbf{Q}, \quad \mathbf{V}_o = \mathbf{0}, \quad \mathbf{b}_o = \mathbf{0},$

we find that

$$\psi(\mathbf{X},t) = \Psi[\boldsymbol{\kappa}', \theta'] \tag{3.7}$$

$$\Psi[\mathbf{Q}\boldsymbol{\kappa}', \theta'] = \Psi[\boldsymbol{\kappa}', \theta'] \tag{3.8}$$

where

$$\boldsymbol{\kappa}' = \mathbf{x}(\mathbf{X}') - \mathbf{x}(\mathbf{X}) \tag{3.9}$$

is the relative motion of the material point $\mathbf{X}'$ with respect to $\mathbf{X}$. For brevity, henceforth we do not indicate explicitly the dependence on t and $\mathbf{X}$ in the argument functions of the response functionals. Thus, for example, $\mathbf{x}(\mathbf{X}')$ and $\mathbf{x}$ shall mean $\mathbf{x}(\mathbf{X}',t)$ and $\mathbf{x}(\mathbf{X},t)$, respectively.

The invariance requirement (3.7) states that ψ must be an isotropic functional in $\boldsymbol{\kappa}'$. Hence it will depend on $\boldsymbol{\kappa}'$ only through its invariants

$$\chi'' \equiv \boldsymbol{\kappa}(\mathbf{X}') \cdot \boldsymbol{\kappa}(\mathbf{X}''), \quad \mathbf{X}', \mathbf{X}'' \in V \tag{3.10}$$

i.e.,

$$\psi(\mathbf{X},t) = \Psi[\chi(\mathbf{X}',\mathbf{X}''), \theta(\mathbf{X}')] \tag{3.11}$$

Note that $\chi'' = |\boldsymbol{\kappa}'||\boldsymbol{\kappa}''| \cos(\boldsymbol{\kappa}', \boldsymbol{\kappa}'')$ and thus ψ at $\mathbf{X}$ not only depends on the distances $|\boldsymbol{\kappa}'|$ and $|\boldsymbol{\kappa}''|$ but also the angle between any two rays $\boldsymbol{\kappa}'$ and $\boldsymbol{\kappa}''$ from the point $\mathbf{x}$ in the deformed body $\mathcal{V}$. This implies an influence, between the material points, of non-central nature. In Newtonian mechanics the central force attraction being a main postulate, here too we assume that the attraction is central. Consequently, the dependence on the relative motions is only through the distances of pairs of points. With this assumption we have

$$\psi(\mathbf{X},t) = \Psi[\chi(\mathbf{X}'), \theta(\mathbf{X}')] \tag{3.12}$$

where

$$\chi' \equiv \chi(\mathbf{X}') \equiv \boldsymbol{\kappa}(\mathbf{X}') \cdot \boldsymbol{\kappa}(\mathbf{X}'). \tag{3.13}$$

For a class of nonlinear functionals, that are relevant to a wide class of physical problems, a representation theorem was provided by Friedman and Katz [6]. Let C denote the class of bounded real-valued continuous functions defined in V. Define the norm of a function $f(\underset{\sim}{X}) \subset C$ by

$$||f|| \equiv \max_{\underset{\sim}{X} \in V} |f(\underset{\sim}{X})| . \tag{3.14}$$

Suppose that the functional $\Psi(f)$ is continuous and bounded with the norm (3.14). Moreover, it is additive in the sense that if $(f,g) \subset C$ and $\Psi_g(f) \equiv \Psi(f+g) - \Psi(g)$, then for two function f_1 and f_2 have <u>disjoint supports</u>, i.e.,

$$\Psi_g(f_1 + f_2) = \Psi_g(f_1) + \Psi_g(f_2) . \tag{3.15}$$

Under these considerations Ψ can be represented in the form

$$\Psi(f) = \int_V K[f(\underset{\sim}{X}'), \underset{\sim}{X}'] dV(\underset{\sim}{X}') \tag{3.16}$$

where K is a continuous function of its arguments. We identify $f = (\chi', \theta')$ and recall that ψ is calculated at $\underset{\sim}{X}$ and therefore it must also be a function of $f(\underset{\sim}{X})$ at $\underset{\sim}{X}$. Hence

$$\psi(\underset{\sim}{X}, t) = \int_{V-\Sigma} K\left[\chi(\underset{\sim}{X}'), \theta(\underset{\sim}{X}'), \underset{\sim}{X}'; \theta, \underset{\sim}{X}\right] dV(\underset{\sim}{X}') \tag{3.17}$$

since $\chi(\underset{\sim}{X}) = 0$ and $\theta(\underset{\sim}{X}) = \theta$. Here Σ is the image of σ (discontinuity surface sweeping V).

Next we make the following important observation: Any functional of $\chi(\underset{\sim}{X}') = (\underset{\sim}{x}' - \underset{\sim}{x}) \cdot (\underset{\sim}{x}' - \underset{\sim}{x})$ can equivalently be considered to be a functional of the Green deformation tensor

$$C_{KL}(\underset{\sim}{X}') = \underset{\sim}{x}_{,K}(\underset{\sim}{X}') \cdot \underset{\sim}{x}_{,L}(\underset{\sim}{X}') . \tag{3.18}$$

This is clear from the fact that distance between every pair of points $\underset{\sim}{x}'$ and $\underset{\sim}{x}$ in the deformed body $\mathcal{V}$ is fully determined by the knowledge of $C_{KL}(\underset{\sim}{X}')$ at all points $\underset{\sim}{X}'$ of the body at the reference state. Thus we may replace (3.17) equivalently by*

---

* Of course, the norm and disjointness are modified accordingly.

$$\psi(\underset{\sim}{X},t) = \int_{V-\Sigma} \Psi\left[C_{KL}(\underset{\sim}{X}'),\theta(\underset{\sim}{X}'),\underset{\sim}{X}';C_{KL},\theta,\underset{\sim}{X}\right] dV(\underset{\sim}{X}'). \qquad (3.19)$$

Representations of this form are valid for the other members of (3.1).

4. THERMODYNAMIC RESTRICTIONS

The second law of thermodynamics places restrictions on the constitutive response functionals. According to (2.10) we have for the entropy production:

$$-\rho J(\dot{\psi}+\dot{\theta}\eta) + \tfrac{1}{2}T_{KL}\dot{C}_{KL} + \frac{1}{\theta}Q_K\theta_{,K} - \hat{\rho}J(\psi - \tfrac{1}{2}v^2)$$

$$- \rho J\hat{\underset{\sim}{f}}\cdot\underset{\sim}{v} + \rho J(\hat{h} - \theta\hat{b}) \geqq 0. \qquad (4.1)$$

Clearly the entropy production should not change if the spatial frame of reference is given a constant velocity translation, i.e.,

$$\overline{\underset{\sim}{x}}' = \underset{\sim}{x}' + \underset{\sim}{V}_o t, \quad \overline{\underset{\sim}{x}} = \underset{\sim}{x} + \underset{\sim}{V}_o t.$$

We posit that the C-D inequality is form-invariant under Galilean transformations. None of the response functionals and $\rho, J, \dot{C}_{KL}, \theta$ and $\theta_{,K}$ are affected by this change but the terms containing $\hat{\rho}$ and $\hat{\underset{\sim}{f}}$ are modified by the extra terms

$$\hat{\rho}J\left(\frac{1}{2}V_o^2 + \underset{\sim}{v}\cdot\underset{\sim}{V}_o\right) - \rho J\hat{\underset{\sim}{f}}\cdot\underset{\sim}{V}_o.$$

For arbitrary variations of V_o these terms must vanish. Hence

$$\hat{\rho} = 0, \quad \hat{\underset{\sim}{f}} = \underset{\sim}{0}. \qquad (4.2)$$

With this $\rho J = \rho_o$, and (4.1) reduces to

$$-\rho_o(\dot{\psi}+\dot{\theta}\eta) + \tfrac{1}{2}T_{KL}\dot{C}_{KL} + \frac{1}{\theta}Q_K\theta_{,K} + \rho_o(\hat{h} - \theta\hat{b}) \geqq 0. \qquad (4.3)$$

For simplicity, henceforth we confine our attention to non-heat conducting materials. In this case the temperature is uniform throughout the body. Thus response functionals depend on θ at one point; i.e., they are not functionals but functions of θ. From (3.19) we calculate $\dot{\psi}$ and substitute into (4.3) to obtain

$$-\rho_o\left(\int_{V-\Sigma}\frac{\partial\Psi}{\partial\theta}\,dV'+\eta\right)\dot{\theta}+\left(\tfrac{1}{2}T_{KL}-\rho_o\int_{V-\Sigma}\frac{\partial\Psi}{\partial C_{KL}}\,dV'\right)\dot{C}_{KL}$$

$$-\rho_o\int_{V-\Sigma}\frac{\partial\Psi}{\partial C'_{KL}}\dot{C}'_{KL}dV'+\frac{1}{\theta}Q_K\theta_{,K}+\rho_o(\hat{h}-\theta\hat{b})\geqq 0. \tag{4.4}$$

This inequality is linear in $\dot{\theta}$ and $\theta_{,K}$. If it is not to be violated for all possible values of these quantities we must have

$$\eta=-\int_{V-\Sigma}\frac{\partial\Psi}{\partial\theta}\,dV',\quad Q_K=0. \tag{4.5}$$

The inequality (4.4) may now be put into the form

$$\left\{\tfrac{1}{2}T_{KL}-\int_{V-\Sigma}\left[\rho_o\frac{\partial\Psi}{\partial C_{KL}}+\left(\rho_o\frac{\partial\Psi}{\partial C'_{KL}}\right)^*\right]dV'\right\}\dot{C}_{KL}$$

$$+\int_{V-\Sigma}\left[\left(\rho_o\frac{\partial\Psi}{\partial C'_{KL}}\right)^*\dot{C}_{KL}-\rho_o\frac{\partial\Psi}{\partial C'_{KL}}\dot{C}'_{KL}\right]dV'+\rho_o(\hat{h}-\theta\hat{b})\geqq 0 \tag{4.6}$$

where an asterisk placed on quantities indicate the interchange of $\underset{\sim}{X}$ by $\underset{\sim}{X}'$, e.g.,

$$[F(\underset{\sim}{X}',\underset{\sim}{X})]^*=F(\underset{\sim}{X},\underset{\sim}{X}'). \tag{4.7}$$

If we now integrate (4.6) over the volume on $\underset{\sim}{X}$, the second term vanishes on account of being skew-symmetric in $\underset{\sim}{X}$ and $\underset{\sim}{X}'$ and the last terms containing residuals vanish because of conditions (2.6) on residuals, and we are left with

$$\int_{V-\Sigma}\left\{\tfrac{1}{2}T_{KL}-\int_{V-\Sigma}\left[\rho_o\frac{\partial\Psi}{\partial C_{KL}}+\left(\rho_o\frac{\partial\Psi}{\partial C'_{KL}}\right)^*\right]dV'\right\}\dot{C}_{KL}dV\geqq 0. \tag{4.8}$$

This integral is linear in $\dot{C}_{KL}$. For arbitrary variations of this tensor throughout V, it cannot be maintained in one sign, unless the coefficient of $\dot{C}_{KL}$ vanishes. Hence we obtain

$$T_{KL}=2\int_{V-\Sigma}\left[\rho_o\frac{\partial\Psi}{\partial C_{KL}}+\left(\rho_o\frac{\partial\Psi}{\partial C'_{KL}}\right)^*\right]dV'. \tag{4.9}$$

For the non-heat conducting elastic solids no further need arises for the use of the energy equation and the entropy inequality. Thus we leave $\hat{h}$ and $\tilde{b}$ undetermined. The relevant members of (3.1) are now fully determined in terms of the free energy functional Ψ. The constitutive equations obtained are valid for the nonlinear nonlocal elastic materials.

5. LINEAR THEORY

To obtain the constitutive equations that are linear in the strain measure, we introduce the Lagrangian strain tensor E_{KL} by

$$E_{KL} = \tfrac{1}{2}(C_{KL} - \delta_{KL}) \tag{5.1}$$

and approximate the function ψ by a polynomial quadratic in E'_{KL} and E_{KL}, i.e.,

$$\begin{aligned}\Psi = \overset{o}{\psi} + \overset{o}{\psi}_{KL}E_{KL} + \tfrac{1}{2}\overset{o}{\psi}_{KLMN}E_{KL}E_{MN} + \overset{1}{\psi}_{KL}E'_{KL} \\ + \overset{1}{\psi}_{KLMN}E_{KL}E'_{MN} + \overset{2}{\psi}_{KLMN}E'_{KL}E_{MN} + \tfrac{1}{2}\overset{3}{\psi}_{KLMN}E'_{KL}E'_{MN}\end{aligned} \tag{5.2}$$

where $\overset{o}{\psi}, \overset{o}{\psi}_{KL}, \ldots, \overset{3}{\psi}_{KLMN}$ are functions of θ, $\underset{\sim}{X}'$ and $\underset{\sim}{X}$. Substituting (5.2) into (4.9) we obtain

$$T_{KL} = A_{KL} + A_{KLMN}E_{MN} + \int_{V-\Sigma} A'_{KLMN}(\underset{\sim}{X}', \underset{\sim}{X})E_{MN}(\underset{\sim}{X}')dV' \tag{5.3}$$

where

$$\begin{aligned}A_{KL} &\equiv \int_{V-\Sigma}\left(\rho_o\overset{o}{\psi}_{KL} + \overset{*}{\rho}_o\overset{*1}{\psi}_{KL}\right)dV', \\ A_{KLMN} &\equiv \int_{V-\Sigma}\left(\rho_o\overset{o}{\psi}_{KLMN} + \overset{*}{\rho}_o\overset{*3}{\psi}_{KLMN}\right)dV', \\ A'_{KLMN} &\equiv \rho_o\left(\overset{1}{\psi}_{KLMN} + \overset{2}{\psi}_{MNKL}\right) + \overset{*}{\rho}_o\left(\overset{*1}{\psi}_{MNKL} + \overset{*2}{\psi}_{KLMN}\right).\end{aligned} \tag{5.4}$$

The material moduli A_{KL} and A_{KLMN} are functions of θ and $\underset{\sim}{X}$ and

A'_{KLMN} are functions of θ, $\underset{\sim}{X}$ and $\underset{\sim}{X}'$. The following symmetry regulations are obvious from (5.2) and (5.4)

$$A_{KL} = A_{LK}, \quad A_{KLMN} = A_{LKMN} = A_{KLNM} = A_{MNKL},$$

$$A'_{KLMN}(\underset{\sim}{X}',\underset{\sim}{X}) = A'_{LKMN}(\underset{\sim}{X}',\underset{\sim}{X}) = A'_{KLNM}(\underset{\sim}{X}',\underset{\sim}{X}) = A'_{MNKL}(\underset{\sim}{X},\underset{\sim}{X}'). \tag{5.5}$$

Thus the number of A_{KL} is six, the number of A_{KLMN} is 21 and the number of A'_{KLMN} is also 21. If the natural state is stress free, then the stress must vanish with the strain, i.e., $A_{KL}=0$. Henceforth we assume that this is the case.

The linear constitutive equations (5.3) is valid for inhomogeneous anisotropic bodies. For homogeneous solids A'_{KLMN} will depend on $\underset{\sim}{X}'$ and $\underset{\sim}{X}$ only through $\underset{\sim}{X}'-\underset{\sim}{X}$. For the isotropic solids we have

$$A_{KLMN} = \lambda\delta_{KL}\delta_{MN} + \mu\left(\delta_{KM}\delta_{LN} + \delta_{KN}\delta_{LM}\right) \tag{5.6}$$

where λ and μ are the classical Lamé constants that depend on θ. An isotropic fourth order tensor that is a function of a vector $\underset{\sim}{X}'-\underset{\sim}{X}$ has the form

$$\begin{aligned} A'_{KLMN} &= \lambda'\delta_{KL}\delta_{MN} + \mu'\left(\delta_{KM}\delta_{LN} + \delta_{KN}\delta_{LM}\right) \\ &+ A'_1\left(X'_K-X_K\right)\left(X'_L-X_L\right)\delta_{MN} + A'_2\left(X'_K-X_K\right)\left(X'_M-X_M\right)\delta_{LN} \\ &+ A'_3\left(X'_K-X_K\right)\left(X'_N-X_N\right)\delta_{LM} + A'_4\left(X'_L-X_L\right)\left(X'_M-X_M\right)\delta_{KN} \\ &+ A'_5\left(X'_L-X_L\right)\left(X'_N-X_N\right)\delta_{KM} + A'_6\left(X'_M-X_M\right)\left(X'_N-X_N\right)\delta_{KL} \\ &+ A'_7\left(X'_K-X_K\right)\left(X'_L-X_L\right)\left(X'_M-X_M\right)\left(X'_N-X_N\right) \end{aligned} \tag{5.7}$$

where λ', μ' and A_1' to A_7' are functions of θ and $|\underset{\sim}{X}'-\underset{\sim}{X}|$. From (5.6) and (5.7) it is clear that the linear nonlocal, homogeneous and isotropic elastic solids are characterized by eleven <u>nonlocal moduli</u> $\lambda, \mu, \lambda', \mu', A_1'$ to A_7'. Here A_1' to A_7' indicate local symmetry regulations which may be interpreted as the symmetry due to "molecular orientations." Such molecular properties are, of course, absent in classical elasticity.

It is well-known that the intermolecular and interatomic attractions die out quickly with the distance. If a is the internal characteristic length beyond which the nonlocal attractions, at the center of a sphere with radius a, are negligible then we expect that the terms containing products of $(X'_K - X_K)$ in (5.7) may be neglected as compared to λ' and μ'. We call such materials as <u>macro-isotropic nonlocal elastic solids</u>. In this case the linear stress constitutive equations read:

$$T_{KL} = \lambda E_{MM}\delta_{KL} + 2\mu E_{KL} + \int_{V-\Sigma} \Big[\lambda'(|\underset{\sim}{X}'-\underset{\sim}{X}|)E'_{MM}\delta_{KL} + 2\mu'(|\underset{\sim}{X}'-\underset{\sim}{X}|)E'_{KL}\Big]dV'. \tag{5.8}$$

Upon carrying this into $(2.9)_1$ and further linearizing, we obtain the spatial form of the stress constitutive equations

$$t_{k\ell} = \lambda e_{rr}\delta_{k\ell} + 2\mu e_{k\ell} + \int_{V-\sigma} \Big[\lambda'(|\underset{\sim}{x}'-\underset{\sim}{x}|)e_{rr}(\underset{\sim}{x}')\delta_{k\ell} + 2\mu'(|\underset{\sim}{x}'-\underset{\sim}{x}|)e_{k\ell}(\underset{\sim}{x}')\Big]\,dv(\underset{\sim}{x}') \tag{5.9}$$

where we wrote $\rho/\rho_0 \cong 1$, $x_{k,K} \cong \delta_{kK}$, $u_k = U_K\delta_{kK}$. $e_{k\ell}(\underset{\sim}{x},t)$ is the linear spatial strain tensor defined by

$$e_{k\ell} \equiv \tfrac{1}{2}(u_{k,\ell} + u_{\ell,k}). \tag{5.10}$$

The elastic moduli λ, μ and the nonlocal elastic moduli λ', μ' depend on θ. If we extend the class of functions—that λ' and μ' belong to—to distributions, then the classical terms involving λ and μ in (5.9) can be incorporated to the volume integral so that we may alternatively write

$$t_{k\ell} = \int_{V-\sigma} \Big[\lambda'(|\underset{\sim}{x}'-\underset{\sim}{x}|)e_{rr}(\underset{\sim}{x}')\delta_{k\ell} + 2\mu'(|\underset{\sim}{x}'-\underset{\sim}{x}|)e_{k\ell}(\underset{\sim}{x}')\Big]dv(\underset{\sim}{x}'). \tag{5.11}$$

In this case, however, to include the classical continuum limit we employ the normalization condition

$$\int_{V-\sigma} \left\{\lambda'(|\underset{\sim}{x}'-\underset{\sim}{x}|)/\lambda,\ \mu'(|\underset{\sim}{x}'-\underset{\sim}{x}|)/\mu\right\} dv(\underset{\sim}{x}') = 1. \tag{5.12}$$

Note that if (5.9) is used the classical continuum limit is obtained by setting $\lambda' = \mu' = 0$. The constitutive equations (5.11) are found to be simpler for the treatment of certain classes of problems.

Upon substituting (5.9) and (5.10) into Cauchy's second law (2.2), we obtain the field equations

$$(\lambda + 2\mu)\nabla\nabla\cdot\underset{\sim}{u} - \mu\nabla x\nabla x\underset{\sim}{u}$$

$$+ \int_{V-\sigma} \left[(\lambda' + 2\mu')\nabla'\nabla'\cdot\underset{\sim}{u}' - \mu'\nabla' x\nabla' x\underset{\sim}{u}'\right] dv'$$

$$- \int_{\partial V-\sigma} \underset{\sim}{T}'_k da'_k + \int_{\sigma} [\underset{\sim}{T}'_k] da'_k + \rho(\underset{\sim}{f} - \ddot{\underset{\sim}{u}}) = \underset{\sim}{0} \tag{5.13}$$

where

$$\underset{\sim}{T}'_k = T'_{k\ell}\underset{\sim}{i}_\ell, \quad T'_{k\ell} = \lambda' e'_{rr}\delta_{k\ell} + 2\mu' e'_{k\ell} \tag{5.14}$$

are the surface tractions and surface stress densities, respectively. Here operators carrying a prime are on $\underset{\sim}{x}'$. In deriving (5.13) we used the identity

$$\frac{\partial\lambda'}{\partial x_k} F' = -\frac{\partial}{\partial x'_k}(\lambda' F') + \lambda'\frac{\partial F'}{\partial x'_k}$$

and Green-Gauss theorem to convert the first term on the right to a surface integral.

It is interesting that (5.13) contains the effect of <u>surface stresses</u> which are not included in the classical field theories. Thus <u>the nonlocal theory accounts for the surface physics</u>.

These results were derived by Eringen [2], [7]. If in place of (5.9) we use (5.11), the result would be identical to (5.13) excluding the classical Navier terms containing $\lambda + 2\mu$ and μ, outside of integrals, i.e.,

$$\int_{V-\sigma} \left[(\lambda' + 2\mu')\nabla'\nabla'\cdot\underset{\sim}{u}' - \mu'\nabla' x\nabla' x\underset{\sim}{u}'\right] dv' - \int_{\partial V-\sigma} \underset{\sim}{T}'_k da'_k +$$

$$+ \int_{\sigma} [\underset{\sim}{T}'_k] da'_k + \rho(\underset{\sim}{f} - \ddot{\underset{\sim}{u}}) = \underset{\sim}{0}. \tag{5.15}$$

For the linear, homogeneous, anisotropic solids, the corresponding equations to (5.11) and (5.15) are

$$t_{k\ell} = \int_{V-\sigma} a'_{k\ell mn}(\underset{\sim}{x}'-\underset{\sim}{x}) e_{mn}(\underset{\sim}{x}') dv(\underset{\sim}{x}'), \tag{5.16}$$

$$\int_{V-\sigma} a'_{k\ell mn}(\underset{\sim}{x}'-\underset{\sim}{x}) \frac{\partial^2 u_m(\underset{\sim}{x}')}{\partial x'_n \partial x'_k} dv(\underset{\sim}{x}') - \int_{\partial V-\sigma} T'_{k\ell} da_k(\underset{\sim}{x}')$$

$$+ \int_{\sigma} [\underset{\sim}{T}'_{k\ell}] da_k(\underset{\sim}{x}') + \rho(f_\ell - \ddot{u}_\ell) = 0, \tag{5.17}$$

where

$$T'_{k\ell} = a'_{k\ell mn}(\underset{\sim}{x}'-\underset{\sim}{x}) e_{mn}(\underset{\sim}{x}')$$

is the surface stress tensor.

6. DETERMINATION OF NONLOCAL ELASTIC MODULI

The nonlocal moduli $\lambda'(|\underset{\sim}{x}'-\underset{\sim}{x}|)$, $\mu'(|\underset{\sim}{x}'-\underset{\sim}{x}|)$ for isotropic solids and $a_{k\ell mn}(\underset{\sim}{x}'-\underset{\sim}{x})$ for the anisotropic solids may be determined by requiring that the solution of the field equations (5.15) and (5.17) for some typical problems give identical results to their solutions based on the atomic lattice dynamics. To this end Eringen [7], employing the dispersion relations for plane waves, has determined λ' and μ'. Several different functions have been used by other authors (cf. Krumhansl [8], Kunin [9]). It appears that the specific forms of these functions are not as crucial as the general properties that

(a) they acquire maxima at $\underset{\sim}{x}' = \underset{\sim}{x}$,

(b) they attenuate quickly over several internal characteristic lengths from $\underset{\sim}{x}$, and

(c) they possess distribution property in the limit when characteristic length goes to zero (classical continuum limit) satisfying (5.12).

For example, λ' and μ' determined by the one-dimensional plane wave solution have the simple, elegant forms

$$\left\{\lambda'(|\underset{\sim}{x}|)/\lambda,\ \mu'(|\underset{\sim}{x}|)/\mu\right\} = \left\{\begin{matrix} a^{-1}[1-|x|/a], & x \leqq a \\ 0 & x > a \end{matrix}\right\} \tag{6.1}$$

where a is the lattice parameter. This is the triangular moduli illustrated in Figure 1. With λ' and μ' given by (6.1), the dispersion curves of plane elastic waves coincide with those of Born-Von Kármán model of the lattice dynamics in the entire Brillouin zone. This result is certainly remarkable in that no continuum theory hitherto has been able to penetrate thus far into the atomic scale phenomena.

A more useful candidate for the nonlocal elastic moduli is the Gaussian function

$$\left\{\lambda'/\lambda,\ \mu'/\mu\right\} = A \exp\left[-(k/a)^2 (x'_k - x_k)(x'_k - x_k)\right] \tag{6.2}$$

where k and A are constants, a is the internal characteristic length (e.g., the lattice parameter). The constant A is determined by the normalization condition (5.12). If the dimension of the space is denoted by N, then we have

$$A = \pi^{-N/2} (k/a)^N . \tag{6.3}$$

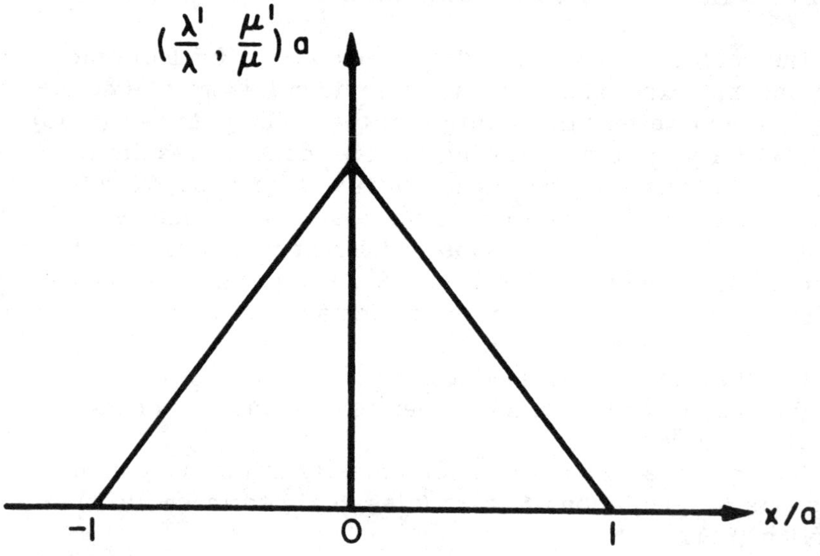

Figure 1. Nonlocal elastic moduli.

It remains to determine k. This may be done by requiring that the dispersion curves for the plane elastic waves obtained by use of (6.2) in (5.15) approximate those of the lattice dynamics as closely as possible. For various values of k the comparison of the dispersion curves is shown on Figure 2. For example, for $k = 1.65$ the dispersion curves based on Born-Von Kármán model of lattice dynamics and (5.15) is indistinguishable in the entire Brillouin zone. The maximum error is less than 0.2 percent. However, note that (6.2) does not have finite support to be considered as a distribution. Nevertheless as $a \to 0$ it gives Dirac delta measure so that it satisfies all three conditions (a) to (c). We note that clearly other criteria more crucial to the problem at hand may be used to determine k, including experimental curve fitting. For example, if the attenuation reaches its 1 percent value at n lattice parameter from the origin using (6.2), we find that

$$k = 2.146/n. \tag{6.4}$$

If we take $n = 3$ and 5, we find, respectively, $k_3 = 0.71$, $k_5 = 0.43$. This flexibility of the theory should be considered an asset rather than a defect when we consider that the materials are never perfect and the pure crystals are merely man-made idealizations, generally for the purpose of simplicity and understanding. Whatever the exact forms of the nonlocal moduli for the homogeneous anisotropic materials, we may express them in the forms

$$a_{k\ell mn} = c_{k\ell mn}\,\alpha(\underset{\sim}{x}'-\underset{\sim}{x}) \tag{6.5}$$

where $c_{k\ell mn}$ are the classical elastic constants and $\alpha(\underset{\sim}{x}'-\underset{\sim}{x})$ is the attenuation function which is subject to

$$\int_{\mathcal{V}-\sigma} \alpha(\underset{\sim}{x}'-\underset{\sim}{x})\,dv(\underset{\sim}{x}') = 1; \tag{6.6}$$

the form (6.5) is, of course, an approximation in the sense discussed above. The attenuation factor $\alpha(\underset{\sim}{x}'-\underset{\sim}{x})$ brings the influence of the strain fields in the neighboring directions. For isotropic solids, this range is spherical with center $\underset{\sim}{x}$ [as is clear from (6.1)] but for anisotropic materials it may, in general, attenuate at different rates in different directions.

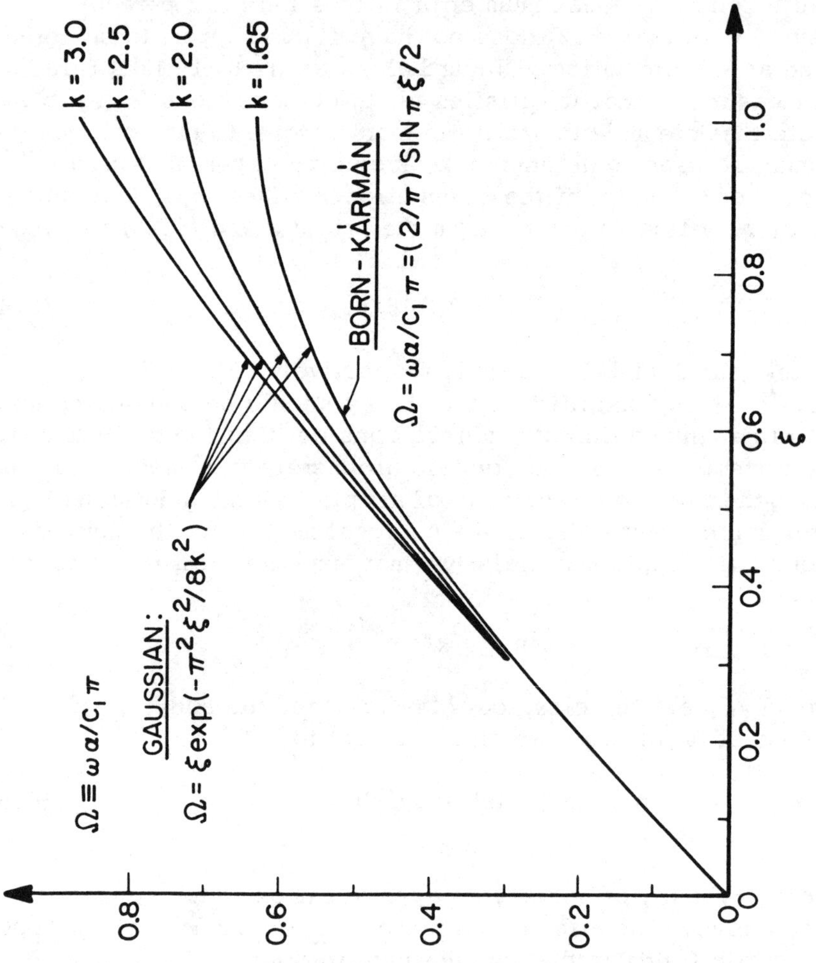

Figure 2. Dispersion curves for Gaussian-type nonlocal moduli.

7. SURFACE WAVES

In this section we discuss the propagation of surface waves in an isotropic, elastic half-space occupying the region $x_2 \geqq 0$. We examine the plane strain problem so that the displacement field is independent of x_3 and the problem is reduced to determination of the plane components of the displacement field $u_1(x_1, x_2, t)$ and $u_2(x_1, x_2, t)$. For simplicity, we assume that the nonlocal effects in the x_2-direction is negligible. This is permissible since a solid sliced along $x_2 = 0$ plane stiffens near the surface of the half-space $x_2 \geqq 0$ on account of the elimination of intermolecular attractions from the other half-space occupying $x_2 < 0$. With this assumption, we may write

$$t_{k\ell} = \int_{-\infty}^{\infty} \alpha_1(|x_1' - x_1|)\sigma_{k\ell}(x_1', x_2)dx_1' \tag{7.1}$$

where

$$\alpha_1(|x_1' - x_1|) = (k/a\pi^{\frac{1}{2}}) \exp\left[-(k/a)^2(x_1' - x_1)^2\right] \tag{7.2}$$

and $\sigma_{k\ell}$ is the classical (local) stress which is given by the Hookes law:

$$\sigma_{k\ell} = \lambda e_{rr}\delta_{k\ell} + 2\mu e_{k\ell} \tag{7.3}$$

$$e_{k\ell} = \tfrac{1}{2}(u_{k,\ell} + u_{\ell,k}). \tag{7.4}$$

Equations of linear momentum read

$$t_{k\ell,k} - \rho\ddot{u}_\ell = 0. \tag{7.5}$$

We must determine u_k by solving (7.1) to (7.5) subject to the boundary conditions

$$\begin{aligned} t_{22} = t_{21} &= 0, \quad x_2 = 0 \\ u_k &= 0, \quad x_2 \to \infty. \end{aligned} \tag{7.6}$$

The solution is effected by means of the Fourier transform defined by

$$\bar{F}(\xi, x_2, \omega) = \frac{1}{2\pi} \int\int_{-\infty}^{\infty} F(x_1, x_2, t) \exp(i\xi x_1 + i\omega t) dx_1 dt. \quad (7.7)$$

Upon taking the Fourier transforms of (7.5) and (7.1) and eliminating $\bar{t}_{k\ell}$, we obtain

$$\begin{aligned} -i\xi\bar{\sigma}_{11} + \bar{\sigma}_{21,2} + (\rho\omega^2/\bar{\alpha}_1)\bar{u}_1 &= 0 \\ -i\xi\bar{\sigma}_{21} + \bar{\sigma}_{22,2} + (\rho\omega^2/\bar{\alpha}_1)\bar{u}_2 &= 0. \end{aligned} \quad (7.8)$$

The Fourier transforms of the boundary conditions read

$$\begin{aligned} \bar{\sigma}_{22} = \bar{\sigma}_{21} &= 0, \quad x_2 = 0 \\ \bar{u}_k &= 0, \quad x_2 \to \infty. \end{aligned} \quad (7.9)$$

A comparison of (7.8) and (7.9) with the Rayleigh surface wave problem in classical elasticity shows that if in the classical solution we replace ρ by $\rho/\bar{\alpha}_1$, we obtain the transforms $\bar{\sigma}_{k\ell}$ and $\bar{u}_k$ of the nonlocal fields. Hence we can invert them to obtain $\sigma_{k\ell}$ and u_k. The stress field $t_{k\ell}$ then follows from (7.1).

In the surface wave problem the great interest lies in the surface wave velocity. According to the foregoing consideration, this is given by

$$c_R/c_2 = \omega/\xi c_2 = \gamma(\bar{\alpha}_1)^{\frac{1}{2}} \quad (7.10)$$

where γ is a constant that depends on the Poisson's ratio. This is the same constant as listed in Table 7.5.1 of Eringen and Suhubi [10], e.g.,

$\gamma = 0.9194$ for $\nu = 0.25$ $(\lambda = \mu)$

$\gamma = 0.9553$ for $\nu = 0.5$ (incompressible solids).

The dispersion curve based on (7.10) is plotted in Figure 3. The calculations based on atomic lattice dynamics as given in Maraddudin et al. [11] are also indicated on this figure by heavy dots. The agreement is remarkable, considering the fact that no curve-fitting is made by adjustment of any constant.

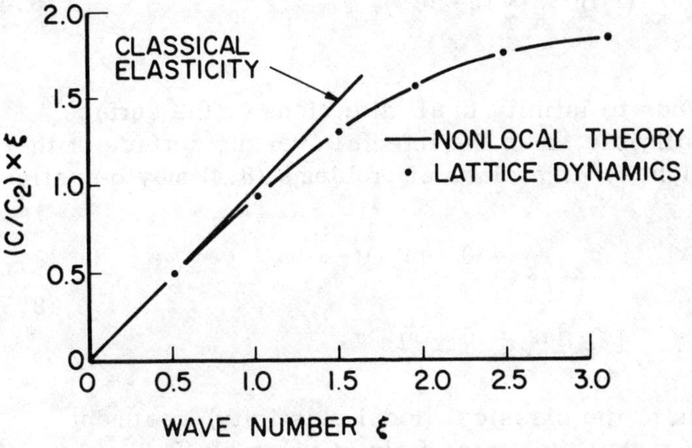

Figure 3. Dispersion of Surface Waves (c = phase velocity, c_2 = equivol. wave velocity).

8. SCREW DISLOCATION

In the case when the inertial effects are negligible and the body force is absent we have

$$t_{k\ell,k} = 0 \tag{8.1}$$

where the stress tensor is given by

$$t_{k\ell} = \int_{V-\sigma} \alpha(\underset{\sim}{x}'-\underset{\sim}{x})\sigma_{k\ell}(\underset{\sim}{x}')dv(\underset{\sim}{x}') \tag{8.2}$$

in which $\sigma_{k\ell}$ is related to the strain tensor $e_{k\ell}$ by the Hooke's law:

$$\begin{aligned} \sigma_{k\ell}(\underset{\sim}{x}') &= c_{k\ell mn}e_{mn}(\underset{\sim}{x}') \\ e_{k\ell}(\underset{\sim}{x}') &= \tfrac{1}{2}(u'_{k,\ell} + u'_{\ell,k}). \end{aligned} \tag{8.3}$$

Substituting (8.2) into (8.1), in the same way as before, we obtain

$$\int_{V-\sigma} \alpha(\underset{\sim}{x}'-\underset{\sim}{x})\sigma_{k\ell,k}(\underset{\sim}{x}')dv(\underset{\sim}{x}') - \int_{\partial V-\sigma} \alpha(\underset{\sim}{x}'-\underset{\sim}{x})\sigma_{k\ell}(\underset{\sim}{x}')da_k(\underset{\sim}{x}') +$$

$$+ \int_{\sigma} \alpha(\underset{\sim}{x}'-\underset{\sim}{x})[\sigma_{k\ell}(\underset{\sim}{x}')]da_k(\underset{\sim}{x}') = 0. \tag{8.4}$$

If the body extends to infinity in all directions or the surface stresses are negligible, then the integral over the surface of the body $\partial\mathcal{V}-\sigma$ vanishes. In a class of problems (8.4) may be satisfied by taking

$$\begin{aligned} \sigma_{k\ell,k} &= 0 \quad \text{in} \quad \mathcal{V}-\sigma \\ [\sigma_{k\ell}]n_k &= 0 \quad \text{on} \quad \sigma. \end{aligned} \tag{8.5}$$

This corresponds to the classical (local elasticity) treatment with the exception that the stress field is given by (8.2). Thus if the solution of (8.5) does not violate the boundary conditions involving the stress field $t_{k\ell}$ or if the problem is a displacement boundary value problem not involving any boundary conditions on $t_{k\ell}$, all we need is to borrow the displacement field u_k satisfying (8.3) and (8.5). This turns out to be the case for various Volterra dislocation problems and crack problems (cf. Eringen [12], [13], and Eringen et al. [14]).

A screw dislocation is obtained if the lower face of a radial plane ($r>0$, $\theta=2\pi$) of a circular cylinder is given a constant relative displacement b in the axial (z) direction, Figure 4. Here b is known as the Burgers vector. The classical elasticity solution contains singularities in both stress field and the elastic energy. We shall soon see that the nonlocal theory eliminates these singularities.

The displacement field satisfying (8.5), and (8.3) for the isotropic solids, is given by the classical solution

$$u_z = b\theta/2\pi, \quad u_r = u_\theta = 0 \tag{8.6}$$

where u_r, u_θ and u_z are the components of the displacement vector $\underset{\sim}{u}$ in cylindrical coordinates (r, θ, z). The classical stress field $\sigma_{k\ell}$ is given by

$$\sigma_{z\theta} = \mu b/2\pi r, \quad \text{all other } \sigma_{k\ell} = 0. \tag{8.7}$$

It is now clear that the integral over the surface σ in (8.4) is satisfied identically since

$$\sigma_{\theta r} = \sigma_{\theta\theta} = 0 \quad \text{and} \quad [\sigma_{\theta z}] = 0$$

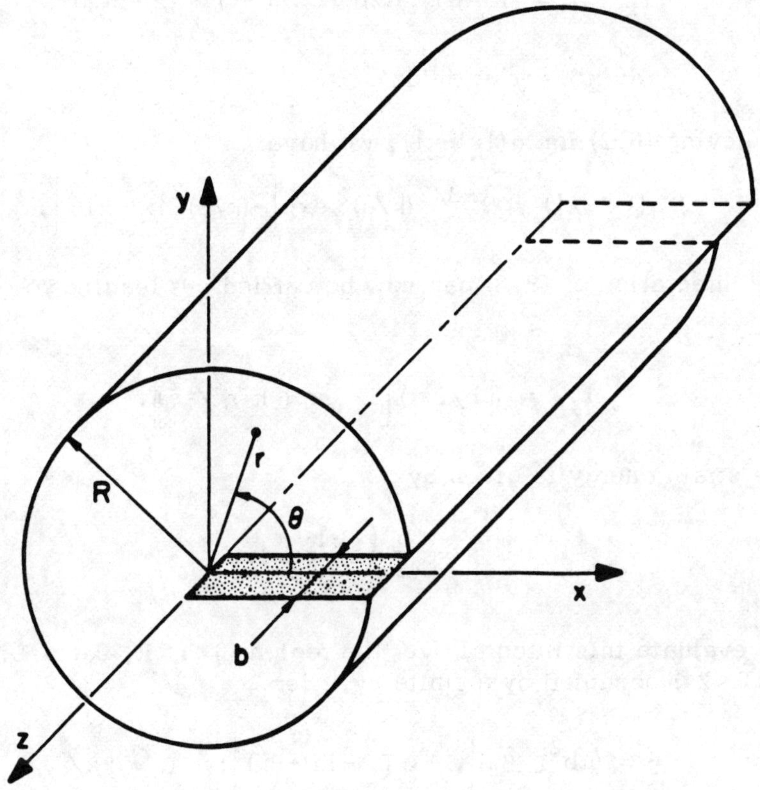

Figure 4. Screw dislocation.

across the discontinuity surface $\theta = 0, 2\pi$. Thus far the displacement field (8.6) satisfies (8.4) for a cylinder of infinite (large) radius. The stress field is given by (8.2). In cylindrical coordinates, this reads

$$\begin{aligned} t_{rz} &= -\int_V \alpha(|\underset{\sim}{x}'-\underset{\sim}{x}|)\sigma_{\theta z}(\underset{\sim}{x}')\sin(\theta'-\theta)\,dv(\underset{\sim}{x}') \\ t_{\theta z} &= \int_V \alpha(|\underset{\sim}{x}'-\underset{\sim}{x}|)\sigma_{\theta z}(\underset{\sim}{x}')\cos(\theta'-\theta)\,dv(\underset{\sim}{x}') \end{aligned} \tag{8.8}$$

where

$$|\underset{\sim}{x}'-\underset{\sim}{x}| = \left[r'^2 + r^2 - 2rr'\cos(\theta'-\theta) + (z'-z)^2\right]^{\frac{1}{2}}$$

$$dv(\underset{\sim}{x}') = r'dr'd\theta'dz'. \tag{8.9}$$

Employing (6.2) for $\alpha(|\underset{\sim}{x}'-\underset{\sim}{x}|)$, we have

$$\alpha(|\underset{\sim}{x}'-\underset{\sim}{x}|) = \pi^{-3/2}(k/a)^3 \exp\left[-(k/a)^2|\underset{\sim}{x}'-\underset{\sim}{x}|^2\right]. \tag{8.10}$$

The integration of (8.8) can now be carried out leading to

$$t_{rz} = 0,$$

$$t_{z\theta} = (\mu b/2\pi r)\left[1 - \exp(-k^2r^2/a^2)\right]. \tag{8.11}$$

The strain energy is given by

$$\Sigma = \frac{b}{4\pi}\int_{V} r^{-1} t_{z\theta} dv.$$

We evaluate this integral over the region ($0 < r \leqq R$, $0 \leqq z \leqq L$, $0 \leqq \theta < 2\pi$) occupied by a finite cylinder

$$\Sigma = (\mu b^2 L/8\pi)\left[C + \ln P^2 - \mathrm{Ei}(-P^2)\right]; \quad P = kR/a \tag{8.12}$$

where C is the Euler's constant and $\mathrm{Ei}(x)$ is the exponential integral function

$$C = 0.577216\ldots, \quad \mathrm{Ei}(-x) = -\int_x^{\infty}(e^{-t}/t)dt, \quad x > 0. \tag{8.13}$$

For $R \neq \infty$ and $a \neq 0$ (a = lattice constant), it is clear the Σ is finite.

The shear stress and the strain energy for the screw dislocation, according to classical elasticity, are given by

$$t_{z\theta} = \mu b/2\pi r,$$

$$\Sigma = (\mu b^2 L/4\pi)\ln(R/r_o) \quad \text{as} \quad r_o \to 0 \tag{8.14}$$

where r_o is the inner radius for a hollow cylinder of length L and outer radius R. Clearly the stress is singular at $r = 0$ and the strain energy is unbounded for the full cylinder ($r_o = 0$). This situation is fully realized in the dislocation theory and to remedy it, the atomic lattice theory is brought into play with various considerations regarding the so-called "core radius." The solutions (8.11) and (8.12) based on the nonlocal theory contain no singularity for $a \neq 0$ and $R \neq \infty$. In the limit $a = 0$ they revert to the classical elasticity solution. The maximum shear stress occurs at the root of ρ_m of

$$\exp \rho^2 = 1 + 2\rho^2, \quad \rho \equiv kr/a \tag{8.15}$$

and it is given by

$$t_{z\theta\,\max} = \tau_c = (\mu bk/\pi a)\rho_m (1 + 2\rho_m^2)^{-1}. \tag{8.16}$$

If we equate this to the cohesive shear stress τ_c, holding the bonds, we obtain

$$\tau_c \cong 0.3191\ (\mu bk/\pi a) \tag{8.17}$$

which determines the parameter k when τ_c is known. For a single dislocation of face-centered cubic metals $b = a/\sqrt{6}$ and this gives

$$k \cong 24.1156\,(\tau_c/\mu). \tag{8.18}$$

If we employ the value $k = 1.65$ which matches the dispersion curve of plane waves with that based on the Born Kármán theory of lattice dynamics, we obtain $\tau_c/\mu = 0.068$. If we use (6.4) with $n = 3$, we find $\tau_c/\mu = 0.030$. These results are in the range predicted by other considerations in the atomic theory. For example, for aluminum and copper Kelly ([15], page 19) gives the value $\tau_c/\mu = 0.039$. These values correspond to a value of $k \cong 0.94$. At this value, the attenuation function drops to 1 percent of its maximum at 2.28 atomic distances. Thus it seems that the nonlocal theory is capable of predicting physical phenomena in the atomic scale.

9. FRACTURE MECHANICS

A plate with a line crack, subject to uniform tension at infinity perpendicular to the direction of line crack, is a crucial problem sometimes known as the Griffith problem, in fracture mechanics. Griffith [1] noticed that such a plate can sustain considerable tension before it fractures. Yet classical elasticity solution of this problem predicts an infinite hoop stress at the tip of the crack no matter how small the applied tension is. The major discrepancy that exists between experimental observations and the theory led Griffith to propose his celebrated fracture criterion which results from equating the elastic energy to the "surface tension energy":

$$t_o^2 \ell = C_G, \quad C_G = \frac{2E}{\pi(1-\nu^2)} \gamma \tag{9.1}$$

where t_o is the applied tension, ℓ is the half crack length. Griffith constant C_G is expressed in terms of the Young's modulus E, Poisson's ratio ν and the surface tension energy γ. Accordingly, when the applied stress satisfies (9.1), the crack begins to propagate.

The past half century registered many advances in this field, influenced mainly by this central idea. While attempts exist to eliminate the stress singularity at the crack tip by some artificial means, such as assuming additional cohesive surface stress to close the crack tip (Barenblatt [16], Zheltov and Khristianovich [17], Dougdale [18]), nonlinear springs along the tips of the crack (Goodier and Kanninen [19]), the problem basically remained unresolved until recently. Eringen et al. [14] gave a solution of this problem by solving nonlocal field equations. Here we outline this solution employing, however, for the attenuation function (6.2) in place of (6.1) used in their solutions.

Consider a plate in $(x_1 = x, x_2 = y)$-plane weakened by a line crack of length 2ℓ along the x-axis. The plate is subjected to a constant tension t_o at $y = \pm\infty$ perpendicular to the line of crack, Figure 5. For the plane strain problem, (8.4) takes the form

$$\displaystyle{\not}\!\!\int_R \alpha(|\underset{\sim}{x}'-\underset{\sim}{x}|)\sigma_{k\ell,k}(x',y')dx'dy' + \\ + \int_{-\ell}^{\ell} \alpha(|\underset{\sim}{x}'-\underset{\sim}{x}|)\underset{\sim}{[}\sigma_{2\ell}(x',0)\underset{\sim}{]}dx' = 0 \tag{9.2}$$

where the integral with a slash is over two-dimensional infinite space excluding the crack line $(|x| < \ell, y = 0)$. The displacement fields u(x,y) and v(x,y) possess the following symmetry regulations

$$u(x,-y) = u(x,y), \quad v(x,-y) = -v(x,y). \tag{9.3}$$

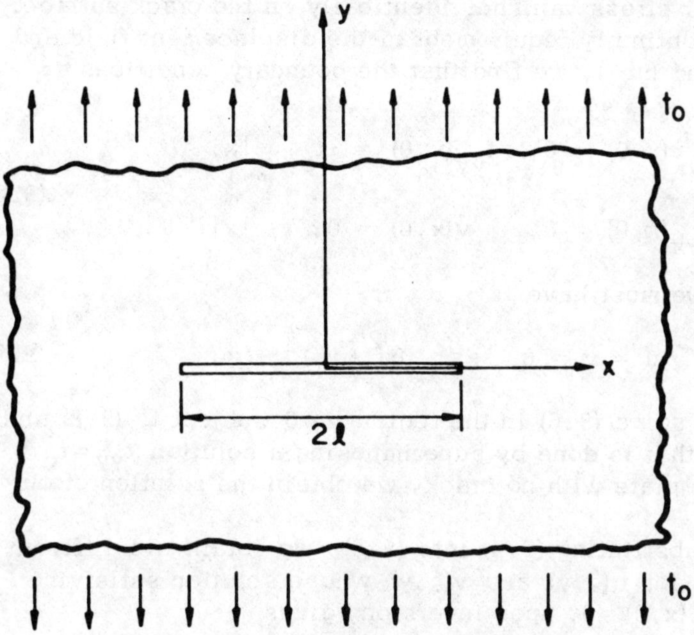

Figure 5. Sharp crack subject to tension.

Employing this in Hooke's law

$$\sigma_{k\ell} = \lambda u_{r,r}\delta_{k\ell} + \mu(u_{k,\ell} + u_{\ell,k}), \tag{9.4}$$

we find that

$$[\underset{\sim}{\sigma}_{2\ell}(x,0)]\underset{\sim}{} = 0, \quad |x| > \ell. \tag{9.5}$$

Hence the limits $(-\ell,\ell)$ in the line integral in (9.1) can be replaced by $(-\infty,\infty)$.

By the use of double Fourier transform, one can show that (9.5) is also valid for $|x| < \ell$. By the use of the Fourier transform of (9.2) with respect to x (indicated by a superimposed bar), we then find that the solution of (9.2) is equivalent to that of

$$-i\xi\bar{\sigma}_{1\ell} + \frac{d\bar{\sigma}_{2\ell}}{dy} = 0, \quad \ell = 1, 2 \tag{9.6}$$

$$[\underset{\sim}{\sigma}_{2\ell}(x,0)]\underset{\sim}{} = 0. \tag{9.7}$$

Since $\sigma_{21}(x,-y) = -\sigma_{21}(x,y)$, we also have $t_{21}(x,0) = 0$ for $|x| < \ell$. Thus the shear stress vanishes identically on the crack surface. Through the continuity requirement of the displacement field and using (9.5) and (9.7), we find that the boundary conditions at $y = 0$ are

$$\begin{aligned} \sigma_{yx}(x,0) &= 0, \quad t_{yy}(x,0) = -t_0, \quad |x| < \ell \\ \sigma_{yx}(x,0) &= 0, \quad v(x,0) = 0, \quad |x| > \ell. \end{aligned} \tag{9.8}$$

In addition, we must have

$$u = v = 0 \quad \text{as} \quad (x^2 + y^2)^{\frac{1}{2}} \to \infty. \tag{9.9}$$

Thus we must solve (9.6) in the region $y \geqq 0$ subject to (9.8) and (9.9). Once this is done by superimposing a solution $t_{yy} = t_0 =$ const. for the plate with no crack, we obtain the solution of our problem.

Upon substituting (9.4) into (9.6), we obtain two differential equations for $\bar{u}(\xi, y)$ and $\bar{v}(\xi, y)$ whose solution satisfying (9.9) and $\sigma_{yx}(x,0) = 0$, upon inversion, gives

$$\begin{aligned} u(x,y) &= -(2/\pi)^{\frac{1}{2}} \int_0^\infty A(\xi)(1 - 2\nu - \xi y) e^{-\xi y} \sin(\xi x) d\xi \\ v(x,y) &= (2/\pi)^{\frac{1}{2}} \int_0^\infty A(\xi)(2 - 2\nu + \xi y) e^{-\xi y} \cos(\xi x) d\xi \end{aligned} \tag{9.10}$$

where we wrote $\lambda = 2\mu\nu/(1-2\nu)$ which introduces Poisson's ratio ν. The unknown function $A(\xi)$ is to be determined by using the remaining two boundary conditions in (9.8). To this end, we calculate $\sigma_{yy}(x,y)$ from (9.4) and (9.10)

$$\begin{aligned} \sigma_{yy}(x,y) &= \sigma_{yy}(x,-y) = \\ &= -2\mu(2/\pi)^{\frac{1}{2}} \int_0^\infty \xi A(\xi)(1 + \xi y) e^{-\xi y} \cos(\xi x) d\xi . \end{aligned} \tag{9.11}$$

The stress $t_{k\ell}$ now follows from (8.2) in two-dimensions. Here we calculate only t_{yy}. The other components of the stress tensor can be evaluated in the same way

$$t_{yy}(x,y) = \int_0^\infty dy' \int_{-\infty}^\infty \sigma_{yy}(x',y') \Big[\alpha(|x'-x|, |y'-y|) + \alpha(|x'-x|, |y'+y|)\Big] dx' \tag{9.12}$$

where for α we employ

$$\alpha = \frac{1}{\pi}\left(\frac{k}{a}\right)^2 \exp\left\{-\left(\frac{k}{a}\right)^2 \left[(x'-x)^2 + (y'-y)^2\right]\right\}. \tag{9.13}$$

Substituting this and (9.11) into (9.12) after some lengthy manipulations to perform the integrations, we obtain

$$\begin{aligned} t_{yy}(x,y) = -\mu\left(\frac{2}{\pi}\right)^{\frac{1}{2}} \int_0^\infty \xi A(\xi)\cos(\xi x) \Big\{ & 2\xi(\pi p)^{-\frac{1}{2}} \exp\left(-py^2 - \frac{\xi^2}{4p}\right) \\ & + \left[1 - \frac{\xi}{2p}(\xi - 2py)\right] \exp(-\xi y)\,\mathrm{erfc}\left[(\xi - 2py)/2\sqrt{p}\right] \\ & + \left[1 - \frac{\xi}{2p}(\xi + 2py)\right] \exp(\xi y)\,\mathrm{erfc}\left[(\xi + 2py)/2\sqrt{p}\right]\Big\} d\xi \end{aligned} \tag{9.14}$$

where $p \equiv (k/a)^2$ and erfc(z) is the error function complementary defined by

$$\mathrm{erfc}(z) = 1 - 2\pi^{-\frac{1}{2}} \int_0^z \exp(-t^2)dt. \tag{9.15}$$

The boundary conditions $v(x,0) = 0$, $|x| > \ell$ and $t_{yy}(x,0) = -t_o$ for $|x| < \ell$ now take the forms

$$\begin{aligned} &\int_0^\infty A(\xi)\cos(\xi x)d\xi = 0, && |x| > \ell \\ &(2/\pi)^{\frac{1}{2}} \int_0^\infty \xi A(\xi) K(\xi/2\sqrt{p})\cos(\xi x)d\xi = t_o/2\mu, && |x| < \ell \end{aligned} \tag{9.16}$$

where

$$K(\xi/2\sqrt{p}) \equiv \xi(\pi p)^{-\frac{1}{2}} \exp(-\xi^2/4p) + \left[1 - (\xi^2/2p)\right] \mathrm{erfc}(\xi/2\sqrt{p}). \tag{9.17}$$

Dual integral equations (9.16) will have to be solved to determine $A(\xi)$. The exact solution of these equations is unknown. However, approximate solutions can be effected to a desirable accuracy for the small values of a/ℓ (the ratio of lattice parameter to half crack length). To this end we set

$$
\begin{gathered}
x/\ell = z, \quad \xi\ell = \zeta, \quad A(\xi) = \zeta^{-\frac{1}{2}} C(\zeta) \\
T_o = t_o \ell^2 / 2\mu, \qquad \epsilon = a/2k\ell \qquad (9.18) \\
K(\epsilon\zeta) = 1 + k(\epsilon\zeta) = 2\pi^{-\frac{1}{2}} \epsilon\zeta \exp(-\epsilon^2\zeta^2) + (1 - 2\epsilon^2\zeta^2)\,\mathrm{erfc}(\epsilon\zeta)
\end{gathered}
$$

and recall the expression

$$
\cos(\zeta z) = (\pi\zeta z/2)^{\frac{1}{2}} J_{-\frac{1}{2}}(\zeta z) \qquad (9.19)
$$

where $J_\nu(z)$ is the Bessels function of order ν. With these, (9.16) takes the form

$$
\begin{aligned}
&\int_0^\infty C(\zeta) J_{-\frac{1}{2}}(\zeta z)\, d\xi = 0, && z > 1 \\
&\int_0^\infty \zeta C(\zeta)[1 + k(\epsilon\zeta)] J_{-\frac{1}{2}}(\zeta z)\, d\zeta = T_o z^{-\frac{1}{2}}, && z < 1.
\end{aligned} \qquad (9.20)
$$

It is now possible to reduce the problem to the solution of a Fredholm equation (cf. Sneddon [20], Section 4.6).

$$
h(x) + \int_0^1 h(u) L(x,u)\, du = \tfrac{1}{2} (\pi x)^{\frac{1}{2}} T_o \qquad (9.21)
$$

for the function $h(x)$ where

$$
L(x,u) = (xu)^{\frac{1}{2}} \int_0^\infty t k(\epsilon t) J_o(xt) J_o(ut)\, dt. \qquad (9.22)
$$

Once (9.21) is solved then $C(\zeta)$ is calculated by

$$
C(\zeta) = (2\zeta)^{\frac{1}{2}} \int_0^1 x^{\frac{1}{2}} J_o(\zeta x) h(x)\, dx . \qquad (9.23)
$$

Of course, numerical solution may be found to (9.21). However, we observe that for $\epsilon = 0$ the system (9.20) reduces to dual integral equations obtained in classical elasticity. For this case

$k(0) = 0$ and the classical result follows from (9.21) and (9.23)

$$C_o(\zeta) = (\pi/2)^{\frac{1}{2}} T_o \zeta^{-\frac{1}{2}} J_1(\zeta). \tag{9.24}$$

Note that $\epsilon = a/2k\ell$ is a small number. For example, for a crack length of 100 atomic distances (an extremely small submicroscopic length!) $\epsilon \cong 1/100$. In such cases $k(\epsilon\zeta)$ is very small as compared to unity and (9.24) suffices to calculate the stress field. Of course, calculations by use of (9.24) lead to a hoop stress at $y = 0$, for all $z = x/\ell$

$$t_{yy}/t_o = -\int_0^\infty [1 + k(\epsilon\zeta)] J_1(\zeta)\cos(\zeta z)d\zeta. \tag{9.25}$$

This integral for $0 < z < 1$ gives $t_{yy}/t_o = -1$ since in this interval it converges even for $\epsilon = 0$. For $z > 1$ the integral converges for all $\epsilon > 0$ and it is permissible to ignore $k(\epsilon\zeta)$ as compared to unity. However, for $z = 1$ this is no longer the case and we cannot ignore $k(\epsilon\zeta)$ as compared to unity.

Numerical calculations of t_{yy} given by (9.25) were carried out for $\epsilon = 1/20$, $1/50$, $1/100$ and $1/200$ and the hoop stress $\bar{t}_{yy} = t_o + t_{yy}$ are plotted in Figure 6. For a crack length of 20 atomic distances ($\epsilon \cong 1/20$), the results are not very good since $\bar{t}_{yy}$ does not vanish all along the crack length. However, for a crack length of 100 or more atomic distances, the surface of the crack is fully cleared of tractions. In fact, the relative error is less than 1 percent for $\epsilon \leqq 1/100$.

The stress concentration occurs at the crack tip and it is given by

$$\bar{t}_{yy}(\ell, 0)/t_o = c_1/\sqrt{\epsilon}, \quad \epsilon = a/2k\ell, \quad c_1 \cong 0.55 \tag{9.26}$$

where c_1 converges to about 0.55. We now make the following significant observations:

(i) The maximum stress occurs at the crack tip, and it is finite [Equation (9.26)].

(ii) The hoop stress at the crack tip becomes infinite as the atomic distance $a \to 0$. This is the classical continuum limit of square root singularity.

(iii) <u>When $\bar{t}_{yy}(\ell, 0) = t_c$</u> (= cohesive bond stress), <u>fracture will occur</u>. In this case (9.26) gives

$$t_o^2 \ell = C_G \tag{9.27}$$

where

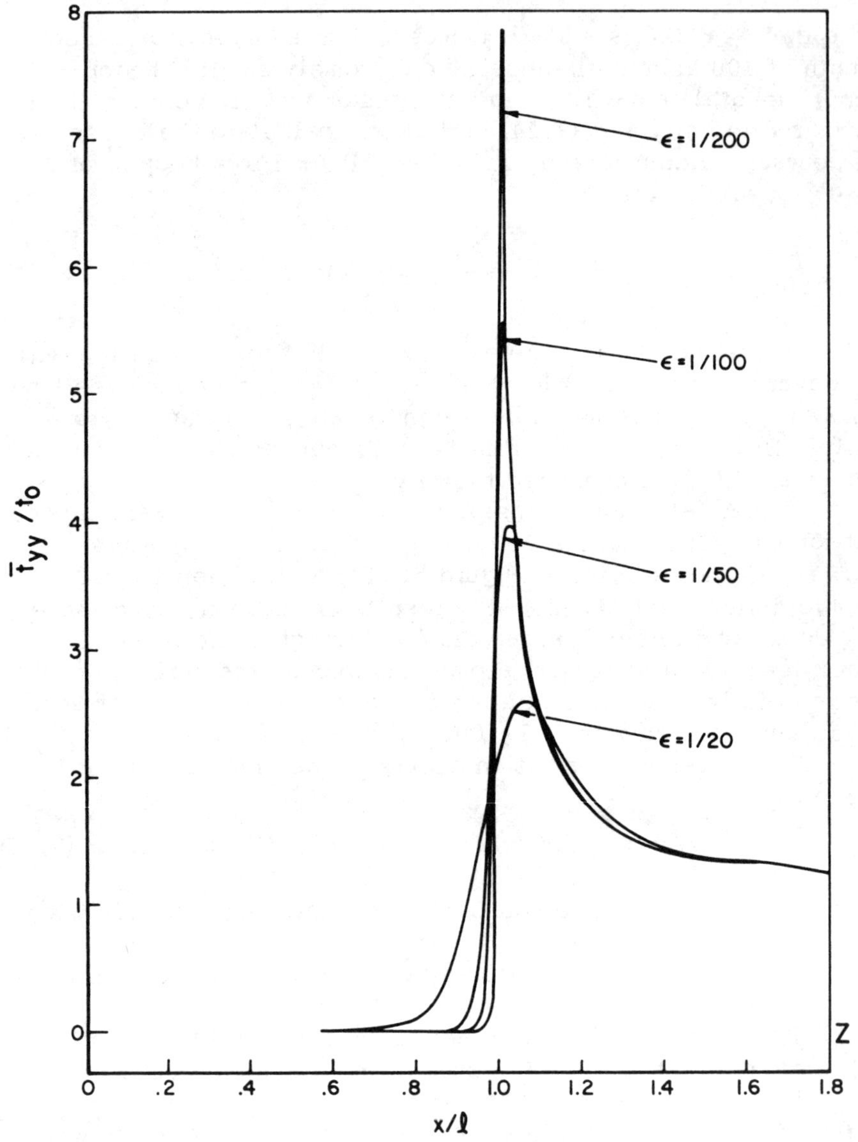

Figure 6. Hoop stress distribution along crack line.

$$C_G = (a/2kc_1^2)t_c^2. \tag{9.28}$$

Hence we have arrived at the Griffith criterion (9.1) by a direct method of equating the maximum stress to the stress holding atomic bonds together. This, of course, is not possible through the classical elasticity solution where $\bar{t}_{yy}(\ell, 0) = \infty$. Note that we have also determined the Griffith constant.

(iv) Equation (9.28) may be used to predict the cohesive bond stress for perfect crystals. To this end one may employ the Griffith definition of C_G as given by (9.1). Equating (9.28) to $(9.1)_2$, we obtain

$$t_c^2 a = k_c \gamma, \quad k_c = 8\mu c_1^2 k/\pi(1-\nu). \tag{9.29}$$

Employing the known values of γ and the elastic constants for steel

$$\begin{aligned} \gamma &= 1975 \text{ CGS}, \quad \mu = 6.92 \times 10^{11} \text{ CGS} \\ \nu &= 0.291, \quad a = 2.48 \text{ A}^\circ, \quad k = 1.65, \end{aligned} \tag{9.30}$$

we find that

$$t_c/E = 0.176 \tag{9.31}$$

where E is the Young's modulus defined by $E = 2\mu(1+\nu)$. This result is in the right range as predicted by physicists and metallurgists based on other considerations. For example, Lawn and Wilshaw [21] give

$$t_c/E = 0.18.$$

10. NONLOCAL FLUID MECHANICS AND TURBULENCE

The balance laws of nonlocal fluid mechanics are as given in Section 2, with $\hat{\underset{\sim}{f}} = \underset{\sim}{0}$ and $\hat{\rho} = 0$. However, the constitutive equations must now start with

$$\psi(\underset{\sim}{x}, t) = \Psi[\underset{\sim}{v}(\underset{\sim}{x}'), \theta(\underset{\sim}{x}')], \quad \underset{\sim}{x}' \in \mathcal{V} \tag{10.1}$$

instead of (3.3), and similar equations written for η, $t_{k\ell}$, $\hat{h}$ and $\hat{b}$. The invariance under Galilean invariance (3.4), (10.1) reduces to

$$\psi = \Psi[\underset{\sim}{\nu}', \theta'] \tag{10.2}$$

where

$$\underset{\sim}{\nu}' = \underset{\sim}{\nu}(\underset{\sim}{x}') \equiv \underset{\sim}{v}(\underset{\sim}{x}') - \underset{\sim}{v}(\underset{\sim}{x}) \tag{10.3}$$

is the relative velocity field and Ψ is subject to

$$\Psi[\underset{\sim}{Q}\underset{\sim}{\nu}', \theta'] = \Psi(\underset{\sim}{\nu}', \theta'). \tag{10.4}$$

For simplicity, we consider non-heat conducting fluids so that θ is uniform throughout.

Employing the theorem of Friedmann and Katz, we have the representation

$$\psi(\underset{\sim}{x}, t) = \int_{\mathcal{V}} K[\underset{\sim}{\nu}(\underset{\sim}{x}'), \underset{\sim}{x}'; \theta] dv(\underset{\sim}{x}'). \tag{10.5}$$

We again consider only central attractions. The Galilean invariance requirement (10.4) applied to (10.5) now leads to

$$\psi(\underset{\sim}{x}, t) = \int_{\mathcal{V}} \Psi(I'_\alpha; \theta) dv(\underset{\sim}{x}') \tag{10.6}$$

where I_α are the invariants of $\underset{\sim}{\nu}'$ and $\underset{\sim}{\kappa}' \equiv \underset{\sim}{x}' - \underset{\sim}{x}$, i.e.,

$$I'_1 = \underset{\sim}{\kappa}' \cdot \underset{\sim}{\kappa}', \quad I'_2 = \underset{\sim}{\kappa}' \cdot \underset{\sim}{\nu}', \quad I'_3 = \underset{\sim}{\nu}' \cdot \underset{\sim}{\nu}'. \tag{10.7}$$

A similar representation for $t_{k\ell}$ reads

$$\underset{\sim}{t} = \int_{\mathcal{V}} \left[T_0 \underset{\sim}{I} + T_1 \underset{\sim}{\kappa}' \otimes \underset{\sim}{\kappa}' + T_2 (\underset{\sim}{\kappa}' \otimes \underset{\sim}{\nu}' + \underset{\sim}{\nu}' \otimes \underset{\sim}{\kappa}') + T_3 \underset{\sim}{\nu}' \otimes \underset{\sim}{\nu}' \right] dv' \tag{10.8}$$

where T_α are functions of the invariants (10.7) and θ, and

$$(\underset{\sim}{a} \otimes \underset{\sim}{b})_{k\ell} \equiv a_k b_\ell . \tag{10.9}$$

If the stress is to reduce to a hydrostatic pressure when $\underset{\sim}{\nu}' = \underset{\sim}{0}$, then

$$p = -\int_{\mathcal{V}} T_0 dv', \quad T_1 = 0 \quad \text{when } \underset{\sim}{\nu}' = \underset{\sim}{0}. \tag{10.10}$$

If the incompressibility is defined by $I'_1 = \text{const.}$ but $\underset{\sim}{\nu}' \neq \underset{\sim}{0}$, then $I'_2 = 0$ and T_α for the incompressible fluids will depend on $I'_3 = \underset{\sim}{\nu}' \cdot \underset{\sim}{\nu}'$ only.

By using the power series expressions of κ' about $\underset{\sim}{X}$ and letting $\underset{\sim}{X} \to \underset{\sim}{x}$ (in view of the fact that every configuration of the body leaving density unchanged must be a reference frame for fluids), one can show that the independent variables $\underset{\sim}{\kappa}'$ and $\underset{\sim}{\nu}'$ may also be replaced by

$$d_{k\ell} \equiv \tfrac{1}{2}(v_{k,\ell} + v_{\ell,k}), \quad d_{k\ell}(\underset{\sim}{x}') = \tfrac{1}{2}(v'_{k,\ell} + v'_{\ell,k}) \quad (10.11)$$

$$\beta_k(\underset{\sim}{x}') = \tfrac{1}{2}(x'_m - x_m)\left[v_{m,k}(\underset{\sim}{x}') - v_{m,k}(\underset{\sim}{x})\right] + v_k(\underset{\sim}{x}') - v_k(\underset{\sim}{x}) \quad (10.12)$$

as originally arrived at by Eringen [22]. Of these $d_{k\ell}$ is the classical deformation rate tensor, but the vector $\beta_k(\underset{\sim}{x}')$ crucial to turbulence is new.

To simplify the matter, we consider incompressible fluids and employ a very simple special form for $t_{k\ell}$ suitable for the study of generation of the secondary flow.

$$t_{k\ell} = -p\delta_{k\ell} + 2\mu d_{k\ell} + \int_V \mu'(|\underset{\sim}{x}' - \underset{\sim}{x}|)\Big\{(x'_k - x_k)\left[d_{\ell m}(\underset{\sim}{x}') - d_{\ell m}(\underset{\sim}{x})\right]\beta_m(\underset{\sim}{x}') + (x'_\ell - x_\ell)\left[d_{km}(\underset{\sim}{x}') - d_{km}(\underset{\sim}{x})\right]\beta_m(\underset{\sim}{x}')\Big\}\, dv(\underset{\sim}{x}') \quad (10.13)$$

where $\mu'(|\underset{\sim}{x}'-\underset{\sim}{x}|)$ is the nonlocal attenuation function. Here we have separated the local effects (the terms outside the integral) from the nonlocal effects. Note that (10.13) does not violate the second law of thermodynamics

$$\int_V \frac{1}{\theta}(t_{k\ell} + p\delta_{k\ell}) d_{k\ell}\, dv \geqq 0 \quad (10.14)$$

for all possible motions.

Employing (10.13) and the equations of motion

$$t_{k\ell,k} - \rho\dot{v}_\ell = 0, \quad v_{k,k} = 0 \quad (10.15)$$

for a constant pressure gradient $p_{,k} = -G\delta_{k3}$ in the axial direction of a long cylinder of rectangular cross section (Figure 7), we have determined the velocity field

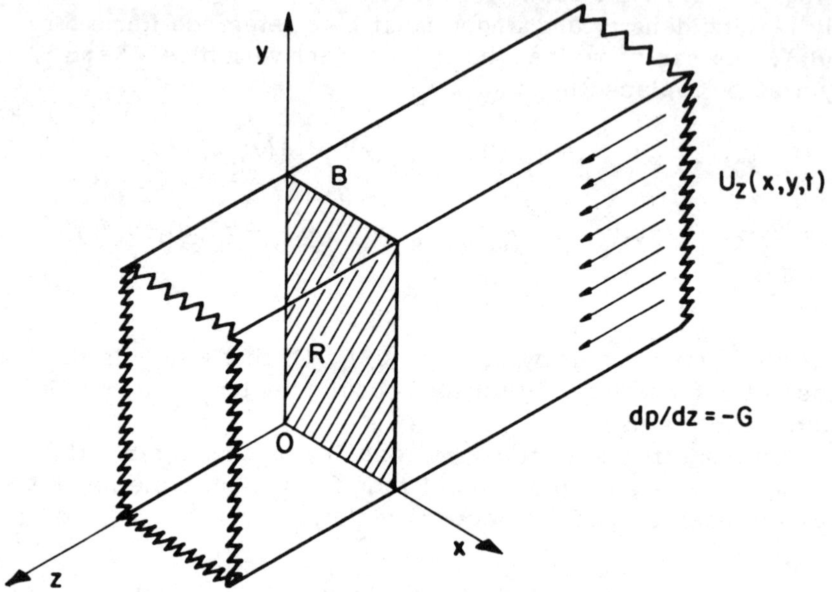

Figure 7. Flow of fluid in rectangular pipe under impulsively applied pressure gradient—G.

$$\underset{\sim}{v} = \left\{ u(x,y,t), \quad v(x,y,t), \quad w(x,y,t) \right\} \tag{10.16}$$

by solving the nonlinear integro-partial differential equations resulting from (10.13) and (10.15). For the nonlocal modulus μ', we chose

$$\mu'(|\underset{\sim}{x}'-\underset{\sim}{x}|) = \mu_o \exp(-k_o|\underset{\sim}{x}'-\underset{\sim}{x}|) \tag{10.17}$$

where μ_o and k_o are constants. In general, k_o can be chosen so that μ' attenuates properly over a characteristic length of the problem (e.g., eddy size, mixing length, etc.). It is possible, in fact, to relate k_o to a universal length which produces proper limits in the two extremes of the turbulence (separation and final stages). In a future paper such results will be discussed in full detail together with applications to the shear flow problem. Here we give some results just to demonstrate the power and potential of the nonlocal theory. The constant μ_o is a normalization factor which is not important as far as mechanism of the secondary flow is concerned.

By means of a finite difference technique, computer calculations* were carried out for a 2×1 pipe containing water with the physical properties $\mu/\rho = 1.1 \times 10^{-5} \text{ft}^2/\text{sec}$, $\rho = 1.94$ slugs/ft^3. The pressure gradient $G = 0.002$ lb/ft^3 which yielded a Reynolds number $R_e = 2500$ which is in the transition region to turbulence. For flows with velocities 100 m/sec., the smallest eddies are of the order of 2mm (cf. Hinze [23]). In order for μ' to attenuate to 1 percent of its maximum value within this characteristic length, we take $k_o = 1000$ ft^{-1}. μ_o can be chosen so that the secondary flow has the correct size relative to the main flow ($\mu_o/\rho \cong 10^6 \text{ft}^{-2}$).

On Figures 8 to 14 results of these calculations are plotted. Small eddies form at the corners gradually diffusing to the interior of the pipe. The steady state solution (Figures 13 and 14) is in remarkably good agreement with experimental results (Figures 15 and 16). Unfortunately, experimental work is not available in the transition region $R_e = 2500$. Available profiles are for $R_e = 50{,}000$ and they are relatively flat at the pipe wall. Nevertheless, profiles compare favorably. Later calculations using $\mu_o/\rho = 10^5$ and 10^7ft^{-2} produced some changes in the magnitude of the secondary flow, but the streamline patterns were not affected. The change in the value of k_o to $k_o = 500$ ft^{-1} gave again similar streamlines although larger and faster secondary flow. Thus it can be concluded that the streamlines are not very sensitive to the choice of μ_o and k_o. This is reasonable since the level of turbulence and the streamlines of steady secondary flow are known not to depend on the Reynolds number (cf. Gessner and Jones [24]). Considering the fact that Navier-Stokes theory predicts no secondary flow, the secret for the mechanism seems to lie in the nonlocal effects.

Turbulence has been speculated to be a nonlocal phenomena by many research workers in this field. The fact that various available approximate theories (e.g., mixing length, Reynold stresses) one way or another make use (in highly crude forms even often violating basic principles, such as objectivity) of the nonlocal effects and produce better curve fitting is an indication of the nonlocality. Nevertheless, until the introduction of nonlocal fluid mechanics by us [22], there was no rational approach incorporating nonlocality to fluid mechanics excluding, of course, polar theories which represent "extremely short nonlocality."

---

* Computer calculations were carried out several years ago on the basis of (10.13) rather than (10.8) since the present approach was not available then. I am indebted to Mr. C. Speziale for the calculations.

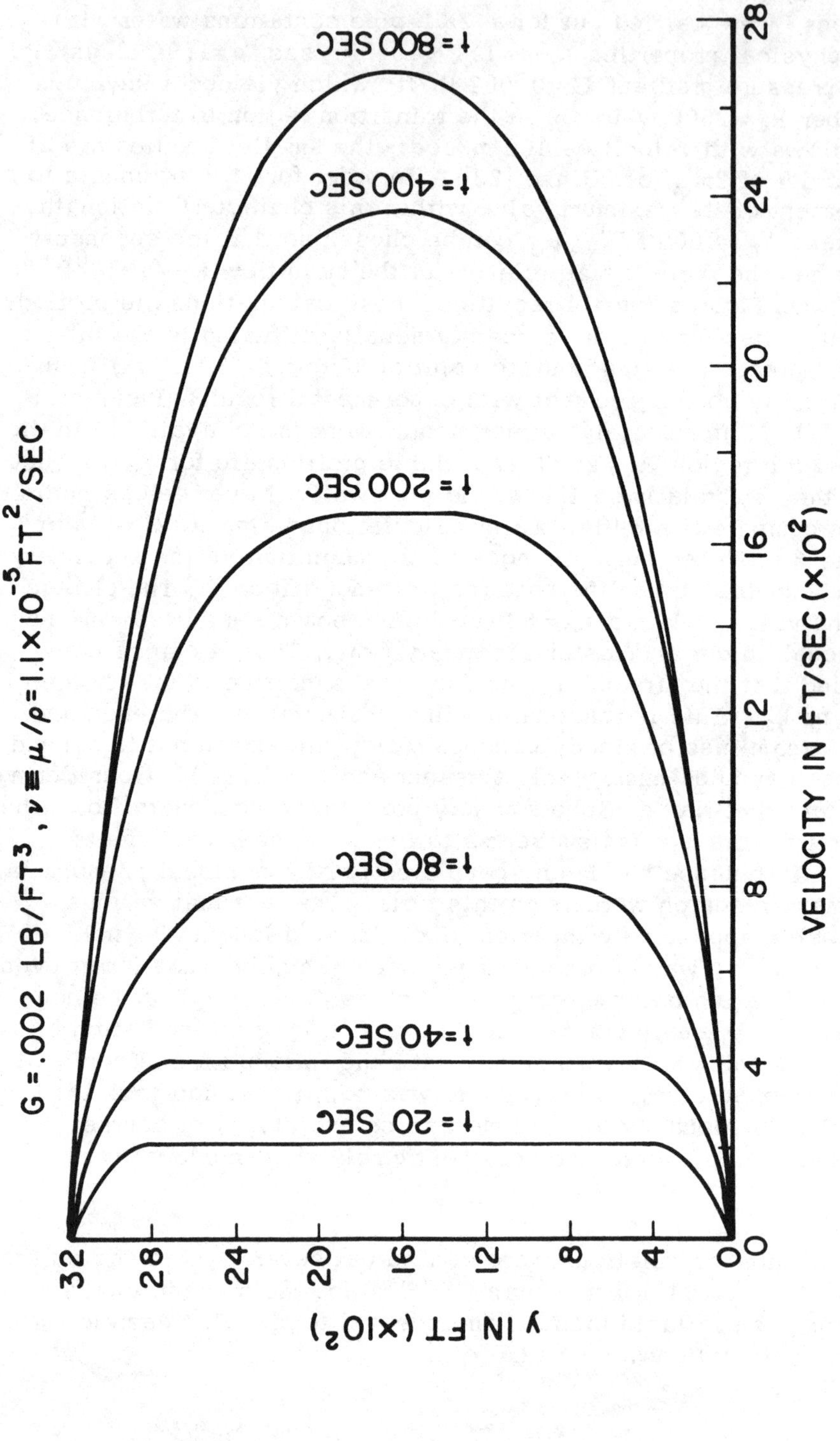

Figure 8. Velocity profiles along centerline of 1×2 rectangular pipe.

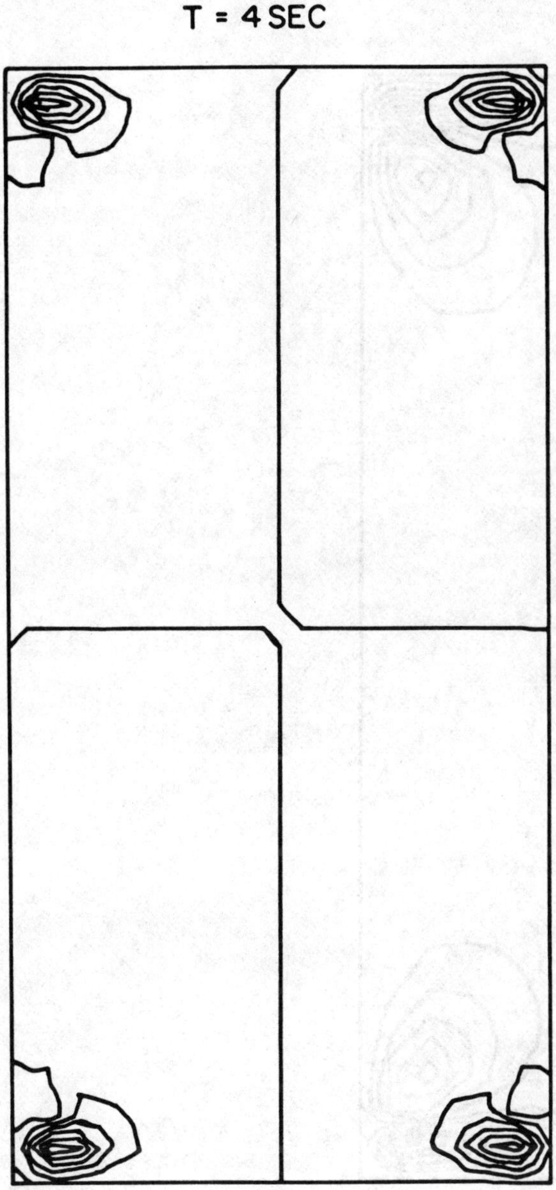

Figure 9. Development of streamlines of the secondary flow in time.

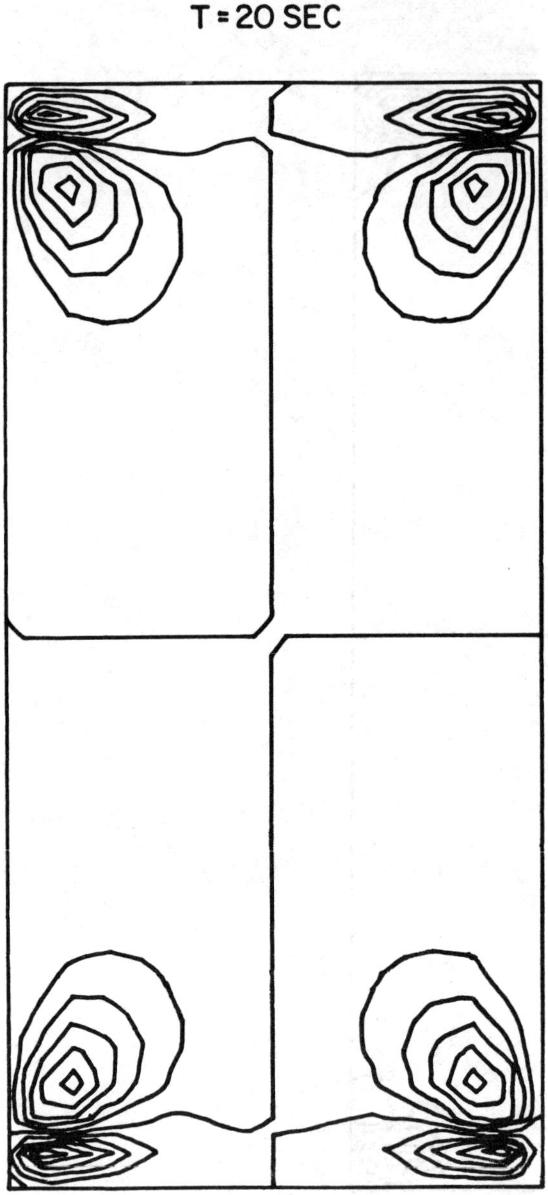

Figure 10. Development of streamlines of the secondary flow in time.

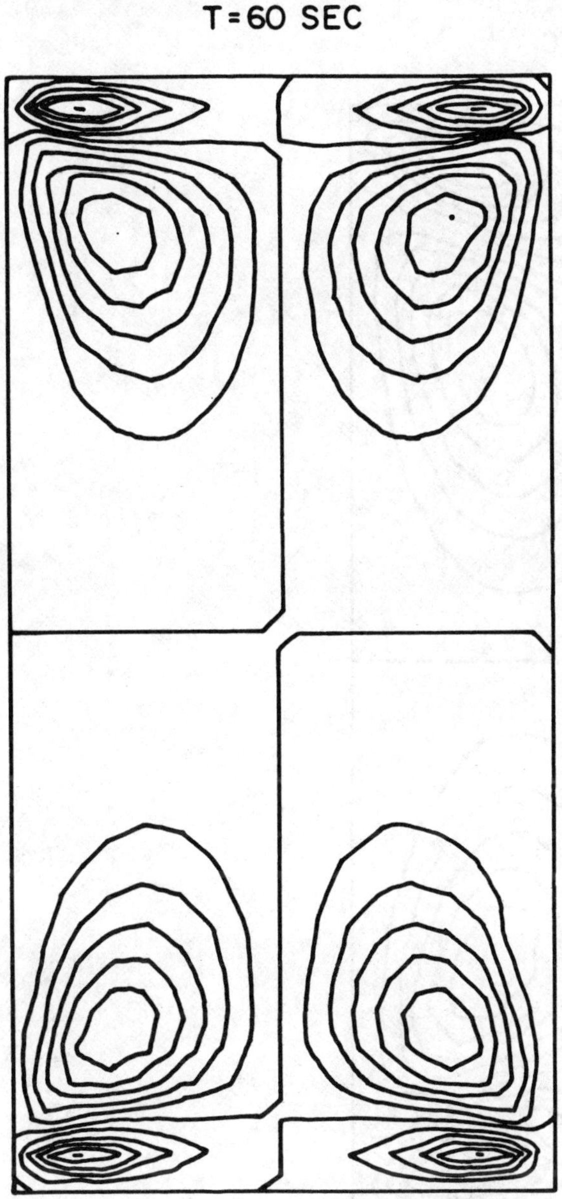

Figure 11. Development of streamlines of the secondary flow in time.

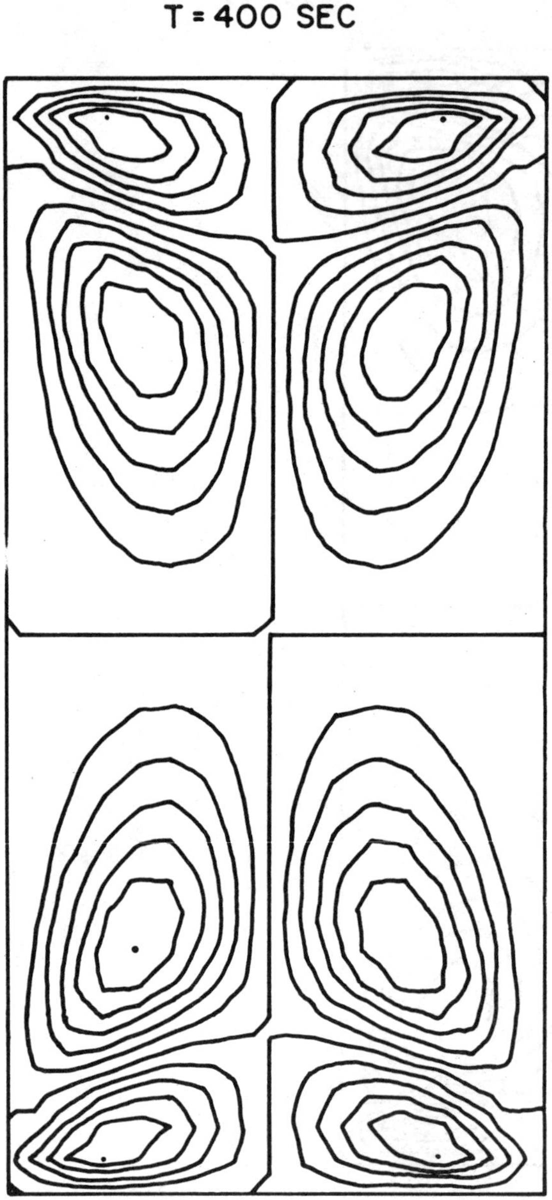

Figure 12. Development of streamlines of the secondary flow in time.

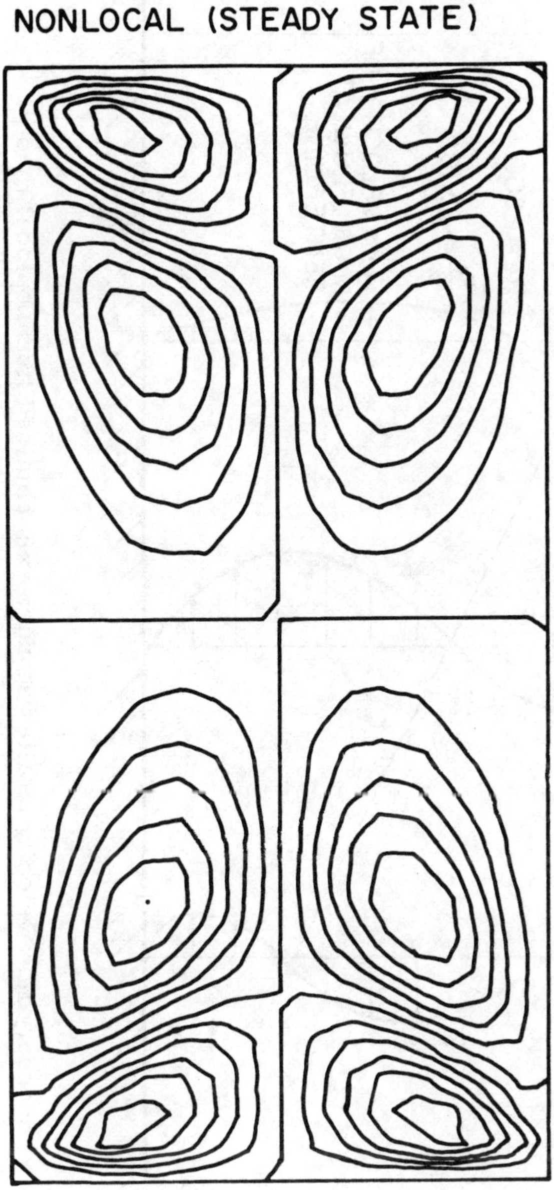

Figure 13. Steady streamlines of the secondary flow.

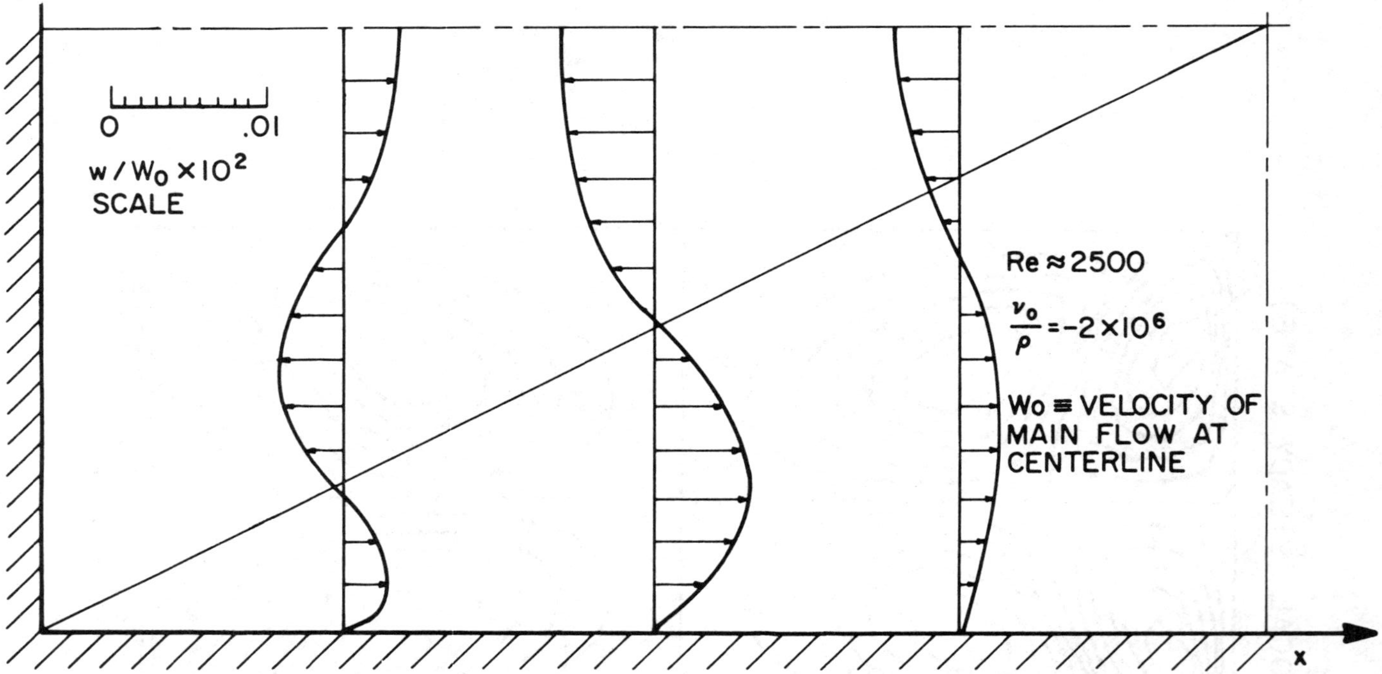

Figure 14. Secondary flow velocity profiles along x-axis obtained from nonlocal theory.

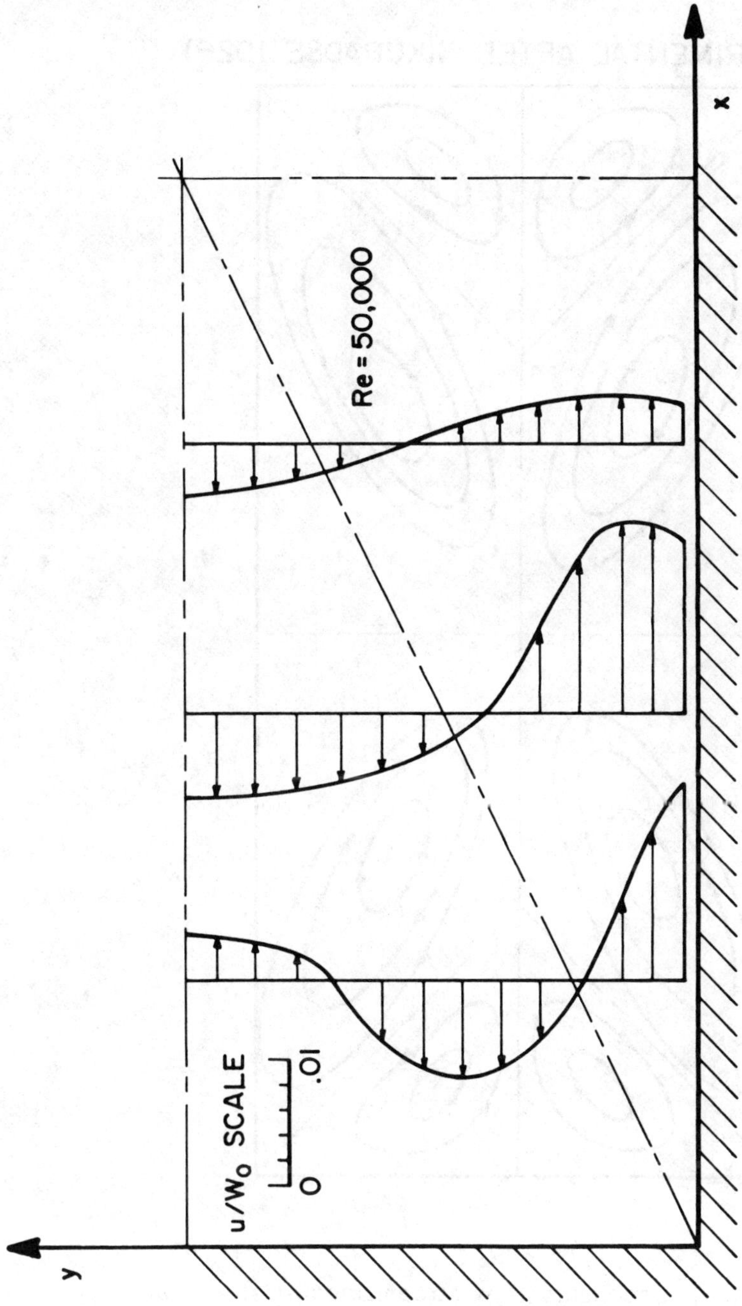

Figure 15. Secondary flow velocities (experiments of Gessner and Jones [24]).

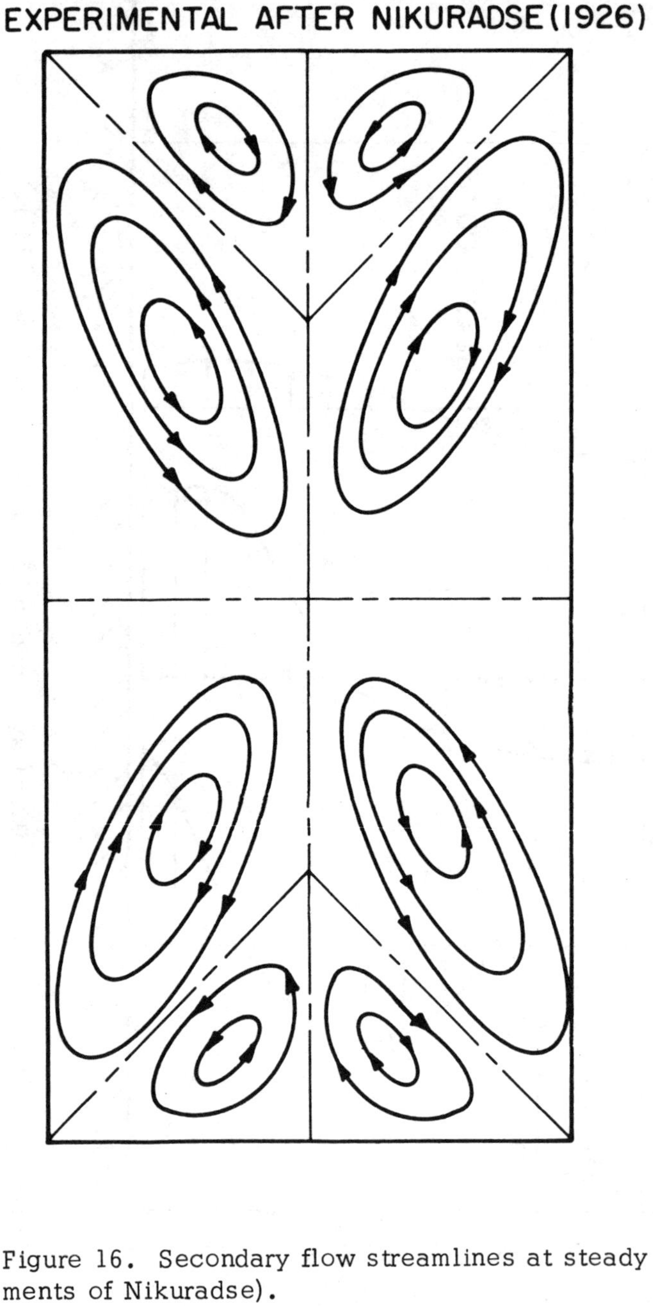

Figure 16. Secondary flow streamlines at steady state (experiments of Nikuradse).

ACKNOWLEDGEMENT

The research contained here was partially supported by the Army Research Office at Durham and the Office of Naval Research. The author is indebted to Dr. F. Balta for proofreading.

REFERENCES

1. A. A. Griffith, The Phenomena of Rupture and Flow in Solids, Phil. Trans. Roy. Soc. (London) A221, 163-198 (1920).
2. A. C. Eringen, Continuum Physics, Acad. Press 4, 274 (1976).
3. A. C. Eringen and D. G. B. Edelen, On Nonlocal Elasticity, Int. J. Engng. Sci. 10, 233-248 (1972).
4. A. C. Eringen, Mechanics of Continua, John Wiley & Sons (1967).
5. A. C. Eringen, Unified Theory of Thermomechanical Materials, Int. J. Engng. Sci. 41, 179-202 (1966).
6. N. Friedmann and M. Katz, A Representation Theorem for Additive Functions, Arch. Ration. Mech. Anal., 49-57 (1966).
7. A. C. Eringen, Linear Theory of Nonlocal Elasticity and Dispersion of Plane Waves, Int. J. Engng. Sci. 10, 425-435 (1972).
8. J. Krumhansl, Generalized Continuum Field Representations for Lattice Vibrations," Lattice Dynamics, edited by P. F. Wallis, Pergamon Press, Oxford (1964).
9. I. A. Kunin, Theories of Elastic Continua with Microstructure, Proceedings of Vibration Problems (Warsaw) 3 9, 323-336 (1968).
10. A. C. Eringen and E. S. Suhubi, Elastodynamics 2, 521 (1975).
11. A. A. Maraddudin, E. W. Montroll, G. H. Weiss, and I. P. Ipatova, Theory of Lattice Dynamics in Harmonic Approximation, Second Edition, Acad. Press, 531 (1971).
12. A. C. Eringen, Screw Dislocation in Nonlocal Elasticity, J. of Phys. D: Applied Physics 10, 671-678 (1977).
13. A. C. Eringen, Edge Dislocation in Nonlocal Elasticity, Int. J. Engng. Sci. 15, 177-183 (1977).
14. A. C. Eringen, C. G. Speziale, and B. S. Kim, Crack Tip Problem in Nonlocal Elasticity, J. Mech. and Phys. of Solids 25, 339-355 (1977).

15. A. Kelly, Strong Solids, Oxford (1966).
16. G. E. Barenblatt, Stresses in a Plate Due to the Presence of Cracks and Sharp Corners, Proc. Inst. Naval Architects (1962).
17. Iu. P. Zheltov and S. A. Khristianovich, Izv. Acad. Nauk., SSSR Otd Tech. Nauk. 5, 3-41 (1955).
18. D. S. Dougdale, Yielding of Steel Sheets Containing Slits, J. Mech. Phys. of Solids 8, 100-104 (1960).
19. J. N. Goodier and M. Kanninen, Crack Propagation in a Continuum Model with Nonlinear Atomic Separation Laws, Tech. Rep. 165 (ONR), Div. of Engineering Mechanics, Stanford University (1966).
20. I. N. Sneddon, Mixed Boundary Value Problems in Potential Theory, North Holland Publishing Company, Amsterdam (1966).
21. B. R. Lawn and T. R. Wilshaw, Fracture of Brittle Solids, Table 7.1, page 160, Cambridge University Press (London) (1975).
22. A. C. Eringen, On Nonlocal Fluid Mechanics, Int. J. Engng. Sci. 10, 561-575 (1972).
23. J. O. Hinze, Turbulence (Second Edition), McGraw Hill, New York (1975).
24. F. B. Gessner and J. B. Jones, On Some Aspects of Fully Developed Turbulent Flow in Rectangular Channels, J. Fluid Mech. 23(4), 689 (1965).

PART IV

NONLINEAR FIELD THEORIES AND QUANTIZATION

QUANTIZATION OF A NONLINEAR FIELD EQUATION†

L. Castell

Max-Planck-Institut
D-813 Starnberg, Germany

The physical interpretation of the manifold of solutions of a classical nonlinear field equation is so far an unsolved problem of elementary particle physics. For some equations [1] ($\lambda\varphi^4$ theory), it has been proved that the solutions develop no singularities for sufficiently smooth initial data. However, it is an open question if these solutions form an infinite dimensional Riemannian space, whose tangent space is an quantum mechanical Hilbert space. Once these questions are solved one can hope to understand the quantum theory of a nonlinear field equation much better. (N.B. Each interacting, fully relativistic field theory is nonlinear.)

If a relativistic field theory has a bigger symmetry group than the Poincaré group, some of the above questions can be more easily approached. Classical solutions can be given explicitly [2]. The most important group, which contains the Poincaré group as a subgroup, is the conformal group $SO_0(4,2)/C_2$. Usually it is easier to consider field theories not in four space-time dimensions, but in two- or one-dimension (with the corresponding symmetry group $SO_0(2,2)/C_2$ or $SO_0(2,1)$), and such a theory we should like to consider in greater detail.

Scalar field theories which have besides the kinematical invariance also dilatational invariance of their action integrals $\int L\,dx$ are determined by unique field equations, which we shall list now.

---

† Lecture given at NATO Advanced Study Institute on Nonlinear Equations in Physics and Mathematics, Istanbul, August 1977.

A. O. Barut (ed.), Nonlinear Equations in Physics and Mathematics, 321–333.

Field Equations	Real Solutions
(1) $\partial_t\partial_t\varphi - (\lambda/\varphi^3) = 0,$	$\varphi = \pm\lambda^{\frac{1}{4}} a\sqrt{(t-t_0)^2 + (1/a^4)},$
(2) $(\partial_x\partial_x - \partial_t\partial_t)\varphi = 0,$	$\varphi = f(x-t) + g(x+t)$
(3) $(\partial_x\partial_x + \partial_y\partial_y - \partial_t\partial_t)\varphi - \lambda\varphi^5 = 0,$	$\varphi = \lambda^{-\frac{1}{4}}\varphi(x,y,t),$
(4) $(\partial_x\partial_x + \partial_y\partial_y + \partial_z\partial_z - \partial_t\partial_t)\varphi - \lambda\varphi^3 = 0,$	$\varphi = \frac{a(\xi,\xi^*)}{2\sqrt{\lambda}}\left[\frac{(\xi-\xi)^2}{(y-\xi)^2(y-\xi^*)^2}\right]^{\frac{1}{2}} \cdot \mathrm{cn}\left[(1+a^2)^{\frac{1}{2}}(\alpha-\alpha_0),\, k\right]$

$$\alpha = \frac{1}{2i}\,\ell n\frac{(y-\xi)^2}{(y-\xi^*)^2}, \qquad k^2 = \frac{1}{2(1+1/a^2)},$$

ξ complex, Im ξ in the forward cone, cn elliptical cosine.

We have chosen $\lambda > 0$ in order to have a theory with positive energy. The last eight-parameter manifold of solutions is generated from the SO(4)-invariant solutions by all symmetry transformations of Equation (4).

Spinor equations with dilatational symmetry are, for example, the massless Thirring model in two dimensions and the massless Q.E.D. in four dimensions.

There may not be a continuous limit $\lambda \to 0$ between the free (linear) theory $\lambda = 0$ and the interacting theory $\lambda > 0$. One obtains instead a pseudo-free theory. This is the so-called Klauder phenomenon, and we shall exhibit it explicitly for Equation (1). So we first consider the classical theory and then quantize the field equation. It will turn out that the quantization of the nonlinear field equation depends crucially on the magnitude of the coupling constant λ.

Moreover the quantization is not unique. So the conjecture of W. Heisenberg that the nonlinearities of the equation determine the quantization uniquely is not fulfilled in this case.

1. THE CLASSICAL THEORY

Consider the equation $\ddot{\varphi} - \lambda\varphi^{-3} = 0$. The total energy T is given by $T = \frac{1}{2}(\dot{\varphi}^2 + \lambda/\varphi^2)$ which is positive definite for $\lambda > 0$. The solutions are given by

$$\varphi(t) = \pm\sqrt{2Tt^2 + 4Dt + 2K},$$

with $TK - D^2 = \lambda/4$, and

$$T = \tfrac{1}{2}(\dot{\varphi}(0)^2 + \lambda/\varphi(0)^2), \quad D(0) = \tfrac{1}{2}\varphi(0)\dot{\varphi}(0), \quad K(0) = \tfrac{1}{2}\varphi(0)^2.$$

The Lie algebra of the conserved quantities is the Lie algebra of $SO_0(2,1)$:

$$\{T,D\} = \frac{\partial T}{\partial \dot{\varphi}(0)}\frac{\partial D}{\partial \varphi(0)} - \frac{\partial T}{\partial \varphi(0)}\frac{\partial D}{\partial \dot{\varphi}(0)} = T,$$

$$\{KD\} = -K, \qquad \{T,K\} = 2D.$$

The coupling constant $\lambda/4$ determines the Casimir invariant of this algebra. For $\lambda \to 0$ we obtain the pseudo-free solution

$$\varphi = \pm|\dot{\varphi}(0)t + \varphi(0)|.$$

In the following figure we compare the different solutions with the same initial conditions at $t=0$, i.e., with the same $D(0)$ $K(0)$.

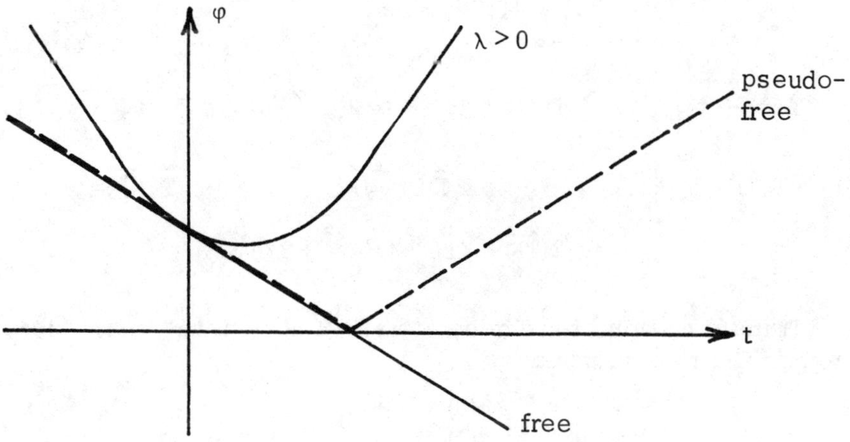

The field equation for the pseudo-free solutions is nonlinear

$$\ddot{\varphi} \mp 2\dot{\varphi}^2\delta(\varphi) = 0.$$

2. SYMMETRY TRANSFORMATIONS

Besides the translational symmetry (kinematical symmetry), there exist two further dynamical symmetry transformations.

$$\varphi'(t') = \varphi(t), \quad t' = t + \tau, \quad -\infty < \tau < +\infty, \quad T = \text{const.},$$

$$\varphi'(t') = 1/\sqrt{\rho}\ \varphi(t), \quad t' = t/\rho, \quad 0 < \rho < \infty, \quad D(t) - tT = D(0).$$

$$\varphi'(t') = \varphi(t)/(1-bt), \quad -\infty < b < +\infty, \quad K(t) - 2tD(t) + t^2T = K(0),$$

$$\text{or} \quad K(t) = K(0) + 2tD(0) + t^2T.$$

The symmetry group $SO_0(2,1)$ is the projective group of the line. The solution $\varphi(t)$ of the field equation is connected with the conserved quantities by

$$\varphi(t) = \pm\sqrt{2K(t)}.$$

3. QUANTUM MECHANICS OF THE FREE FIELD EQUATION

The generators of the symmetry for the free theory become linear operators on a dense subset of the Hilbert space

$$\int_{-\infty}^{+\infty} |f|^2 dt < \infty.$$

$$\hat{T} = -\frac{1}{i}\partial_t, \qquad \hat{D} = \frac{1}{i}(-\tfrac{1}{2} + t\partial_t) \qquad \hat{K} = \frac{1}{i}(t - t^2\partial_t),$$

$$[\hat{T},\hat{D}] = -i\hat{T}, \qquad [\hat{K},\hat{D}] = i\hat{K}, \qquad [\hat{T},\hat{K}] = -i2\hat{D},$$

$$C_{II} = 1/2(\hat{T}\hat{K} + \hat{K}\hat{T}) - \hat{D}^2 = +3/4.$$

Transformation of the generators in "momentum" space, i.e., action on the free solutions

$$\begin{pmatrix} \dot{\varphi}(0) \\ \varphi(0) \end{pmatrix}$$

is given by

$$\Delta(\hat{T}) = \frac{1}{i}\begin{pmatrix} 0 & 0 \\ -1 & 0 \end{pmatrix}, \quad \Delta(\hat{D}) = \frac{1}{i}\begin{pmatrix} \frac{1}{2} & 0 \\ 0 & -\frac{1}{2} \end{pmatrix}, \quad \Delta(\hat{K}) = \frac{1}{i}\begin{pmatrix} 0 & 1 \\ 0 & 0 \end{pmatrix}.$$

The two-dimensional Hilbert space has the norm: $\psi^*\sigma_2\psi$.

4. COMPACTIFICATION OF TIME

Consider the special projective transformation $t' = t/(1-bt)$, which is singular on $-\infty < t < +\infty$. The real line R is not a homogeneous space of the full symmetry group. However, if we introduce

$$t = \tan \alpha/2, \qquad \phi(\alpha) = \sqrt{1+\cos\alpha}\,\varphi(\tan\alpha/2),$$

we obtain as field equation

$$\ddot{\phi} + \phi/4 - \lambda/\phi^3 = 0.$$

The points at infinity, $t=\pm\infty$, correspond to $\alpha=\pm\pi$. The free solution, however, has a period of 4π:

$$\phi(\alpha) = \phi(0)\cos\alpha/2 + 2\dot{\phi}(0)\sin\alpha/2.$$

So for the free solution we cannot identify $-\pi$ and π, but -2π and 2π. Therefore, φ can be extended only to the double covering of the compactified line.

The interacting solution

$$\phi(\alpha) = \pm\sqrt{4H + 4\sqrt{H^2-\lambda/4}\,\cos(\alpha-\alpha_0)},$$

or

$$\phi(\alpha) = \pm\sqrt{4H - 2E_+e^{i\alpha} - 2E_-e^{-i\alpha}},$$

$$E_+ = E_-^*, \quad H = \frac{1}{2}\left(\dot{\phi}^2 + \frac{\phi^2}{4} + \frac{\lambda}{\phi^2}\right),$$

$$H^2 - E_+E_- = \frac{\lambda}{4}, \quad H = \frac{T+K}{2}, \quad E_\pm = \frac{T-K}{2} \pm iD,$$

has the period 2π.

For the pseudo-free solutions $\phi_{ps.free} = \pm|\phi_{free}|$, we obtain the equations

$$\ddot{\phi} + \frac{1}{4}\phi \mp 2\dot{\phi}^2\delta(\phi) = 0.$$

The next figure shows the function ϕ.

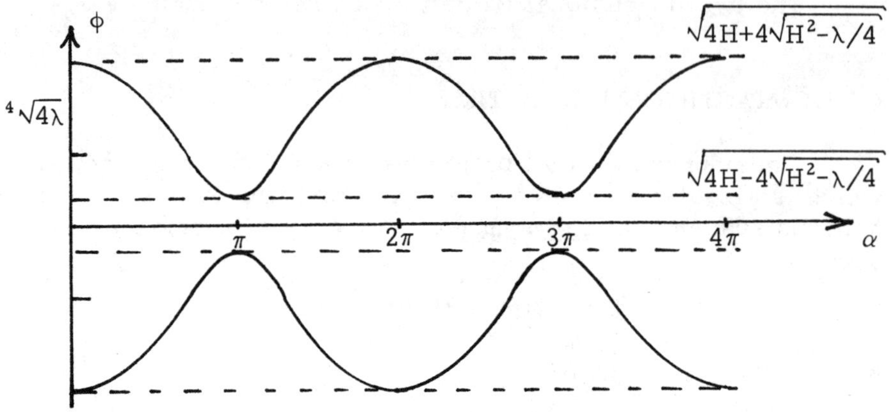

We obtain the minimal energy $H_{min} = \frac{1}{2}\sqrt{\lambda}$ for constant $\phi = \sqrt[4]{4\lambda}$.

5. PERTURBATION EXPANSION

It is interesting to compare the exact solution with the usual perturbation expansions. It is clear that one cannot start the perturbation series with ϕ_{free}, as the frequency 1/2 is not contained in the interacting solution, but only the frequencies 0, 0, 1, 2, 3 . . . . Thus the perturbation series for small λ leads immediately to singular equations and solutions.

$$\phi = a \cos \alpha/2 + \frac{2\lambda}{a^3} \frac{1}{\cos \alpha/2} + \cdots .$$

However, if we start the perturbation series with the constant <u>interacting</u> solutions, we obtain in first order

$$\phi = \sqrt[4]{4\lambda} + a \cos \alpha, \qquad |a| \ll \sqrt[4]{4\lambda},$$

which one usually interprets as a Goldstone excitation. In second order we obtain

$$\phi = \sqrt[4]{4\lambda} + a \cos \alpha + \frac{a^2}{3\sqrt[4]{4\lambda}} - \frac{a^2}{9\sqrt[4]{4\lambda}} \cos 2\alpha .$$

This exhibits a property of the perturbation series of a nonlinear equation: the higher order corrections change the lower frequencies' coefficients.

We have to compare this last expression with the expansion of the exact solution for large λ. We obtain

$$\phi = 2\sqrt{H} - \sqrt{H - \lambda/4H}\cos\alpha + \cdots .$$

Here the two coefficients $2\sqrt{H}$ and $\sqrt{H-\lambda/4H}$ are directly coupled!

Concerning the interpretation of the classical equation, we should like to point out that the manifold of solutions is determined by three variables which obey the equation

$$\left(\frac{T+K}{2}\right)^2 - \left(\frac{T-K}{2}\right)^2 - D^2 = \frac{\lambda}{4} .$$

This is a two-dimensional Riemannian space with negative curvature. The free case is obtained here as a direct product.

$$\begin{pmatrix}\dot\varphi\\ \varphi\end{pmatrix} \times \begin{pmatrix}\dot\varphi\\ \varphi\end{pmatrix} = \begin{pmatrix}\dot\varphi^2\\ \frac{1}{2}(\varphi\dot\varphi+\dot\varphi\varphi)\\ \varphi^2\\ \frac{1}{2}(\varphi\dot\varphi-\dot\varphi\varphi)\end{pmatrix} = \begin{pmatrix}2T\\ 2D\\ 2K\\ 0\end{pmatrix},$$

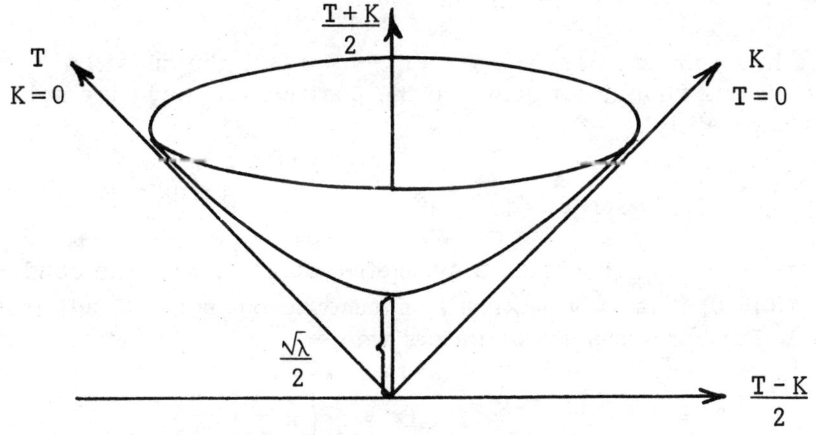

6. THE CLASSICAL SCATTERING THEORY

The $\varphi_{\substack{in\\out}}$ states, with $\ddot\varphi_{\substack{in\\out}} = 0$, are determined by the following equation:

$$\lim_{t\to\mp\infty} |\varphi(t) - \varphi_{\substack{in\\out}}(t)| = 0.$$

For the interacting solution $\varphi = \sqrt{2T}\sqrt{(t + (D/T))^2 + \lambda/4T^2}$, we find

$$\varphi_{\substack{in\\out}} = \mp\sqrt{2T}\,(t + (D/T)),$$

with the same D and T as in the interacting case. Now the initial conditions at $-\infty$ or $+\infty$ are the same. For the scattering operator [3], $\varphi_{out} = S\varphi_{in}$, we obtain $S = -1$.

7. THE QUANTIZATION

The canonical quantization procedure [6] requires that φ and $\dot{\varphi}$ are self-adjoint operators, which obey the commutation relations

$$[\varphi(t),\ \dot{\varphi}(t)] = i.$$

The only integrable realization up to equivalence is given in the Hilbert space of square integrable functions $\int |\psi|^2 dx < \infty$:

$$\varphi = x, \quad \dot{\varphi} = \frac{1}{i}\frac{\partial}{\partial x}, \qquad -\infty < x < +\infty.$$

As we have classically $0 < \varphi < \infty$ or $-\infty < \varphi < 0$, the question arises if one should not quantize the positive and negative solutions separately.

$$\varphi = x, \quad \dot{\varphi} = \frac{1}{i}\frac{\partial}{\partial x}, \qquad 0 < x < +\infty, \qquad \int_0^\infty |\psi|^2 dx < \infty.$$

The operator $\frac{1}{i}\frac{\partial}{\partial x}$ becomes a symmetric operator with the condition $\psi(0) = 0$; it is only maximally symmetric but not self-adjoint $\dot{\varphi} \neq \dot{\varphi}^+$. For our symmetry operators we get

$$\hat{T} = \frac{1}{2}\left(-\frac{\partial^2}{\partial x^2} + \frac{\lambda}{x^2}\right), \qquad \hat{D} = \frac{1}{2i}\left(x\frac{\partial}{\partial x} + \frac{1}{2}\right)$$

$$\hat{K} = \tfrac{1}{2}x^2.$$

The eigenfunctions of T are given by

$$\sqrt{x}\,J_{\sqrt{\lambda + \frac{1}{4}}}(x\sqrt{2e}), \qquad \lambda \geqq 3/4, \qquad 0 < e < \infty,$$

$$\sqrt{x}\, J_{\pm\sqrt{\lambda+\frac{1}{4}}}(x\sqrt{2e}), \qquad 0 \leqq \lambda < 3/4.$$

As the energy operator H (with respect to α) has discrete eigenvalues, it is of advantage to quantize in the α-picture. We have to solve ($\phi(x) = x$, $\dot{\phi}(\alpha) = (1/i)(\partial/\partial x)$),

$$\frac{1}{2}\left(-\frac{\partial^2}{\partial x^2} + \frac{x^2}{4} + \frac{\lambda}{x^2}\right)\psi(x) = h\psi(x).$$

The eigenfunctions are

$$\psi_{n,\ell_\pm} = N_{n,\ell_\pm} e^{-x^2/4} x^{\ell_\pm - \frac{1}{2}} L_n^{\ell_\pm - 1}(x^2/2),$$

$$N_{n,\ell} = \sqrt{\frac{2^{1-\ell} n!}{\Gamma(n+\ell)}},$$

$$h = n + \ell/2, \quad n = 0, 1, 2, \ldots, \quad \ell_\pm = 1 \pm \sqrt{\lambda + \tfrac{1}{4}}.$$

The eigenfunction ψ_{n,ℓ_-} are square-integrable only for $0 \leqq \lambda < 3/4$. They also form a complete system and a basis of an irreducible unitary representation of $SO_0(2,1)$

$$C_{II} = \frac{\lambda}{4} - \frac{3}{16} = \frac{\ell_+}{2}\left(\frac{\ell_+}{2} - 1\right) = \frac{\ell_-}{2}\left(\frac{\ell_-}{2} - 1\right).$$

The discrete representations $\mathcal{D}^+$ of $SO_0(2,1)$ are determined by the minimal eigenvalue of H, $H_{min} = \ell/2$, $\ell > 0$. This was derived by V. Bargmann and L. Pukánszky [4].

The connection [5] between the weights of the finite dimensional representation and the $\mathcal{D}^+$ and $\mathcal{D}^-$ series of $SU_0(1,1)$ is given in the next figure.

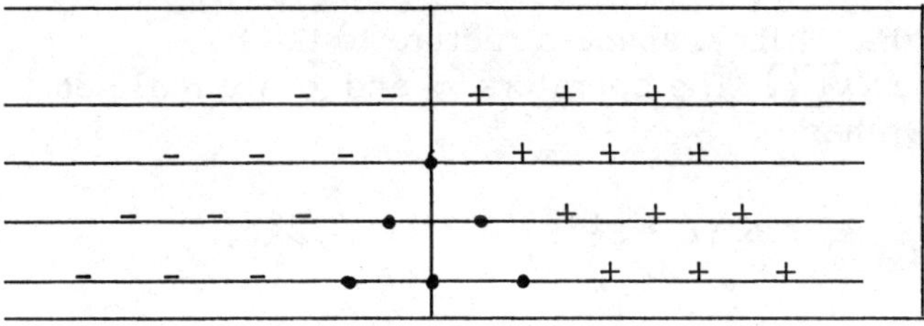

C_{II}	j	ℓ
-1/4	-	1
0	0	2
3/4	1/2	3
2	1	4

• weights of the finite dimensional representation $\mathcal{D}^j$ of $SU_0(1,1)$
+ weights of the $\mathcal{D}^+_{\ell/2}$ series of $SU_0(1,1)$
- weights of the $\mathcal{D}^-_{\ell/2}$ series of $SU_0(1,1)$

Before quantizing the interacting field, we split the Hilbert space of the free field, $\lambda = 0$, into odd and even eigenfunctions. They span the representations $\ell_+ = 3/2$ and $\ell_- = 1/2$, respectively. This is expressed by the relations

$$H_{2n+1}\left(\frac{x}{\sqrt{2}}\right) = (-1)^n n!\, 2^{2n+1} \frac{x}{\sqrt{2}} L_n^{\frac{1}{2}}\left(\frac{x^2}{2}\right)$$

$$H_{2n}\left(\frac{x}{\sqrt{2}}\right) = (-1)^n n!\, 2^{2n} L_n^{-\frac{1}{2}}\left(\frac{x^2}{2}\right).$$

The operators

$$\frac{1}{2}\left(-\frac{\partial^2}{\partial x^2} + \frac{x^2}{4}\right),\quad x,\quad \frac{1}{i}\frac{\partial}{\partial x}$$

in the domain $-\infty < x < +\infty$ transform into the operators

$$\frac{1}{2}\left(-\frac{\partial^2}{\partial x^2} + \frac{x^2}{4}\right)\begin{pmatrix}1 & 0\\ 0 & 1\end{pmatrix},\quad x\begin{pmatrix}0 & 1\\ 1 & 0\end{pmatrix},\quad \frac{1}{i}\frac{\partial}{\partial x}\begin{pmatrix}0 & 1\\ 1 & 0\end{pmatrix},$$

respectively, for $0 < x < \infty$. The two subspaces of the Hilbert space

$$\psi = \begin{pmatrix}\psi^{3/2}\\ \psi^{1/2}\end{pmatrix}$$

are defined by the following boundary conditions

$$\begin{pmatrix}\psi^{3/2}\\ 0\end{pmatrix}:\ \psi(x)\Big|_0 = 0, \qquad \begin{pmatrix}0\\ \psi^{1/2}\end{pmatrix}:\ \frac{\partial}{\partial x}\psi(x)\Big|_0 = 0.$$

Now we generalize this Hilbert space structure to the case $0 < \lambda < 3/4$, $\ell_\pm = 1 \pm \sqrt{\lambda + \frac{1}{4}}$. The operators ϕ and $\dot{\phi}$ are realized as before. For H we have

$$H = \frac{1}{2}\left(-\frac{\partial^2}{\partial x^2} + \frac{x^2}{2} + \frac{\lambda}{x^2}\right)\begin{pmatrix}1 & 0\\ 0 & 1\end{pmatrix}, \qquad 0 < x < \infty.$$

The subspaces of the Hilbert space spanned by

$$\psi = \begin{pmatrix} \psi^{\ell+} \\ \psi^{\ell-} \end{pmatrix}$$

are defined by

$$\begin{pmatrix} \psi^{\ell+} \\ 0 \end{pmatrix}: \left.\frac{\psi}{x^{\frac{1}{2}-\sqrt{\lambda+\frac{1}{4}}}}\right|_0 = 0, \quad \begin{pmatrix} 0 \\ \psi^{\ell-} \end{pmatrix}: x^{1-2\sqrt{\lambda+\frac{1}{4}}} \cdot \frac{\partial}{\partial x} \left.\frac{\psi}{x^{\frac{1}{2}-\sqrt{\lambda+\frac{1}{4}}}}\right|_0 = 0.$$

For $\lambda \geqq 3/4$ there exists only the quasicanonical quantization. So we obtain the following energy levels

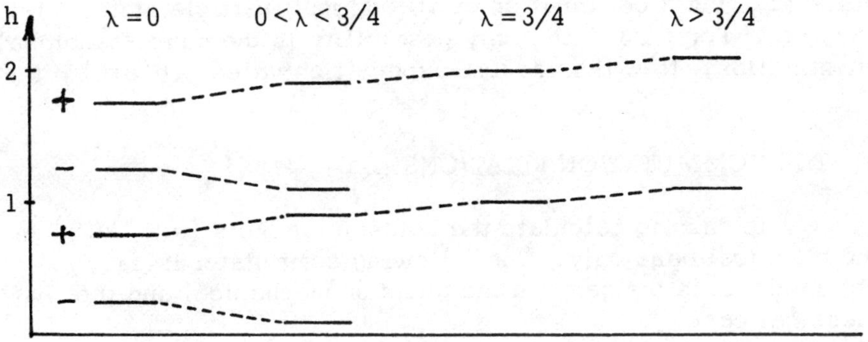

In the free case $\lambda = 0$, $\Delta h = 1/2$, and we have free multi-particle states with energy per particle $h = 1/2$. For $0 < \lambda < 3/4$, $\Delta h = \sqrt{\lambda+\frac{1}{4}}$ or $\Delta h = 1 - \sqrt{\lambda+\frac{1}{4}}$. The one-particle state has the energy $h = \sqrt{\lambda+\frac{1}{4}}$. The two-particle state has the energy $h = 1 = 2\sqrt{\lambda+\frac{1}{4}} + B$ with the binding energy $B = -(2\sqrt{\lambda+\frac{1}{4}} - 1)$. The three-particle state consists of a two-particle bound state and of a third free particle. The four-particle state consists of two pairs, etc. For $\lambda = 3/4$, the one- and two-particle states become identical and form the new ground state ($\Delta h = 1$). The picture is not altered essentially for $\lambda > 3/4$, as $\Delta h = 1$ is independent of λ; only the vacuum energy becomes larger.

Where does the attractive potential come from? Note that the Casimir operator is given quantum mechanically by $C = \lambda/4 - 3/16$; whereas, classically we have $C_{class} = \lambda/4$. To $C_{class} < 0$ corresponds quantum-mechanically $C < 0$ or $\lambda < 3/4$. So all values $0 < \lambda < 3/4$ may lead to an attraction.

The matrix element for the field operator

$$\phi = \begin{pmatrix} 0 & x \\ x & 0 \end{pmatrix}$$

$$\int \begin{pmatrix} 0 \\ \psi_{n'}^{\ell_-} \end{pmatrix}^+ \begin{pmatrix} 0 & x \\ x & 0 \end{pmatrix} \begin{pmatrix} \psi_n^{\ell_+} \\ 0 \end{pmatrix} dx = NN \int e^{-x^2/2} L_n^{\ell_- - 1}\left(\frac{x^2}{2}\right) L_{n'}^{\ell_+ - 1}\left(\frac{x^2}{2}\right) dx,$$

$$\neq 0 \quad \text{for all} \quad n, n' \quad \text{if} \quad 0 < \lambda < 3/4,$$

$$\neq 0 \quad \text{for} \quad n' = n,\ n' = n-1, \ \text{if} \ \lambda = 0.$$

So ϕ creates from the vacuum $|0\rangle$, not just the one-particle state $|1\rangle$ but a combination of all odd multiparticle states $|3\rangle$, $|5\rangle$, etc. For $\lambda \geqq 3/4$ the only possibility is the quasi-canonical quantization. Now the vacuum-expectation value $\langle 0|\phi|0\rangle \neq 0$.

8. THE COMMUTATION RELATIONS

It is easy to calculate the Poisson brackets $\{\varphi(t), \varphi(t')\}$ in the classical case only. The following commutator is easily obtained. It is the same in the quantum mechanical and the classical case.

$$\begin{aligned} [\varphi^2(t), \varphi^2(t')] &= 4[Tt^2 + 2Dt + K,\ Tt'^2 + 2Dt' + K] \\ &= -8i(t-t')(Ttt' + D(t+t') + K) \\ &= \left[\varphi^2_{free}(t), \varphi^2_{free}(t')\right] - \frac{4i(t-t')tt'\lambda}{\varphi^2(0)} . \end{aligned}$$

9. QUANTIZATION IN ANALOGY TO THE THIRRING MODEL

In the mass 0 Thirring model one can show that the two currents obey a free mass 0 equation. And one postulates for these operators the canonical commutation relations. In analogy we introduce the field

$$\Xi = \phi^2 - 4H, \qquad H = \frac{1}{2}\left(\dot{\phi}^2 + \frac{\phi^2}{4} + \frac{\lambda}{\phi^2}\right),$$

which obeys the equation

$$\ddot{\Xi} + \Xi = 0.$$

The canonical Poisson brackets (for simplicity) for Ξ and $\dot{\Xi}$ are given by

$$\{\Xi, \dot{\Xi}\} = -1, \qquad \dot{\Xi} = 2\phi\dot{\phi}.$$

For the fields ϕ, $\dot{\phi}$, and H we obtain

$$\phi^2 = \Xi + \sqrt{\dot{\Xi}^2 + \Xi^2 + 4\lambda},$$

$$\dot{\phi}^2 = \frac{\dot{\Xi}^2}{4(\Xi + \sqrt{\dot{\Xi} + \Xi^2 + 4\lambda}},$$

$$H = \frac{1}{4}\sqrt{\dot{\Xi}^2 + \Xi^2 + 4\lambda} = \frac{\sqrt{2}}{4}\sqrt{H_{\Xi} + 2\lambda}.$$

Now the eigenvalues of H_{Ξ} quantum mechanically are $n+\frac{1}{2}$. H, however, is not the translation operator any more (which is given by H_{Ξ}) and thus has lost its physical significance. The noncanonical Poisson brackets for the fields ϕ and $\dot{\phi}$ are given by

$$\{\phi, \dot{\phi}\} = -\frac{1}{16H}.$$

REFERENCES

The present lecture is based on results derived by the author in 1974 and on the Diploma thesis "Klassische und quantentheoretische Aspekte einer nichtlinearen skalaren Feldgleichung," H. Zinnecker, München, 1976.

1. M. C. Reed, "Abstract nonlinear wave equations," Lecture Notes Math., Vol. 507, Springer, 1976.
2. L. Castell, Phys. Rev. D6, 536 (1972).
3. L. Castell and W. P. Renz, Phys. Rev. D7, 1264 (1973).
4. V. Bargmann, Annals of Math. 48, 568 (1947); L. Pukánszky, Math. Annalen 156, 96 (1964).
5. A. O. Barut, "Dynamical Groups and Generalized Symmetries in Quantum Theory," University of Cantury, Christchurch, New Zealand, 1972.
6. V. de Alfaro, S. Fubini, and G. Furlan, CERN preprint Ref. TH 2115-CERN, 1976.

CHARACTERISTIC "QUANTA" OF NONLINEAR FIELD EQUATIONS†

A. O. Barut

Department of Physics
The University of Colorado
Boulder, Colorado 80309

ABSTRACT. Periodic solutions of nonlinear field equations depending on a single degree of freedom are quantized canonically. The quantized excitations of nonlinear fields generalize the concepts of "photon" or "phonon" of linear fields, and satisfy characteristic equations. For example, the "quanta" of the sine-Gordon equation are Mathieu functions with a nonlinear energy spectrum. They take the place of the ordinary plane waves. It is suggested that these nonlinear quanta should be taken as building blocks for the quantization of nonlinear fields. Spinor-type quantized wave functions arise. Soliton solution is a limiting case. These quantized excitations may be observable.

1. LINEAR FIELDS

There is a straightforward generalization of the procedure of field quantization from linear to nonlinear fields. We first reformulate the standard free field quantization to point out this generalization.

Consider the classical free scalar real field

$$(\Box + m^2)\phi(x) = 0. \tag{1}$$

---

† Lectures given at NATO Advanced Study Institute on Nonlinear Equations in Physics and Mathematics, Istanbul, August 1977.

A. O. Barut (ed.), Nonlinear Equations in Physics and Mathematics, 335–345. *All Rights Reserved.*

We look for special solutions of (1) depending on a single variable $\tau = k^{\mu}x_{\mu}$, i.e., $\phi = \phi(\tau)$. These special solutions satisfy the ordinary differential equation:

$$\phi_{\tau\tau} + \frac{m^2}{k^2}\phi = 0; \quad k^2 = k_{\mu}k^{\mu}. \tag{2}$$

This is a classical equation of motion of one mechanical degree of freedom of the type $\ddot{q} = -(1/\mu)(\partial V/\partial q)$: τ plays the role of "time" and ϕ is the "coordinate." Hence we write

$$\phi_{\tau\tau} + \frac{1}{\mu}\frac{\partial V(\phi)}{\partial\phi} = 0, \quad V(\phi) = \phi^2, \quad \mu = \frac{2k^2}{m^2}. \tag{2'}$$

We now quantize Equation (2) by the standard canonical formalism: The Hilbert space is $L^2(\mathbb{R})$ of functions $\psi(\phi)$, $-\infty < \phi < \infty$, and the Schrödinger equation becomes

$$i\hbar\frac{\partial\psi(\phi)}{\partial\tau} = \left[-\frac{\hbar^2}{2\mu}\frac{\partial^2}{\partial\phi^2} + V(\phi)\right]\psi(\phi). \tag{3}$$

Note again that ϕ plays the role of x ("coordinate") and τ the role of "time," and μ the role of "mass"; $\psi(\phi)$ is the wave function for the value of the field or potential ϕ, $|\psi(\phi)|^2$ the probability of finding the field (depending on τ only) to have the value ϕ. Equation (13) shows the evolution of $\psi(\phi,\tau)$ as a function of τ.

The stationary solutions of (3) for which

$$\psi(\phi,\tau) = u_n(\phi)e^{-(i/\hbar)\mathcal{E}_n\tau} \tag{4}$$

satisfy

$$-\frac{\hbar^2}{2\mu}u_n'' + V(\phi)u_n = \mathcal{E}_n u_n. \tag{5}$$

The quantity $\mathcal{E}$ is conjugate to τ. Extremely large values of ϕ are improbable; hence we may put the boundary condition $\psi(\phi) \to 0$, as $\phi \to \pm\infty$. The solutions of (5) are then clearly the usual harmonic oscillator states, but in the variable ϕ:

$$\mathcal{E}_n = \lambda(n+\tfrac{1}{2}), \quad u_n(\phi) = N_n e^{-(\mu/\hbar^2)\phi^2} H_n\left(\sqrt{\frac{\mu}{\hbar^2}}\,\phi\right), \tag{6}$$

$$\lambda = \hbar\sqrt{\frac{2}{\mu}}.$$

The limiting case $m^2 = 0$, in Equation (1), fits also into the above scheme if we let $m^2 \to 0$, $k^2 \to 0$ such that $\mu = \mu_0$ finite in Equation (2').

The energy-momentum of the classical field

$$P^{\mu} = \int d\sigma_{\nu} T^{\mu\nu} = \int d\sigma_{\nu}\left(\phi'^{\mu}\phi'^{\nu} - \tfrac{1}{2}g^{\mu\nu}(\phi'^{\sigma}\phi_{,\sigma} - m^2\phi^2)\right) \tag{7}$$

becomes, for the special solutions obeying (2),

$$P^{\mu} = \int d\sigma_{\nu}\left[\left(k^{\mu}k^{\nu} - \tfrac{1}{2}g^{\mu\nu}k^2\right)\dot{\phi}^2 + \tfrac{1}{2}m^2\phi^2\right]. \tag{7'}$$

Or, choosing the space-like surface as $t=0$, with $\sigma_1 = \vec{k}\cdot\vec{x}$,

$$P^{0} = \int\left[\tfrac{1}{2}\vec{k}^2\left(\frac{d\phi}{d\sigma_1}\right)^2 + \tfrac{1}{2}m^2\phi^2(\sigma_1)\right] d\sigma_1 d\sigma_2 d\sigma_3 , \tag{8}$$

where $\sigma_1, \sigma_2, \sigma_3$ are three orthogonal coordinates on the surface $t=0$. The quantity in parentheses is a constant of the motion of (2). Hence (8) is precisely the classical energy in the one-degree of freedom multiplied by the volume that the field occupies at constant τ. If the field extends to infinity this volume is infinite. But it is a multiplicative constant and can be renormalized by the energy of the ground state. It is clear that a medium or a field of infinite extension that oscillates collectively must have infinite-energy. In practice, our Ansatz leading to Equation (2) is valid in a localized region, so that the field can be assumed to be zero outside a finite volume. The ratios of the masses of the excited states to the ground states are finite in any case, and we are interested in this finite mass spectrum of the field.

Thus for each k^{μ} we have an energy spectrum (6). Then we can count the possible number of linearly independent k-vectors in the usual manner.

This version of field quantization shows why and how oscillators and the corresponding Schrödinger equation (3) or (5) enter into a manifestly covariant theory—our "Schrödinger equation" is fully covariant because τ and ϕ are Lorentz scalars.

It is possible to quantize the field, instead of plane waves, in terms of spherical waves, or other invariant combinations of variables; one can easily classify the oscillations of one degree of freedom in terms of various variables replacing τ [4].

2. NONLINEAR FIELDS

We now generalize the same procedure to nonlinear equations. This is best explained again in terms of a general class of wave equations. We consider now a nonlinear scalar field $\phi(\vec{x},t)$ obeying the equation

$$\Box\phi = F(\phi) \tag{9}$$

and, again, look for solutions depending on the single variable $\tau = k^\mu x_\mu$. These special excitations satisfy

$$\phi_{\tau\tau} = \frac{1}{k^2} F(\phi) \equiv -\frac{1}{\mu}\frac{\partial V(\phi)}{\partial\phi}, \tag{10}$$

which is of the form (2'), a mechanical equation for one degree of freedom "ϕ". The quantization of (10) again leads to the Schrödinger equation (3). The only difference is in the form of the "potential" $V(\phi)$. The stationary excitations are given by

$$-\frac{\hbar^2}{2\mu} u'' + V(\phi)u = Eu. \tag{11}$$

Depending on the nonlinear term $F(\phi)$, or potential $V(\phi)$, the quantized excitations may have a discrete or continuous spectrum, or both.

The expression for energy is the same as (8) with $m^2\phi^2$ replaced by the potential energy $V(\phi)$.

Example 1. $F(\phi) = -\lambda \sin\phi$. Then $V(\phi) = +\lambda(1-\cos\phi)$, normalized such that $V(0) = 0$. The stationary quantized oscillations now satisfy the Mathieu equation:

$$-\frac{\hbar^2}{2k^2} u'' + \lambda(1-\cos\phi)u = Eu. \tag{12}$$

Thus the fundamental oscillations of our self-coupled nonlinear field are no longer oscillators with a linear energy spectrum. It is therefore interesting to study the nature and spectrum of these fundamental excitations.

The Mathieu equation is written in the standard form

$$u''(x) + (4\alpha + 16q\cos 2x)u(x) = 0. \tag{13}$$

Hence,

$$\phi = 2x, \quad 4\alpha = \frac{8k^2}{\hbar^2}(-\lambda+E), \quad 16q = +\frac{8\lambda k^2}{\hbar^2}, \quad 4\alpha + 16q = E\frac{8k^2}{\hbar^2}.$$

The periodic solutions of (13) are the Mathieu functions $ce_n(x,q)$ and $se_n(x,q)$ which have a period of 2π in ϕ for n even, and a period of 4π in ϕ for n odd. The "odd" solutions change sign at $\phi \to \phi + 2\pi$, and thus behave like "spinor" solutions. Mathieu functions go over to cosnx and sinnx, respectively, as $q \to 0$ $(\lambda = 0)$.

The spectrum of the energy levels as a function of q or λ (the strength of the self-coupling) is rather complicated but well known [6].

In the limiting case $\lambda \to 0$ $(q \to 0)$, we get a spectrum

$$\alpha = n^2, \quad \text{or} \quad E_n = n^2 \cdot \frac{\hbar^2}{2k^2} \tag{14}$$

with eigenfunctions, $e^{in\phi}$ and $e^{-in\phi}$. There is a discrete topological quantum number indicating the sign of the exponent and distinguishing the two degenerate eigenstates (except when $n=0$). [In the analogy to pendulum equation (5), this number is the direction of rotation of the pendulum in full rotation.]

For the other extreme of infinite coupling, $\lambda \to \infty$, $q \to \infty$, the spectrum of the Mathieu equation is given by

$$4\alpha + 16q \longrightarrow (2n+1)(32q)^{\frac{1}{2}},$$

or, (15)

$$E_n \longrightarrow \frac{\hbar^2}{8k^2}\sqrt{\frac{16\lambda k^2}{\hbar^2}}\,(2n+1) = \hbar\sqrt{\lambda/k^2}\,(n+\tfrac{1}{2})$$

and is nondegenerate.

The one-soliton solution of the classical equation (9) corresponds to the limiting case $\alpha = 4q$. In this case the σ_1-integration in (8) is finite, but we have renormalized the σ_2, σ_3-integration as discussed in Section 1.

It is interesting that the solutions $e^{\pm in\varphi}$ for $\alpha > 4q$ do not join continuously the Hermite polynomial oscillatory solutions for $\alpha < 4q$. But the linear combination

$$\phi = ce_n(x,q)e^{(i/\hbar)Et} + ise_n(x,q)e^{-(i/\hbar)Et}$$

joins continuously e^{inx}. The quantum-mechanical current associated with ϕ is proportional to $(E_1 - E_2)t/\hbar$. For small q, $E_1 - E_2 \to 0$, hence the current changes sign.

It is interesting to exhibit the nature of these simple quantized excitations in the case of some other common nonlinear field equations.

Example 2. ϕ^4-theory. The field equations are

$$\Box\phi - m^2\phi - \lambda^2\phi^3 = 0. \tag{16}$$

For excitations of the type $\phi = \phi(\tau)$, $\tau = k^\mu x_\mu$, we have

$$\ddot{\phi} - \frac{m^2}{k^2}\phi - \frac{\lambda^2}{k^2}\phi^3 = 0.$$

The stationary quantized wave equation now becomes

$$-\frac{\hbar^2}{2k^2}u''(\phi) - \left(\tfrac{1}{2}m^2\phi^2 + \tfrac{1}{4}\lambda^2\phi^4\right)u(\phi) = Eu(\phi),$$

or,

$$u''(\phi) + \frac{k^2}{\hbar^2}\left(m^2\phi^2 + \tfrac{1}{2}\lambda^2\phi^4\right) = -\frac{2k^2}{\hbar^2}Eu(\phi). \tag{17}$$

The "potential" for these excitations is now of the type often considered in recent investigations, $V(\phi) = a\phi^2 + b\phi^4$, with two local minima at $\phi_m = \pm\sqrt{-a/2b}$, $V(\phi_m) = -a^2/2b$, and a local maximum at $\phi = 0$, $V(0) = 0$, depending on the sign of the coupling constant λ^2.

Thus we have anharmonic oscillator levels, and the phenomenon of tunneling between the two local minima for time-dependent problems: The "field ϕ" may tunnel from one value ϕ_1 to another value ϕ_2.

Example 3. Korteweg-de Vries Field. This is a field in one-space, one-time dimensions satisfying

$$\phi_t + \alpha\phi\phi_{,x} + \phi_{xxx} = 0. \tag{18}$$

The periodic excitations $\phi(x,t) = f(\tau)$, $\tau = t - vx$ satisfy:

$$\dot{f} = \alpha v f\dot{f} - v^3\dddot{f} = 0$$

or

$$\ddot{f} + \frac{1}{2}\frac{\alpha}{v^2}f^2 - \frac{1}{v^3}f = K$$

where K is a constant of integration.

We now quantize this mechanical equation in one-degree of freedom as before and obtain the wave equation:

$$-\frac{\hbar^2}{2\mu}u''(f) + \left[\frac{1}{6}\frac{\alpha^2}{v^2}f^3 - \frac{1}{2v^2}f^2 - Kf\right]u(f) = Eu(f). \tag{19}$$

The "potential" now increases from $-\infty$ at $f = -\infty$ to $+\infty$ at $f = +\infty$, with one possible local maximum and minimum. We have thus possible unstable quantized states (resonances), otherwise a continuous spectrum, just like a one-dimensional atom in an external electric field.

3. MULTICOMPONENT FIELDS

Multicomponent scalar fields, ϕ^i, vector fields A_μ^i, or higher order fields, where i is an internal symmetry index, have the interesting property that the reduced equations for characteristic excitations is a multidimensional Schrödinger equation in $\mathbb{R}^n$. So far the reduced equation for a single scalar nonlinear field was always a one-dimensional Schrödinger equation.

Consider the set of coupled nonlinear scalar fields obeying

$$\Box\phi^i = 2\phi^i F(\phi^2), \qquad i = 1, 2, \ldots, N, \tag{20}$$

$$\phi^2 = \sum_{i=1}^{N} \phi^{i2}$$

where we assumed, for simplicity, a special form of nonlinearity. The reduced equation for proper oscillations

$$\ddot{\phi}^i = \frac{2}{k^2}\phi^i F(\phi^2) \tag{21}$$

is now a dynamical problem in N-degrees of freedom. Hence we obtain the stationary Schrödinger equation

$$-\frac{\hbar^2}{2\mu}\nabla u(\vec{\phi}) - G(\phi^2)u(\vec{\phi}) = Eu(\vec{\phi}) \tag{22}$$

where $G' = F$, and ∇ is the N-dimensional Laplacian. Thus the problem is reduced to the spectral theory of an N-dimensional Schrödinger operator.

The problem of nonlinear vector fields has been discussed in a recent publication [4], and we refer the reader to this work.

4. NONLINEAR CHIRAL FIELDS

Consider (n+1) scalar fields ψ^i, $i = 1, 2, \ldots, n+1$, such that

$$\sum_{i=1}^{n+1} \psi^{i2} = R^2.$$

Let $\psi^j = \alpha(\phi)\phi^j$, $j = 1, \ldots, n$, $\psi^{n+1} = (R^2-\alpha^2\phi^2)^{\frac{1}{2}}$ with

$$\phi^2 = \sum_{i=1}^{n} \phi^{i2}.$$

Then a general class of $0(n+1)$-symmetric, so-called chiral, field theories are defined [7] by the metric on the sphere S^n:

$$ds^2 = \alpha^2(\phi)d\phi^i d\phi^i + \beta(\phi)(\phi^i d\phi^i)^2, \tag{23}$$

where

$$\beta(\phi) = \frac{R^2}{R^2-\alpha^2\phi^2}\left[\frac{2\alpha\alpha'}{\phi} + (\alpha')^2 + \frac{\alpha^4}{R^2}\right], \quad \alpha' = \frac{d\alpha}{d\phi},$$

or, alternatively by the Lagrangian density,

$$\mathcal{L} = \frac{1}{2}\left[\alpha^2(\phi)\phi^i_{,\mu}\phi^{i,\mu} + \beta(\phi)(\phi^i\phi^i_{,\mu})(\phi^j\phi^{j,\mu})\right]. \tag{24}$$

We shall now look for the proper oscillations of these fields, which may also be called as the proper oscillations of the Riemannian space (23). Here we consider now a simple special case of (24) with $\alpha(\phi) = 2R^2(R^2+\phi^2)$, $\beta(\phi) = 0$. The proper oscillations of the plane wave type, $\phi^i = \phi^i(\tau)$, are described by a mechanical nonlinear Lagrangian and Hamiltonian for n degrees of freedom

$$L = \frac{R^2}{2}\frac{\dot{\phi}_i\dot{\phi}_i}{(R^2+\phi^2)^2}, \qquad H = \frac{1}{2R^2}\pi^2(R^2+\phi^2)^2 \tag{25}$$

where

$$\pi_i = \frac{\partial L}{\partial\dot{\phi}_i}, \quad \pi^2 = \sum_{i=1}^{n}\pi_i^2,$$

The quantized proper oscillations satisfy still a linear Schrödinger equation, covariant with respect to space-time transformations

$$i\hbar\frac{\partial\psi}{\partial t} = H\psi. \tag{26}$$

The stationary solutions of (26) with (25) can be obtained, for example, by group theoretic arguments. The problem is exactly soluble for an appropriate ordering of operators in H which is dictated by the dynamical group-property of the theory. The spectrum of the quantized oscillations is given by

$$E_n = an^2 + b, \qquad n = 1, 2, \ldots . \tag{27}$$

5. PHYSICAL INTERPRETATION AND APPLICATIONS

It is clear now that we have not quantized the fields in the usual sense by making them operator-valued distributions, but we have only quantized certain simple (periodic) solutions. The aim is, of course, to extend this quantization gradually to more general solutions.

This idea of quantization of solutions of classical field equations has a simple intuitive interpretation in the case of relativistic extended elementary particles [1]-[4]. If we imagine a particle to be a certain continuous medium (a string, membrane, shell, or a ball), then the proper oscillations of such a medium could be quantized and the resultant mass spectrum can be identified with low-lying states of such a medium of matter. Indeed, atomic or nuclear matter could be treated from this point of view. The proper oscillations then correspond to certain simple collective motions of the medium. In such applications the fields represent the deformations of the medium. Not only have the fields ϕ^i finite extensions in space but also bounded magnitudes, and the whole theory is finite.

In the case of electromagnetic and other fields filling the whole space, we can still talk of collective motions of the fields. The field functions $\phi^i, A_\mu \ldots$ do not necessarily have the moaning of displacements of a medium (although A_μ still can be interpreted as displacement of a relativistic ether [2]).

There are elementary proper oscillations of different kinds [4]. Once the parameters of the field equations are fixed, one could immediately decide which of the proper oscillations have lowest energy—for example, the nonlinear plane waves, or radical oscillations.

Motions of the fields which depend on two or more independent variables would correspond to a field theory in two or more dimensions. It is interesting that the knowledge gained by the study of two-dimensional field theories (e.g., soluble Thirring model, etc.) could be considered as realistic special motions of a four-dimensional field.

A problem which is completely open so far is the study of completely localized (in three dimensions) motions of fields—three-dimensional soliton-like solutions of nonlinear field equations and their own oscillations have not been fully studied. This is an outstanding problem in this field.

Finally, we emphasize that the occurrence of Schrödinger-type eigenvalue problems, such as ones used in hadron spectroscopy, could also be interpreted as quantized collective normal modes of the field. They need not be interpreted as point-like constituents (e.g., quarks) of hadrons. This interpretation has the great merit that the Schrödinger-eigenvalue problem is perfectly covariant. In the model with point-like constituents it would be a miracle that a nonrelativistic treatment would give reasonable results at all.

It is remarkable that mass spectra of the type (14), (15) or (27) come qualitatively close to the observed mass spectra of hadrons.

6. CHOICE OF NONLINEAR MODEL

Given a particular nonlinear theory, we can now systematically study the various proper oscillations of the system. For a realistic application (for example, to the problem of hadron spectroscopy), we must have a theory for the choice of the underlying field theory. This is clearly a physical problem.

It is likely that the nonlinear field equations of elementary particles are more complicated than the class of <u>local</u> field equations we have considered. For example, in quantum electrodynamics, we need to study proper oscillations of coupled Maxwell-Dirac equations. Or, if we eliminate the potentials A_μ in these coupled equations, we arrive at a nonlinear, nonlocal integro-differential equation for the matter-field $\psi(x)$:

$$\left(-i\gamma^\mu \partial_\mu - m\right)\psi(x) = e\gamma^\mu \psi(x) \int dy\, D(x-y)\overline{\psi}(y)\gamma_\mu \psi(y) . \qquad (28)$$

Only in the limit when the Green's function $D(x-y)$ is replaced by δ-function, we do obtain a <u>local</u> nonlinear field equation with a ψ^3-nonlinearity (cf. Example 2, Section 2).

There are indications that Equation (28) has various types of localized soliton-like solutions and high energy-narrow resonances [8]. It would certainly be of great value to study the proper oscillations of this basic equation of quantum electrodynamics.

REFERENCES

1. P. A. M. Dirac, An Extensible Model of the Electron, Proc. Roy Soc. (London) A268, 57-67 (1962).
2. A. O. Barut, Relativistic Composite Systems and Extensible Models of Fields and Particles, in Mathematical Physics and Physical Mathematics, edited by K. Maurin and R. Rączka, D. Reidel Publishing Company, 1976.
3. A. O. Barut and R. Rączka, Quantized Excitations of Relativistic Extended Elementary Particles, Nuovo Cimento 31B, 19-31 (1976).
4. A. O. Barut and R. Rączka, Mass Spectrum in Relativistic Extended Models of Elementary Particles, Lett. Math. Phys. 1, 315-322 (1976).
5. E. U. Condon, The Physical Pendulum in Quantum Mechanics, Phys. Rev. 31, 891-894 (1928).
6. M. J. O. Strutt, Lamé-sche-Mathieu-sche und Verwandte Funktionen in Physik und Technik, Berlin, Springer, 1932.
7. A. O. Barut, L. Girardello, and W. Wyss, Nonlinear 0(n+1)-Symmetric Field Theories, Symmetry Breaking and Finite Energy Solutions, Helv. Phys. Acta 49, 807-813 (1976).
8. A. O. Barut and J. Kraus, Form Factor Corrections to Superpositronium and Short Distance Behavior of the Magnetic Moment of the Electron, Phys. Rev. D 16, 161-164 (1977).

NONLINEAR SCHRÖDINGER EQUATION WITH SOURCES: AN APPLICATION OF THE CANONICAL FORMALISM †

L. Girardello‡
Istituto di Fisica dell'Università, Milano
Istituto Nazionale di Fisica Nucleare,
Sezione di Milano

R. Jengo
Istituto di Fisica Teorica dell'Università, Trieste
Istituto Nazionale di Fisica Nucleare,
Sezione di Trieste

1. A GENERAL FIELD THEORETICAL PROBLEM

There are applications in which one naturally comes across nonlinear evolution equations associated with integrable Hamiltonian systems in presence of source or, in general, perturbative terms.

A typical situation is the following: Let $\phi(x,t)$ and $\phi^+(x,t)$ be two canonically complex conjugate fields satisfying the C.C.R.: $[\phi(x,t), \phi^+(x',t)] = \delta(x-x')$ (one-space dimension), with the dynamics specified by the Action

$$S = \int dt\, dx \left[i\phi^+ \delta_t \phi - \mathcal{H} \right] \tag{1}$$

where the Hamiltonian is

$$H \equiv \int dx\, \mathcal{H} = \int dx \left[\partial_x \phi^+ \partial_x \phi - \mu \phi^+ \phi + g^2 (\phi\phi^+)^2 \right]. \tag{2}$$

---

† Abstract of a talk presented at the NATO Advanced Study Institute on "Nonlinear Equations in Mathematics and Physics," Istanbul, August 1977.

‡ Reporting.

A. O. Barut (ed.), Nonlinear Equations in Physics and Mathematics, 347–353. *All Rights Reserved.*

The correlation functions

$$G_{n,m}(x_2-x_1,t_2-t_1) = \langle 0|\left(\phi^+(x_2,t_2)\right)^n \left(\phi^+(x_1,t_1)\right)^m |0\rangle \quad (3)$$

can be obtained from the generating functional

$$Z(J,J^+) = \sum_n \frac{(-JJ^+)^n}{n!^2} G_n(x_2-x_1,t_2-t_1) =$$

$$= \langle 0| e^{iJ^+\phi(x_2,t_2)} e^{iJ\phi(x_1,t_1)} |0\rangle \quad (4)$$

(in this case $G_{n,m} = \delta_{nm} G_n$).

In the path integral approach the "classical" or tree approximation of Z is given by [1]

$$Z = \exp i S_{cl}(J,J^+) =$$

$$= \exp i \int dt\, dx \left[i\phi^+_{cl}\partial_t\phi_{cl} - \mathcal{H}_{cl} + J^+\phi_{cl} + J\phi^+_{cl}\right] \quad (5)$$

where, in the case of δ-like sources, ϕ_{cl} and ϕ^+_{cl} are solutions of the equations

$$i\partial_t\phi_{cl} = -\mu\phi_{cl} - \partial_x^2\phi_{cl} + 2g^2\phi^+_{cl}\phi^2_{cl} - J\delta(t+(T/2))\delta(x-(B/2))$$

$$-i\partial_t\phi^+_{cl} = -\mu\phi^+_{cl} - \partial_x^2\phi_{cl} + 2g^2\phi_{cl}\phi^{+2}_{cl} - J^+\delta(t-(T/2))\delta(x+(B/2)).$$

(6)

We recognize here the well-known repulsive nonlinear Schrödinger equation (N.L.S.E.) to which source and mass terms are added.

In analogy with Q.F.T. the full-fledged functional approach can be given mathematically rigorous meaning in the Euclidean context.

The formal situation of imaginary time (t = -iy) would appear in a quantum statistical treatment of the system. We have another interesting example where the imaginary time is relevant, if we consider this modified N.L.S.E. (6) (t = -iy) as a model for the so-called Regge Gribov field theory (R.G.F.T.) [2]. In this application ϕ and ϕ^+ represent a Reggeon field, describing the propagation of a Pomeron; y is the rapidity variable (roughly, the logarithm of the energy) and x the impact parameter—these being appropriate variables for a high energy small-angle collision. The sources J and J^+ represent the strength with which the Pomeron is coupled to the incident particles where Y = i(-iT) and B are the relative rapidity and impact parameter of the colliding particles. Notice that the mass term can be eliminated in the equations and in the action S by redefining new fields $\phi' = e^{-\mu y}\phi$ and $\phi^{+'} = e^{\mu y}\phi^+$; this amounts to taking as new sources $Je^{\mu(Y/2)}$ and $J^+e^{\mu(Y/2)}$.

In the R.G.F.T. context the generating functional $Z(J, J^+)$ gives the scattering amplitude; i.e., the S-matrix element, at fixed impact parameter.

2. AN APPLICATION OF THE CANONICAL FORMALISM

Having in mind the R.G.F.T. application, we have studied the Equation (6) (-t = iy) under particular "initial" and boundary conditions, the final aim being the evaluation of the classical action functional, i.e., the scattering amplitude in the tree approximation. Actually, the main interest lies in its asymptotic behaviour for large rapidity "delay" Y between the two sources. Since a detailed analysis has been already published [3], here we only outline our approach. Conditions relevant to this situation, leading to a well-posed problem, are

$$\begin{aligned} &(1) \qquad \phi^+\phi \underset{|x| \to +\infty}{\longrightarrow} 0, \quad \text{any } y \\ &(2) \qquad \phi = 0 \quad y < -(Y/2), \qquad \phi^+ = 0 \quad y > (Y/2) \end{aligned} \tag{7}$$

Solutions ϕ_{cl}, ϕ^+_{cl} to Equation (6) can be constructed by considering solutions ϕ_h, ϕ^+_h of the homogeneous, genuine NLSE (no sources) satisfying the conditions:

$$\begin{aligned} \phi_h(x, y = -(Y/2)) &= J\delta(x - (B/2)) \\ \phi^+_h(x, y = (Y/2)) &= J^+\delta(x + (B/2)). \end{aligned} \tag{8}$$

Solutions to Equation (6) will then be

$$\phi_{cl} = \phi_h \theta(y + (Y/2)) \qquad \phi^+_{cl} = \phi^+_h \theta((Y/2) - y). \tag{9}$$

The structure of such solutions can be sketched as in Figure 1.

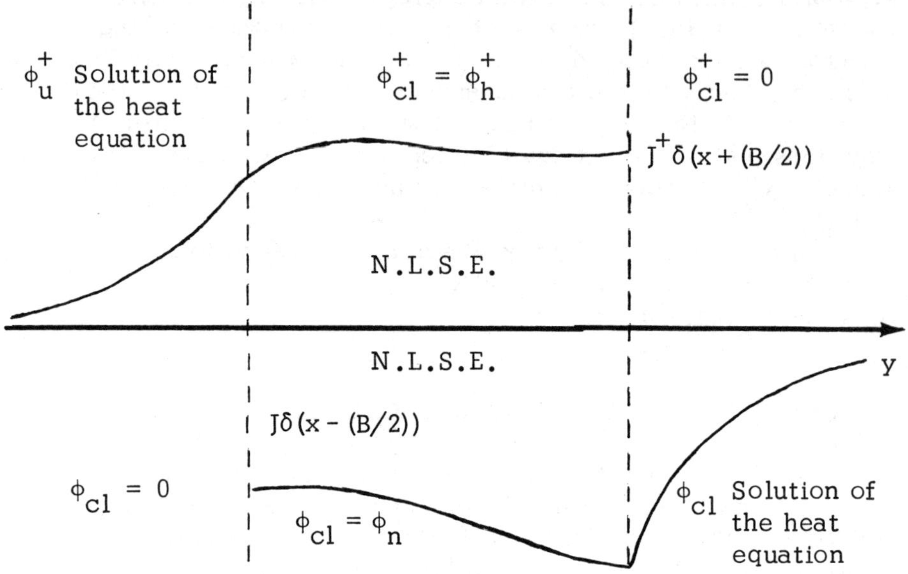

Figure 1

In the strip $|y| \leqq Y/2$, the inverse scattering method [4] (I.S.M.) can still be applied to the "Euclidean" N.L.S.E., the main difference being now the nonunitary "time" evolution of the scattering data.

In order to evaluate the classical action, one does not need to solve the relevant Gelfand-Levitan-Marchenko equations.

Rather, once the direct scattering problem has been solved, in the sense that enough informations on the scattering data are available, our strategy will be based on an unconventional use of the canonical structure of the I.S.M.

We indicate a way of re-expressing the classical action in terms of new canonical variables or, at the very end, of the scattering data, in terms of which calculations are more manageable. Here are the main points:

2.1 Cánonical mapping

We can map the variables ϕ, ϕ^+ into the new canonical variables [4b]:

$$A^{+}(\xi,y) = \frac{1}{g}\sqrt{\frac{\ell n\, a(\xi)\bar{a}(\xi)}{\pi}\;\frac{b(\xi,y)}{\bar{b}(\xi,y)}}$$

$$A(\xi,y) = \frac{1}{g}\sqrt{\frac{\ell n\, a(\xi)\bar{a}(\xi)}{\pi}\;\frac{\bar{b}(\xi,y)}{b(\xi,y)}} \tag{10}$$

where a, $\bar{a}$, b, $\bar{b}$, are the scattering data (with evolution equations)

$$\frac{\partial a}{\partial y} = \frac{\partial \bar{a}}{\partial y} = 0 \qquad \frac{\partial b}{\partial y} = 4\xi^2 \qquad \frac{\partial \bar{b}}{\partial y} = -4\xi^2 . \tag{11}$$

Furthermore, we have

$$a(\xi)\bar{a}(\xi) = 1 + g^2 JJ^{+} e^{-4\xi^2 Y + 2i\xi B} . \tag{12}$$

In terms of the new variables the Hamiltonian and the number of particles $\mathfrak{n}$ (constants of motion) read:

$$H = \frac{4}{\pi g^2}\int_{-\infty}^{+\infty} d\xi\, \xi^2 \ell n\, a(\xi)\bar{a}(\xi) = 4\int_{-\infty}^{+\infty} d\xi\, \xi^2 A^{+}(\xi,y)A(\xi,y) \tag{13}$$

$$\mathfrak{n} \equiv \int_{-\infty}^{+\infty} dx\, \phi^{+}\phi = \frac{1}{\pi g^2}\int_{-\infty}^{+\infty} d\xi\, \ell n\, a(\xi)\bar{a}(\xi)$$

$$= \int_{-\infty}^{+\infty} d\xi\, A^{+}(\xi,y)A(\xi,y) . \tag{14}$$

Notice the absence of the discrete variables (discrete spectrum) [3].

2.2 New expression of the classical action

Formal generalization of classical mechanics for systems of finite degrees of freedom to continuous systems leads us to rewrite our classical action (i.e., the Lagrangian) in terms of the new canonical variables as follows.

$$S_{cl}(y_1,y_2) = \int_{y_1}^{y_2} dy \left(- \int_{-\infty}^{+\infty} dx\, \phi^+ \partial_y \phi - H \right) =$$

$$= \int_{y_1}^{y_2} dy \left(\int_{-\infty}^{+\infty} d\xi\, A(\xi,y) \partial_y A^+(\xi,y) - H \right)$$

$$+ G(y_2) - G(y_1) \tag{15}$$

where $G(y) \equiv G[\phi(y,\cdot), A^+(y,\cdot)]$ is the generating functional (unknown!) of the canonical transformation $\{\phi,\phi^+\} \to \{A,A^+\}$. Here the source terms are included in H. By analyzing the structure of the solutions ϕ_{cl}, ϕ^+_{cl} for $|y| \geqq Y/2$ and the equations defining the scattering data, we find that for $\phi \to 0$

$$A \to 0, \quad A^+ \to \frac{1}{\sqrt{\pi}} \int_{-\infty}^{+\infty} dx\, e^{-2i\xi x} \phi^+ \tag{16}$$

and for $\phi^+ \to 0$,

$$A^+ \to 0, \quad A \to \frac{1}{\sqrt{\pi}} \int_{-\infty}^{+\infty} dx\, e^{-2i\xi x} \phi . \tag{17}$$

When $y_1 = -\infty$ and $y_2 = +\infty$,

$$G(+\infty) - G(-\infty) = G\left(\phi(+\infty), A^+(+\infty)\right) - G\left(\phi(-\infty), A^+(-\infty)\right);$$

by combining the above arguments, we can show that this difference is zero.

2.3 Evaluation of the classical action

The variables $A(\xi,y)$, $A^+(\xi,y)$ are discontinuous at $y = \pm(Y/2)$, so much care must be taken in the y-integration; a further analysis yields the form

$$S_{cl} = J^+\phi\left(y = +\frac{Y}{2}, x = -\frac{B}{2}\right) + J\phi^+\left(y = -\frac{Y}{2}, x = \frac{B}{2}\right) +$$

$$+ \int_{-\infty}^{+\infty} d\xi \left\{ \int_{J^+}^{O} dJ^{+\prime} A \frac{\partial}{\partial J^{+\prime}} A^+ + \int_{O}^{J} dJ' A \frac{\partial}{\partial J'} A^+ \right\} \tag{18}$$

where J, J^+ are the source strengths.

The first two terms are easily recognized to be equal to η (14). The J, J^+ integrations can be performed, once the scattering data are expressed in terms of the initial data and finally S_{cl} can be written as a sum of a finite number of convergent integrals. Their main features of interest can be grasped by inspection and the asymptotic behaviour in Y can be worked out.

For a general discussion of the results we refer once more to our paper [3]; here we limit ourselves to stress that the final result is essentially nonperturbative.

REFERENCES

1. E. S. Abers and B. W. Lee, Physics Reports 9, 1 (1973); S. Coleman, Erice Lectures, 1973.
2. V. N. Gribov, JETP (Sov. Phys.) 26, 414 (1968); H. D. I. Abarbanel, J. D. Bronzan, R. L. Sugar, and R. R. White, Physics Reports 21, 119 (1975) and references therein.
3. L. Girardello and R. Jengo, Nuclear Phys. B120, 23 (1977).

4a. V. E. Zacharov and A. B. Shabat, JETP (Sov. Phys.) 34 (1972).

4b. V. E. Zacharov and S. V. Manakov, Theor. Math. Phys. 19 (1974). (Amer. transl., March 1975.)

4c. M. J. Ablowitz, D. J. Kaup, A. C. Newell, and H. Segur, Studies Appl. Math. 53, 249 (1974).

NONLINEAR FIELD EQUATIONS AND COLLECTIVE PHENOMENA†‡

H. Kleinert

Institut für Theoretische Physik
Freie Universität Berlin

ABSTRACT. The appearance of nonlinear field equations in the description of collective phenomena is discussed using very simple models as examples.

1. INTRODUCTION

Almost all nonlinear equations which were presented at this school arise in the description of collective phenomena in many-body systems [1], [2], independently of whether the system is following classical or quantum statistics. It is the purpose of this lecture to give a simple introduction into the basic properties of collective field theories for the quantum statistical situation. Our representation will focus on a most trivial model of many-body physics: A single degenerate shell of Ω particles, bosons or fermions, with pair and density interactions. Since this model is completely soluble, we can achieve a transparent understanding of how the collective fields manage to render a complete description of a many-body system.

While this model presented is very special, the techniques to be used will be completely general. They can be successfully [3] applied to realistic many-body systems such as superconductors

---

† Lectures presented at NATO Advanced Study Institute on Nonlinear Equations in Physics and Mathematics, Istanbul, August 1977.

‡ Work supported in part by Deutsche Forschungsgemeinschaft under Grant Kl 256.

A. O. Barut (ed.), Nonlinear Equations in Physics and Mathematics, 355–374.

and ^{3}He. These techniques consist in manipulating the path integral representation of quantum statistics in such a way as to produce convenient changes of field variables. Graphically speaking, these manipulations amount to resummations of Feynman diagrams (ladderwise or bubblewise). Since these resummations are trivial in the model at hand, we shall not discuss this aspect any further but refer to the extensive treatment of this subject in References 3 and 4.

We shall show how both, Bose and Fermi systems, can be imbedded completely in another bosonic theory of collective fields. In the new field theory, the original particles become coherent collective states. They take up only a small portion of the collective Hilbert space. They may enter and leave the collective system only via special couplings to external currents. These make sure that the statistics are preserved for any Green's function: In particular, anticommuting currents for Fermions enforce the respectation of Pauli's exclusion principle.

The treatment here is in contradistinction to some recent work in 1+1 dimension [5] where Fermion theories, such as Thirring and Schwinger models, have been transformed completely into a Bose form with purely bosonic expressions taking care of anticommutativity. This situation is peculiar to the two-dimensional world and cannot, in structure, be extrapolated to learn about general properties of collective fields. Thirring and Schwinger model can, however, be treated with the methods presented here. If this is done, the actual calculations of the Fermion Green's functions will be completely equivalent to what can be found in Schwinger's and Klaiber's work [6]. It may be left as a simple exercise to the reader to translate each of the technical manipulations on the path integral of our model to these two 1+1 dimensional examples (or a mixture thereof).

2. A SIMPLE MODEL

Consider a system of Ω bosons or Fermions a_i and b_i ($i=1, \ldots, \Omega$) with a Hamiltonian

$$H = -\frac{V}{4}\left(\sum_{i=1}^{\Omega} a_i^+ a_i \pm b_i b_i^+\right)^2 . \tag{1}$$

Its eigenstates are

$$\prod_{\{i\}} a_i^+ \prod_{\{j\}} b_i^+ |0\rangle \tag{2}$$

with energies

$$E = -\frac{V}{4}\left(N_a + N_b \pm \Omega\right)^2 . \qquad (3)$$

The Lagrangian of the theory is

$$\mathcal{L} = \sum_i a_i^+ i\partial_t a_i + b_i^+ i\partial_t b_i + \frac{V}{4}\left(\sum_i a_i^+ a_i \pm b_i b_i^+\right)^2 \equiv \mathcal{L}_o + \mathcal{L}_{int} . \qquad (4)$$

All Green's functions may be obtained from partially differentiating the generating functional†

$$Z[\eta^+,\eta,\lambda^+,\lambda] \equiv$$

$$\equiv \langle 0 | T e^{i\int dt\{\mathcal{L}_{int}(t) + (\eta^+(t)a(t)^+ + \lambda^+(t)b(t) + h.c.)\}} | 0 \rangle \qquad (5)$$

with respect to the external sources $\eta^+(t), \lambda^+(t), \eta(t), \lambda(t)$. In this expression, a^+, a, b^+, b are in the interaction representation and follow the free field propagators

$$\langle 0 | T \begin{pmatrix} a(t)a^+(t') & a(t)b^+(t') \\ b(t)a^+(t') & b(t)b^+(t') \end{pmatrix} | 0 \rangle \equiv \begin{pmatrix} 1 & 0 \\ 0 & 1 \end{pmatrix} \theta(t-t') \qquad (6)$$

There is another representation for Z which avoids the operator formulation and works only with functional integrals over c-number objects:†

$$Z[\eta^+,\eta,\lambda^+,\lambda] \equiv$$

$$\equiv \int \mathcal{D}a^+\mathcal{D}a\mathcal{D}b^+\mathcal{D}b\, e^{i\int dt\{\mathcal{L}_o(t)+\mathcal{L}_{int}(t)+(\eta^+(t)a(t)+\lambda^+(t)b(t)+h.c.)\}} . \qquad (7)$$

Here, $\mathcal{D}a$, etc. denotes the path integral as introduced by Feynman. It is defined by the infinite product of integrals over all field values from $-\infty$ to $+\infty$ once for every point on the time axis [7]. It is this formulation of Z which allows the introduction of collective fields by mere changes of the integration variables.

† All formulas for Z have to be read modulo an irrelevant normalization factor.

A typical transformation of this type which will be the first example to be studied consists in adding the following complete square to the Lagrangian [4]:

$$\Delta\mathcal{L} = -V\left\{\rho - \tfrac{1}{2}\sum\left(a_i^+ a_i \pm b_i b_i^+\right)\right\}^2 \tag{8}$$

and integrating over the new field ρ in the generating functional Z which certainly does not change Z since the Gaussian integral

$$\int \mathcal{D}\rho\, e^{-iv\int dt\{\rho(t) - \frac{1}{2}\sum_i (a_i^+ a_i(t) \pm b_i b_i^+(t))\}^2} \tag{9}$$

does not depend on a^+, a, b^+, b due to translational invariance of the integral under $\rho \to \rho + f(t)$ [since the individual integrals at each t cover the whole field interval $(-\infty, +\infty)$]. Notice that the new field ρ is equal, at the classical levels, to the density

$$\tfrac{1}{2}\sum_i \left(a_i^+ a_i + b_i^+ b_i \pm \Omega\right).$$

This can be seen immediately by minimizing the Lagrangian with respect to variations in $\delta\rho$ which produces the equation of motion

$$\rho = \tfrac{1}{2}\sum_i \left(a_i^+ a_i + b_i^+ b_i \pm \Omega\right).$$

Now one observes that the complete square (6) can always be chosen as to cancel the quartic interaction in $\mathcal{L}$ leaving

$$\mathcal{L} + \Delta\mathcal{L} = \sum_i \left\{a_i^+ i\partial_t a_i + b_i^+ i\partial_t b_i + V\rho(a_i^+ a_i \pm b_i b_i^+)\right\} - V\rho^2. \tag{10}$$

But this Lagrangian is quadratic also in the fundamental fields a, b such that the path integral over $\mathcal{D}a^+\mathcal{D}a\mathcal{D}b^+\mathcal{D}b$ is Gaussian and can be performed at once in the generating functional

$$Z = \int \mathcal{D}a^+\mathcal{D}a\mathcal{D}b^+\mathcal{D}b\mathcal{D}\rho\, e^{i\int\{\mathcal{L}+\Delta\mathcal{L}+(\eta^+ a+\lambda^+ b+h.c.)\}dt}. \tag{11}$$

Introducing the spinors $\psi = \begin{pmatrix} a \\ b^+ \end{pmatrix}$ and corresponding currents $j = \begin{pmatrix} \eta \\ \lambda^+ \end{pmatrix}$, Z may be written as

$$Z = \int \mathcal{D}\psi^+ \mathcal{D}\psi \mathcal{D}\rho \cdot e^{i\int dt \left\{ \sum_i \psi_i^+(t) \begin{pmatrix} i\partial_t + V\rho & 0 \\ 0 & \mp(i\partial_t - V\rho) \end{pmatrix} \psi_i(t) + j_i^+(t)\psi_i(t) + \psi_i^+(t) j_i(t) \right\}} . \tag{12}$$

If one now denotes the functional matrix between ψ^+ and ψ by $iG_\rho^{-1}(t,t')$, then

$$G_\rho(t,t') = i \begin{pmatrix} i\partial_t + V\rho(t) & 0 \\ 0 & \mp(i\partial_t - V\rho(t)) \end{pmatrix}^{-1} = \begin{pmatrix} \frac{i}{i\partial_t + V\rho(t)} & 0 \\ 0 & \mp\frac{i}{i\partial_t - V\rho(t)} \end{pmatrix} \tag{13}$$

becomes the propagator of the fundamental fields a, b in an external field $\rho(t)$. One may quadratically complete the exponents to

$$i\left[\psi^+ + j^+\left(\frac{1}{i}G_\rho\right)\right] iG_\rho^{-1}\left[\psi + \left(\frac{1}{i}G_\rho\right)j\right] - j^+ G_\rho j .$$

Shifting the $\mathcal{D}\psi$ integration by $(1/i)G_\rho j$, one remains with the purely Gaussian forms:

$$\int \mathcal{D}\psi^+ \mathcal{D}\psi e^{i\int \psi^+(t)\, iG_\rho^{-1}(t,t')\psi(t')dt\,dt'} = \left[\det iG_\rho^{-1}\right]^{\mp 1} . \tag{14}$$

The proof of this formula is trivial for bosons since one may diagonalize $iG_\rho^{-1}(t,t')$ by a unitary transformation and perform the integrals separately for each diagonal field coordinate. The result is the product of the inverse eigenvalues of the matrix $iG_\rho^{-1}(t,t')$ which can be written in a coordinate independent way as a determinant: $[\det iG_\rho^{-1}]^{-1}$. For fermions, the proof is a little more subtle since anti-commuting (Grassmann) field variables have to be used in the path integrals. It may be shown that this changes the sign of the exponent producing $[\det iG_\rho^{-1}]^{+1}$. Using the well-known matrix identity $\det A = \exp \operatorname{tr} \log A$, one obtains for (10)

$$Z = \int \mathcal{D}\rho\, e^{i(\pm i\Omega \operatorname{tr} \log iG_\rho^{-1} - V\int dt\rho^2) - \sum_i \int dt\,dt' j_i^+(t) G_\rho(t,t') j_i(t')} . \tag{15}$$

In this new version, the original fundamental fields a, b have been completely eliminated in favour of the collective density type field $\rho(t)$. All information about a , b is contained in the last external current term.

Up to here, the treatment is extremely general and could be performed in any number of space-time dimensions. The advantage of having only one (time) coordinate lies in the possibility of calculating G_ρ explicitly (the same thing is true also for one space and one time dimensions in Schwinger- and Thirring models): Consider the differential equation for G_ρ:

$$\begin{pmatrix} i\partial_t+V\rho & 0 \\ 0 & \mp(i\partial_t-V\rho) \end{pmatrix} G_\rho(t,t') = i\delta(t-t').$$

Obviously,

$$G_\rho(t-t') = \begin{pmatrix} e^{iV\int_{t'}^{t}\rho(\tau)d\tau}\,\theta(t-t') & 0 \\ 0 & \pm e^{-iV\int_{t'}^{t}\rho(\tau)d\tau}\,\theta(t'-t) \end{pmatrix} \tag{16}$$

is a solution. The boundary conditions have been chosen such as to let G_ρ go to the free Green's function in the vacuum for $\rho=0$:

$$G_0(t-t')_{\alpha\beta} = \langle 0|T\left(\psi_\alpha(t)\psi_\beta^+(t')\right)|0\rangle\Big|_{\rho=0} =$$

$$= \langle 0|T\begin{pmatrix} a(t)a^+(t') & a(t)b(t') \\ b^+(t)a^+(t') & b^+(t)b(t') \end{pmatrix}_{\alpha\beta}|0\rangle\Bigg|_{\rho=0} =$$

$$= \begin{pmatrix} \theta(t-t') & 0 \\ 0 & \pm\theta(t'-t) \end{pmatrix}_{\alpha\beta}. \tag{17}$$

From this G_ρ it is easy to obtain also the $\operatorname{tr}\log iG_\rho^{-1}$. For differentiating with respect to ρ yields

$$\frac{\delta}{\delta\rho(t)}\left(\pm i\operatorname{tr}\log iG_\rho^{-1}\right) = \pm V\operatorname{tr}\left\{\begin{pmatrix} 1 & 0 \\ 0 & \pm 1 \end{pmatrix} G_\rho(t,t')\right\}\Bigg|_{t'=t+0} = \pm V. \tag{18}$$

Here the $t'\to t$ limit has been taken in accordance with the current $a^+a\pm bb^+$ to which ρ couples in the Lagrangian (10):

$$a^+(t)a(t) \pm b(t)b^+(t) = \lim_{t' \to t+\epsilon} T\left(a(t)a^+(t') \pm b^+(t)b(t')\right)$$

$$\hat{=} \lim_{t' \to t+\epsilon} \operatorname{tr}\left\{\begin{pmatrix} 1 & 0 \\ 0 & \pm 1 \end{pmatrix} G_\rho(t,t')\right\} . \tag{19}$$

Equation (18) can now be integrated, yielding

$$\pm i \operatorname{tr} \log i G_\rho^{-1} = \pm V \int_{-\infty}^{\infty} dt\, \rho(t). \tag{20}$$

Thus, the generating functional (14) becomes

$$Z = \int \mathcal{D}\rho$$

$$\cdot e^{i\int dt(-V\rho^2 \pm \Omega V \rho) - \int dt dt' e^{iV\int_{t'}^{t} \rho(\tau) d\tau} \theta(t-t') \sum_i (\eta_i^+(t)\eta_i(t') + \lambda_i^+(t)\lambda_i(t'))} . \tag{21}$$

This expression can be brought to a standard local form by introducing a new field variable

$$\rho \equiv -\frac{1}{V}\dot{\varphi}. \tag{22}$$

Then the exponent becomes

$$A = \int dt\left(-\frac{1}{V}\dot{\varphi}^2 \mp \Omega\dot{\varphi}\right) + i\int dt dt' e^{-i\varphi(t)} e^{i\varphi(t')} \theta(t-t')$$

$$\times \sum_i \left(\eta_i^+(t)\eta_i(t') + \lambda_i^+(t)\lambda_i(t')\right) . \tag{23}$$

The first piece of the action corresponds to a collective Lagrangian

$$\mathcal{L}^c = -\left(\frac{1}{V}\dot{\varphi}^2 \pm \Omega\dot{\varphi}\right) \tag{24}$$

with a Hamiltonian

$$\hat{H} = -\frac{V}{4}(\hat{p} \pm \Omega)^2 = -\frac{V}{4}\left(-i\frac{\delta}{\delta\varphi} \pm \Omega\right)^2 . \tag{25}$$

Thus the collective quantum theory consists of all states $e^{ip\varphi}$ with energy

$$E_p = -V\left(\frac{p\pm\Omega}{2}\right)^2 . \tag{26}$$

The calculation of Green's functions of the original operators a, b proceeds within the collective theory in the following fashion: A Green's function $\langle 0|T(a(t)a^+(t')|0\rangle$ is obtained† by differentiating:

$$-\frac{\delta^2}{\delta\eta^+(t)\,\delta\eta(t')} Z\Bigg|_{\substack{\eta=0\\ \eta^+=0}}$$

$$=\langle 0|Ta(t)a^+(t')|0\rangle = \int \mathcal{D}\varphi e^{i\int dt(-(1/V)\dot\varphi^2 \mp \Omega\dot\varphi)}$$

$$\cdot\, e^{-i\varphi(t)} e^{i\varphi(t')}\,\theta(t-t') . \tag{27}$$

This expression may easily be calculated by quadratic completion and performing the resulting Gaussian integrals. A more transparent way, however, consists in going to the collective quantum theory using again the general rule

$$\int \mathcal{D}\varphi e^{i\int dt\mathcal{L}^c + i\mathcal{A}_{curr}} \equiv \left\{0|Te^{i\mathcal{A}_{curr}}|0\right\} \tag{28}$$

where $\mathcal{A}_{curr}$ is the second piece in the action (23) containing all currents and the right-hand side is calculated by using free field propagators for $\hat\varphi$ which are determined by $\mathcal{L}^c$. The state $|0\}$ denotes the vacuum state of $\mathcal{L}^c$. Hence the fermion propagator becomes

$$\langle 0|Ta(t)a^+(t')|0\rangle = \left\{0|Te^{-i\hat\varphi(t)} e^{i\hat\varphi(t')}|0\right\}\theta(t-t') . \tag{29}$$

We now see that the fermion state corresponds to a state of momentum p = 1 in the collective theory. Its energy is from (23)

$$E_{a^+|0\rangle} \equiv E_1 = -V\left(\frac{1+\Omega}{2}\right)^2 \tag{30}$$

---

† Notice that we have changed from the $\mathcal{D}\rho$ functional integral in (21) to $\mathcal{D}\varphi$. Since $\delta\rho(t)/\delta\varphi(t') = -(1/V)\,\dot\delta(t-t')$, this change produces only an irrelevant factor in front of Z.

in agreement with (1). The expression on the right-hand side of Equation (29) is now evaluated directly by inserting

$$e^{-i\hat{\varphi}(t)} = e^{i\hat{H}t} e^{-i\hat{\varphi}(0)} e^{-i\hat{H}t}$$

and using the known energy values of the states $|0\}$ and $e^{i\hat{\varphi}}|0\}$:

$$\langle 0|Ta(t)a^{+}(t')|0\rangle = e^{-i(E_1-E_0)(t-t')}\theta(t-t').$$

This certainty agrees with the direct evaluation in the fermion Hilbert space based on (1).

The same thing holds for b. If the Green's function contains more than one pair of operators a,a^+ or b,b^+, the product rule of functional differentiation produces several terms corresponding to all possible contractions of pairs as one is used to form Wick's rule. For fermions, a minus sign results from the anticommutativity of the Grassmann variables. Take the simplest nontrivial example of a four-point function:

$$\langle 0|Ta_1(t_1)a_1(t_2)a_1^{+}(t_2')a_1^{+}(t_1')|0\rangle = \left.\frac{\delta^4 Z}{\delta\eta_1(t_1)\delta\eta_1(t_2)\delta\eta_1(t_2')\delta\eta_1(t_1')}\right|_{\substack{\eta=0\\ \eta^+=0}}$$

$$= \int \mathcal{D}\varphi\, e^{i\int dt(-(1/V)\dot{\varphi}^2 \mp \Omega\dot{\varphi})} e^{-i\varphi(t_1)} e^{-i\varphi(t_2)} e^{i\varphi(t_2')} e^{i\varphi(t_1')}$$

$$\times \left(\theta(t_1-t_1')\theta(t_2-t_2') \pm \theta(t_1-t_2')\theta(t_2-t_1')\right) \qquad (31)$$

$$= \left\{0|Te^{-i\hat{\varphi}(t_1)} e^{-i\hat{\varphi}(t_2)} e^{i\hat{\varphi}(t_2)} e^{i\hat{\varphi}(t_1)}|0\right\}$$

$$\times \left(\theta(t_1-t_1')\theta(t_2-t_2') \pm \theta(t_1-t_2')\theta(t_2-t_1')\right). \qquad (32)$$

Inserting again time translation operators $e^{-i\hat{\varphi}(t)} = e^{i\hat{H}t}e^{-i\hat{\varphi}(0)}e^{-iHt}$ and using the energy operator (25), one finds the correct time dependencies. Notice that the two particle state $a^{+2}|0\rangle$ corresponds to momentum 2 and energy $-V((2+\Omega)/2)^2$. However, only for bosons can it contribute to the Green's function. For fermions, such a term is cancelled by the fact that it appears twice with a minus sign arising from the anticommutativity of the external currents.

Notice that the Hilbert space of the collective theory is by far richer than that of the original fundamental fields a^+, a, b^+, b.

It contains states of all momenta p. The fundamental fields, on the other hand, occupy only certain discrete momenta at the integer values $p = 0, 1, 2, \ldots$. For fermions, there is even an additional truncation of states of momentum higher than 1. It is the Grassmann algebra which enforces this truncation while ensuring the respectation of Pauli's exclusion principle.

While the quantum mechanical way of calculating the Green's functions within the collective theory is useful in understanding what happens in the Hilbert space of the collective field theory, it is quite awkward to apply to more than one dimension, in particular to the relativistic situation where the time does not play a special role. A more direct and easily generalizable method for the evaluation of fermion propagators in the collective theory consists in the following procedure: One brings the products of exponentials in (29) or (32) to normal order by using Wick's contraction formula which can be written as

$$\left\{0\left|T e^{i\int dt\, j(t)\hat{\varphi}(t)}\right|0\right\} = e^{+\frac{1}{2}\int dt\,dt'\,(\delta/(\delta\varphi(t)))\,\overbrace{\hat{\varphi}(t)\hat{\varphi}(t')}\,(\delta/(\delta\varphi(t')))} \times \left\{0\left|:e^{i\int dt\, j(t)\hat{\varphi}(t)}:\right|0\right\}$$
$$= e^{-\frac{1}{2}\int dt\,dt'\, j(t)\overbrace{\hat{\varphi}(t)\hat{\varphi}(t')}\, j(t')} \tag{33}$$

where $\overbrace{\hat{\varphi}(t)\hat{\varphi}(t')}$ denotes the propagator of the $\hat{\varphi}$ field. In order to avoid subtleties from the pure divergence term $\mp\Omega\dot{\varphi}$ in the collective Lagrangian, let us transform it to a surface part by rewriting Equations (31) and (32) as

$$\int\mathcal{D}\varphi\, e^{i\int dt(-(1/V)\dot{\varphi}^2)} e^{\mp i\Omega\varphi(\infty)} e^{-i\varphi(t_1)} e^{-i\varphi(t_2)} e^{i\varphi(t_2')} e^{i\varphi(t_1')} e^{\pm i\Omega\varphi(-\infty)}$$
$$\times\left[\theta(t_1-t_1')\theta(t_2-t_2') \pm \theta(t_1-t_2')\theta(t_2-t_1')\right]$$
$$= \left\{0\left|e^{\mp i\Omega\hat{\varphi}(\infty)} T\left(e^{-i\hat{\varphi}(t_1)} e^{-i\hat{\varphi}(t_2)} e^{i\hat{\varphi}(t_2')} e^{i\hat{\varphi}(t_1')}\right) e^{\pm i\Omega\hat{\varphi}(-\infty)}\right|0\right\}$$
$$\times\left[\theta(t_1-t_1')\theta(t_2-t_2') \pm \theta(t_1-t_2')\theta(t_2-t_1')\right]. \tag{34}$$

The field $\hat{\varphi}(t)$ now follows the Lagrangian $-(1/V)\dot{\hat{\varphi}}^2$ <u>without</u> the divergence term. Hence, its propagator is

$$\overbrace{\hat{\varphi}(t)\hat{\varphi}(t')} = -\frac{V}{2}\int\frac{dE}{2\pi}\, e^{-iE(t-t')}\frac{i}{E^2} .$$

This can be made well-defined only by introducing a small regulator mass κ:

$$\overbrace{\hat{\varphi}(t)\hat{\varphi}(t')} = -\frac{V}{2}\int\frac{dE}{2\pi}\, e^{-iE(t-t')}\frac{i}{E^2-\kappa^2} = -\frac{V}{4\kappa} + i\frac{V}{4}|t-t'| + 0(\kappa) . \tag{35}$$

Let us now apply the Wick's rule (33) to (34). The general form of such an expression is

$$\left\{0|Te^{i\sum_i q_i\varphi(t_i)}|0\right\} = \left\{0|Te^{i\int dt(\sum_i q_i\delta(t-t_i))\hat{\varphi}(t)}|0\right\} \tag{36}$$

such that Wick's contraction rule yields

$$\begin{aligned}\left\{0|Te^{i\sum_i q_i\hat{\varphi}(t_i)}|0\right\} &= e^{-\frac{1}{2}\int dtdt'\sum_i q_i\delta(t-t_i)\overbrace{\hat{\varphi}(t)\hat{\varphi}(t')}\sum_j q_j\delta(t'-t_j)} \\ &\times\left\{0|:e^{i\sum_i q_i\hat{\varphi}(t_i)}:|0\right\} \\ &= e^{-\frac{1}{2}\sum_{ij}q_iq_j\overbrace{\hat{\varphi}(t_i)\hat{\varphi}(t_j)}} .\end{aligned} \tag{37}$$

The right-hand side of this equation becomes for $\kappa\to 0$

$$e^{(\sum_i q_i)^2(V/8\kappa)}\, e^{-i(V/8)\sum_{ij}q_iq_j|t_i-t_j|} .$$

If κ approaches zero, this expression explodes unless the sum of all is zero:

$$\sum_i q_i = 0 .$$

But this is always assured due to the equal number of η^+ and η currents (or λ^+ and λ) in (23) which guarantee the (separate) conservation of the fundamental particle numbers

$$N_a = \sum_i a_i^+ a_i \quad \text{and} \quad N_b = \sum_i b_i^+ b_i .$$

Thus one finds the general result

$$\left\{0\left|e^{\pm i\Omega\varphi(\infty)}\, T\left(e^{-i\hat{\varphi}(t_1)},\ldots,e^{-i\hat{\varphi}(t_n)}\, e^{i\hat{\varphi}(t_{n+1})},\ldots,e^{i\hat{\varphi}(t_{2n})}\right) e^{\mp i\Omega\varphi(-\infty)}\right|0\right\}$$

$$= e^{\displaystyle i\frac{V}{4}\sum_{\substack{i=1,\ldots,n\\ j=n+1,\ldots,2n}} |t_i-t_j| \pm i\frac{\Omega V}{2}\left(\sum_{i=1}^{n} t_i - \sum_{j=n+1}^{2n} t_j\right)} \tag{38}$$

In particular, the two-point function (29) agrees with what was calculated in the Schrödinger type of calculation.

3. PRESENCE OF PAIRING FORCE

So far the theory has been free in the collective fields φ. Let us now turn on a standard pairing force used in nuclear physics

$$\mathcal{L}_{pair} = V \sum_{ij} a_i^+ b_i^+ b_j a_j . \tag{39}$$

This term can be made quadratic in the fundamental fields by adding the complete square

$$\Delta\mathcal{L} = -V\left|S^+ - \sum_i b_i a_i\right|^2 . \tag{40}$$

In order to avoid the cumbersome distinction of Fermi and Bose statistics, let us confine attention to fermions from now on. Then the Lagrangian (10) becomes

$$\mathcal{L} = \sum_i a_i^+(i\partial_t + V\rho)a_i + b_i(i\partial_t - V\rho)b_i^+ + (VS^+ a_i^+ b_i^+ + \text{h.c.}) - V|S|^2 - V\rho^2 . \tag{41}$$

As a consequence, the generating functional (12) reads

$$Z = \int \mathcal{D}a^+\mathcal{D}a\mathcal{D}b^+\mathcal{D}b\mathcal{D}S^+\mathcal{D}S\mathcal{D}\rho\; e^{i(\int dt\mathcal{L} + (\eta^+ a + \lambda^+ b + \text{h.c.}))} . \tag{42}$$

Now the calculation of the propagator $G_{S,\rho}(t,t')$ corresponding to Equation (15) is somewhat more complicated:

$$\begin{pmatrix} i\partial_t + V\rho & VS^+ \\ VS & i\partial_t - V\rho \end{pmatrix} G_{s,\rho}(t,t') = i\delta(t-t'). \tag{43}$$

For the purpose of higher symmetry it is useful to introduce $\rho \equiv S_3$. Then

$$(i\partial_t + VS_i\sigma_i)G_{s,\rho}(t,t') = i\delta(t-t'). \tag{44}$$

This is obviously solved by

$$G_{s,\rho}(t,t') = u^{-1}(t)G_o(t,t')u(t'), \tag{45}$$

where G_o is the free propagator (17) if u(t) is made to satisfy

$$i\dot{u}^{-1}(t)u(t) = -iu^{-1}(t)\dot{u}(t) = -VS_i\sigma_i.$$

Parametrizing u(t) in terms of Euler angles gives the equations

$$\begin{aligned} -2VS_1 &= -\dot{\beta}\sin\gamma + \dot{\alpha}\sin\beta\cos\gamma \equiv \omega_1 \\ -2VS_2 &= \dot{\beta}\cos\gamma + \dot{\alpha}\sin\beta\sin\gamma \equiv \omega_2 \\ -2VS_3 &= \dot{\alpha}\cos\beta + \dot{\gamma} \equiv \omega_3. \end{aligned} \tag{46}$$

The collective Lagrangian is particularly simple if the generating functional is constructed not on the vacuum state zero but on the state $|\tilde{0}\rangle \equiv b_1^+ b_2^+ \ldots b^+|0\rangle$. Then the free Green's function has the form

$$G_o(t-t') = \langle\tilde{0}|T\begin{pmatrix} a(t)a^+(t') & a(t)b(t') \\ b^+(t)b(t') & b^+(t)b(t') \end{pmatrix}|\tilde{0}\rangle = \begin{pmatrix} 1 & 0 \\ 0 & 1 \end{pmatrix}\theta(t-t'), \tag{47}$$

and the right-hand side of Equation (19) vanishes in the presence of the external S_1, S_2, S_3 fields (due to the limit $t' = t+\epsilon$). The Lagrangian is therefore simply

$$\begin{aligned} \mathcal{L}^c &= -V\left(S_1^2 + S_2^2 + S_3^2\right) = -\frac{1}{4V}\left(\omega_1^2 + \omega_2^2 + \omega_3^2\right) \\ &= -\frac{1}{4V}\left(\dot{\alpha}^2 + \dot{\beta}^2 + \dot{\gamma}^2 + 2\dot{\alpha}\dot{\gamma}\cos\beta\right). \end{aligned} \tag{48}$$

This Lagrangian is highly nonlinear and coincides with that of the symmetric top. In addition, there is the action involving the external currents

$$A_{curr} = i \sum_i \int dt\, dt'\, j_i^+(t) u^{-1}(t) u(t') j_i(t') \theta(t-t') . \tag{49}$$

The Euler angles have become the collective fields of the degenerate fermion shell with pairing forces (39). The fields can be quantized canonically by means of differential operators on a Hilbert space of functions of $\alpha\beta\gamma$. The corresponding Hamiltonian can be written as

$$H = -V\left(\hat{L}_1^2 + \hat{L}_2^2 + \hat{L}_3^2\right) \tag{50}$$

with L being the differential operator

$$\hat{L}^{\pm} \equiv e^{\pm i\gamma}\left[\pm\partial_\beta + \text{ctf}\ \beta\, i\partial_\alpha - i\frac{1}{\sin\beta}\partial_\gamma\right];\ \hat{L}_3 \equiv -i\partial_\gamma . \tag{51}$$

This result is not surprising at all if one remembers that the Hamiltonian under consideration can be expressed in terms of quasi-spin operators

$$L^+ \equiv \sum_i a_i^+ b_i^+ = (L^-)^+, \quad L_3 = \frac{1}{2}\sum_i a_i^+ a_i - b_i b_i^+, \tag{52}$$

in the alternative form

$$H = -V\left(L_1^2 + L_2^2 + L_3^2\right). \tag{53}$$

Thus the Hamiltonians (50) and (53) coincide with that of a symmetric top with moments of inertia $I_1 = I_2 = I_3 = -1/2V$

$$H = \sum_i \frac{L_i^2}{2I_i} . \tag{54}$$

The Hilbert space of the collective theory defined by Equation (50) consists of all wave functions of the symmetric top $D^J_{KM}(\alpha\beta\gamma)$ (the rotation matrices) with energies depending only upon J. As in the previous simple example, the original fermion theory covers only a very small fraction of this Hilbert space due to the anticommutativity of the external currents.

Let us study the way in which the fermion Green's functions arise in the collective theory. Consider at first

$$G_{mm'}(t-t') = \langle\tilde{0}|T \begin{pmatrix} a_1(t)a_1^+(t') & a_1(t)b_1(t') \\ b_1^+(t)a_1^+(t') & b_1^+(t)b_1(t') \end{pmatrix}_{m,m'} |\tilde{0}\rangle. \quad (55)$$

If we differentiate the generating functional† Z

$$Z[\eta^+\eta\lambda^+\lambda] = \int \mathcal{D}\alpha\, \mathcal{D}\cos\beta\, \mathcal{D}\gamma\, e^{i\int dt\mathcal{L}^c - \sum_i \int dt dt' j_i^+(t) u^{-1}(t) u(t') j_i(t')\theta(t-t')}, \quad (56)$$

accordingly, we find

$$G_{m,m'}(t-t') = \int \mathcal{D}\alpha \mathcal{D}\cos p \mathcal{D}\gamma e^{i\int dt\mathcal{L}^c} (u^{-1}(t)u(t'))_{m,m'} \theta(t-t'). \quad (57)$$

This can be calculated most easily by going to the Schrödinger picture

$$G_{m,m'}(t-t') = \left\{\tilde{0}| D^{\frac{1}{2}*}_{mk}(\alpha\beta\gamma(t))\, D^{\frac{1}{2}}_{km'}(\hat{\alpha}\hat{\beta}\hat{\gamma}(t'))|\tilde{0}\right\}. \quad (58)$$

Since the reference state is a symmetric, it must be associated with the wave function

$$D^{o}_{oo}(\alpha,\beta,\gamma) \equiv \frac{1}{\sqrt{8\pi^2}} \quad (59)$$

with energy from Equation (53)

$$E_{\upsilon} = -V\frac{\Omega^2}{4}. \quad (60)$$

Inserting the time translation operator‡

$$D^{\frac{1}{2}}_{km'}(\alpha\beta\gamma(t)) = e^{i\hat{H}t} D^{\frac{1}{2}}_{km'}(\alpha\beta\gamma(0)) e^{-i\hat{H}t} \quad (61)$$

with the Hamiltonian (50), one finds a phase‡‡

---

† Notice that we have passed from the functional integral over $\mathcal{D}S$ to $\mathcal{D}\alpha\mathcal{D}\cos\beta\mathcal{D}\gamma$. It is straightforward to verify that the Jacobian $\delta(S_1 S_2 S_3)/\delta(\alpha,\cos\beta,\gamma)$ is indeed a constant [4].

‡ The Schrödinger angles α,β,γ coincide with the time dependent angles $\alpha(t), \beta(t), \gamma(t)$ at $t=0$.

‡‡ An infinite phase $e^{iE_0(\infty-(-\infty))}$ from initial and final wave functions has been removed in order to normalize Z to $Z[0,0,0,0]=1$.

$$e^{-i\Delta E(t-t')} \tag{62}$$

where ΔE is the energy difference between the state $D^{\frac{1}{2}}|\tilde{0}\}$ and the reference state $|\tilde{0}\}$

$$\Delta E = E_{J=\frac{1}{2}} - E_{J=0} = -V\frac{3}{4} \tag{63}$$

and an integral

$$\int d\alpha d\cos\beta d\gamma \{\tilde{0}|D^{\frac{1}{2}*}_{mk}(\alpha\beta\gamma)D^{\frac{1}{2}}_{km'}(\alpha\beta\gamma)|\tilde{0}\} = \delta_{mm'} . \tag{64}$$

This coincides exactly with the result one would have obtained from Equation (55) by inserting the Hamiltonian (53). Notice, that the orthogonality relation together with the Grassmann algebra ensures the validity of the anticommutation rules among the operators $a^+, a; b^+, b$. For higher Green's functions the functional derivatives amount again to all possible contractions of pairs, except that now each contraction is associated with

$$\overbrace{\psi_{im}\psi^+_{jm'}} = D^{\frac{1}{2}}_{mm'}(u^{-1}(t)u(t'))\theta(t-t')\delta_{ij} \tag{65}$$

where ψ_{im} stands for $\begin{pmatrix} a_i \\ b_i^+ \end{pmatrix}$. One can now proceed and construct the full Hilbert space by piling up operators a_i^+ or b_i on the reference state $|\tilde{0}\rangle \equiv b_1^+ b_2^+ \ldots b_\Omega^+|0\rangle$. First we shall go to the true vacuum state of a^+, b^+: $|0\rangle$, i.e., we shall calculate $Z[j^+j]$ on this state. For this, we obviously have to bring down successively $b_1^+(\infty)b_2^+(\infty)\ldots b_\Omega^+(\infty)b_\Omega(-\infty)\ldots b_2(-\infty)b_1(-\infty)$ by forming the functional derivatives

$$\frac{\delta^{2\Omega}}{\delta j_{1,-\frac{1}{2}}(\infty)\cdot\ldots\cdot\delta j_{\Omega,-\frac{1}{2}}(\infty)\,\delta j^+_{\Omega,-\frac{1}{2}}(-\infty)\cdot\ldots\cdot\delta j^+_{1,-\frac{1}{2}}(-\infty)} \;{}^{|\tilde{0}\rangle}Z$$

in the functional (56). Of the resulting $n!$ contractions only one combination survives, since all indices i, j are different and the Kronecker δ^{ij} permits only one set of contractions. The result is

$${}^{|0\rangle}Z[0,0,0,0] = \int \mathcal{D}\alpha\mathcal{D}\cos\beta\mathcal{D}\gamma e^{i\int dt\mathcal{L}^C}\left[D^{\frac{1}{2}}_{-\frac{1}{2}-\frac{1}{2}}(u^{-1}(\infty)u(-\infty))\right]^{\Omega} . \tag{66}$$

But from the coupling rules of angular momentum and the group property

$$\left[D^{\frac{1}{2}}_{-\frac{1}{2}-\frac{1}{2}}(u^{-1}(\infty)u(-\infty))\right]^{\Omega} = D^{\Omega/2}_{-(\Omega/2),-(\Omega/2)}(u^{-1}(\infty)u(-\infty))$$

$$= \sum_{k} D^{\Omega/2*}_{k,-(\Omega/2)}(u(\infty))D^{\Omega/2}_{k,-(\Omega/2)}(u(-\infty)). \qquad (67)$$

Going to the Schrödinger picture and inserting the time translation operator, one finds an infinite phase

$$e^{iE_{\Omega/2}2\infty}$$

which can be absorbed in the normalization factor N. Here $E_{\Omega/2}$ is the energy of the ground state $|0\rangle$ which has $J=\Omega/2$. The eigenfunctions

$$D^{\Omega/2*}_{k,-(\Omega/2)} \quad D^{\Omega/2}_{k,-(\Omega/2)}$$

now appear both at $t=0$ and the functional becomes in the Schrödinger picture

$$|0\rangle_{Z[0,0,0,0]} = \int d\alpha\, d\cos\beta\, d\gamma \sum_{k=-(\Omega/2)}^{\Omega/2} \{0k|\alpha\beta\gamma\}\{\alpha\beta\gamma|0k\} \qquad (68)$$

with the vacuum wave functions

$$\{\alpha\beta\gamma|0\} = D^{\Omega/2}_{k,-(\Omega/2)}(\alpha\beta\gamma). \qquad (69)$$

It is easy to verify how an additional unpaired particle a_1^+ added to the vacuum decreases $\Omega/2 \to (\Omega-1)/2$ and raises the third component of quasi-spin by $\frac{1}{2}$ unit. Differentiating (65) by

$$-\frac{\delta^2}{\delta j^+_{1,\frac{1}{2}}(\infty)\,\delta j_{1,\frac{1}{2}}(-\infty)},$$

one finds the following set of contractions. Picturing them within the original fermion language, there are

$$\langle\tilde{0}|T(b_1^+(\infty)\cdots b_\Omega^+(\infty)a_1(\infty)a_1^+(-\infty)b_\Omega(-\infty)\cdots b_1(-\infty))|\tilde{0}\rangle =$$

$$= \langle\tilde{0}|T(b_1^+(\infty)\cdots b_\Omega^+(\infty)a_1(\infty)a_1+(-\infty)b_\Omega(-\infty)\cdots b_1(-\infty))|\tilde{0}\rangle \qquad (70)$$

$$+ \langle\tilde{0}|T(b_1^+(\infty)\cdot\ldots\cdot b_\Omega^+(\infty)a_1(\infty)a_1^+(-\infty)b_\Omega(-\infty)\cdot\ldots\cdot b_1(-\infty))|\tilde{0}\rangle.$$

These render under the functional integral (66)

$$\left[D^{\frac{1}{2}}_{-\frac{1}{2}-\frac{1}{2}}(u^{-1}(\infty)u(-\infty))\right]^{\Omega} D^{\frac{1}{2}}_{\frac{1}{2}\frac{1}{2}}(u^{-1}(\infty)u(-\infty))$$

$$-\left[D^{\frac{1}{2}}_{-\frac{1}{2}-\frac{1}{2}}(u^{-1}(\infty)u(-\infty))\right]^{\Omega-1} D^{\frac{1}{2}}_{-\frac{1}{2}\frac{1}{2}}(u^{-1}(\infty)u(-\infty))D^{\frac{1}{2}}_{\frac{1}{2}-\frac{1}{2}}(u^{-1}(\infty)u(-\infty))$$

$$= D^{(\Omega-1)/2}_{-(\Omega-1)/2,-(\Omega-1)/2}(u^{-1}(\infty)u(-\infty))\Big(D^{\frac{1}{2}}_{-\frac{1}{2}-\frac{1}{2}}(u^{-1}(\infty)u(-\infty))$$

$$\cdot D^{\frac{1}{2}}_{\frac{1}{2}\frac{1}{2}}(u^{-1}(\infty)u(-\infty)) - D^{\frac{1}{2}}_{-\frac{1}{2}\frac{1}{2}}(u^{-1}(\infty)u(-\infty))D^{\frac{1}{2}}_{\frac{1}{2}-\frac{1}{2}}(u^{-1}(\infty)u(-\infty))\Big). \tag{71}$$

Employing the formula

$$D^{J}_{m,m'} = D^{J*}_{-m,-m'}(-)^{m-m'} = \left(D^{J}\right)^{-1}_{-m',-m}(-)^{m-m'}$$

such that

$$D^{\frac{1}{2}}_{\frac{1}{2}\frac{1}{2}} = \left(D^{\frac{1}{2}}\right)^{-1}_{-\frac{1}{2},-\frac{1}{2}}$$

$$D^{\frac{1}{2}}_{\frac{1}{2}-\frac{1}{2}} = -\left(D^{\frac{1}{2}}\right)^{-1}_{\frac{1}{2},-\frac{1}{2}}, \tag{72}$$

the right-hand side of Equation (71) becomes

$$D^{(\Omega-1)/2}_{-(\Omega-1)/2,-(\Omega-1)/2}(u^{-1}(\infty)u(-\infty))\Big\{D^{\frac{1}{2}}_{-\frac{1}{2}-\frac{1}{2}}(u^{-1}(\infty)u(-\infty))$$

$$\cdot\left(D^{\frac{1}{2}}\right)^{-1}_{-\frac{1}{2}-\frac{1}{2}}(u^{-1}(\infty)u(-\infty)) + D^{\frac{1}{2}}_{-\frac{1}{2}\frac{1}{2}}(u^{-1}(\infty)u(-\infty))\left(D^{\frac{1}{2}}\right)^{-1}_{\frac{1}{2}-\frac{1}{2}}(u^{-1}(\infty)u(-\infty))\Big\}$$

$$= D^{(\Omega-1)/2}_{-(\Omega-1)/2,-(\Omega-1)/2}(u^{-1}(\infty)u(-\infty)) \tag{73}$$

and therefore, in analogy to (66), (68):

$$Z[j^+ j]\Big|_{\substack{j^+=0\\ j=0}} = \int \mathcal{D}\alpha\mathcal{D}\cos\beta\mathcal{D}\gamma \cdot D^{(\Omega-1)/2}_{-(\Omega-1)/2\ -(\Omega-1)/2}(u^{-1}(\infty)u(-\infty))e^{i\int \mathcal{L}^c dt} \tag{74}$$

$$= \sum_{k=-(\Omega-1)/2}^{(\Omega-1)/2} \int d\alpha d\cos\beta d\gamma \{a_1,k|\alpha\beta\gamma\}\{\alpha\beta\gamma|a_1^+,k\}$$

with the Schrödinger wave functions

$$\{\alpha\beta\gamma|a_1^+,k\} = D^{(\Omega-1)/2}_{k,-(\Omega-1)/2}(\alpha\beta\gamma). \tag{75}$$

In a similar way, we may work our way through the whole Hilbert space!

4. CONCLUSION

Interacting Bose as well as Fermi theories can be transformed into generally nonlinear collective field theories without loss of information. The transformed theory is much richer than the original one—the fundamental particles occupying only a particular subspace of the full collective Hilbert space. Algebraically, the transition from one language to the other is quite analogous to the two possibilities of representing angular momentum operators: Once via fermionic creation and annihilation operators and once in the standard quantum mechanical form via differential operators on Euler angles—the angles being the new collective fields.

In the model presented here, the a^+, b^+ language provides actually the simpler description of the system. This runs parallel to the treatment of the asymmetric top [11] where the original angular momentum representation is much easier to handle than the Euler angle form. For complicated many-body systems, however, with very pronounced collective phenomena dominating the long-wave and small-frequency excitations the opposite is true: Collective fields do render the more powerful and direct treatment of the low-energy properties of the system.

REFERENCES

1. R. K. Bullough, these Proceedings. Also see R. K. Bullough in Interaction of Radiation with Condensed Matter, Vol. 1, International Atomic Energy Agency, Vienna 1977.
2. See M. D. Kruskal's lecture notes.
3. For a thorough discussion, see H. Kleinert, Erice School on Low-Temperature Physics, Fortschr. Phys. 6 (1978).
4. H. Kleinert, Lectures presented at the 1976 Erice Summer School.
5. S. Coleman, Phys. Rev. D 11, 2088 (1975); S. Mandelstam, Phys. Rev. D 11, 3026 (1975).
6. J. Schwinger, Phys. Rev. 128, 2425 (1962) and Trieste Lectures 1962, p. 89, I.A.E.C., Vienna 1963. B. Klaiber, Lectures in Theoretical Physics, Gordon and Breach, New York, 1968, p. 141.
7. L. P. Feynman and A. R. Hibbs, Path Integrals and Quantum Mechanics, McGraw-Hill, New York, 1968.
8. For a review see: D. R. Bes and R. A. Broglia, Lectures delivered at "E. Fermi" Varenna Summer School, Varenna, Como, Italy, 1976.
   For recent studies see: D. R. Bes, R. A. Broglia, R. Liotta, B. R. Mottelson, Phys. Letters 52B, 253 (1974); 56B, 109 (1975); Nucl. Phys. BA 260, 127 (1976).
   See also: R. W. Richardson, Journ. Math. Phys. 9, 1329 (1968); Ann. Phys. 65, 249 (1971) and N.Y.U. Preprint 1977 as well as references therein.
9. H. Kleinert, Phys. Letters B, June issue 1977.
10. B. S. DeWitt, Rev. Mod. Phys. 29, 377 (1957); K. S. Cheng, J. Math. Phys. 13, 1723 (1972).
11. See, e.g., H. Morgenau and G. M. Murphy, The Mathematics of Physics and Chemistry, Van Norstrand, New York, 1964, p. 368.

NONPERTURBATIVE SELF-INTERACTIONS, SOLITARY WAVES AND OTHERS†

Philip B. Burt
Department of Physics and Astronomy
Clemson University
Clemson, South Carolina 29631

1. INTRODUCTION

The discovery of the existence of exact, collisionally stable, localized solutions of classical field equations—solitons [1]—has inspired a large amount of research in quantum field theory in recent years—some of which has been described at this Institute. The principal analogy is that localized, classical solutions of nonlinear field equations, bags, etc. [2] represent, in some sense, extended objects, i.e., particles with structure. The major problem has been to find suitable methods of quantizing the motion of these systems, for example, by using path integrals, WKB approximations or other semiclassical techniques [3].

The work discussed in these lectures proceeds from a different point of view [4]. Intrinsically nonlinear <u>quantum</u> field theories occupy the center. The motivation is partly empirical—the existence of meson resonances makes the question of persistent self-interactions immediately interesting. In addition, the persistence of the interaction provides a basis for an analogy with intrinsically nonlinear classical theories in which solitary waves or solitons are not perturbative in character; that is, they represent new, nonlinear modes of the system.

In the first lecture the types of self-interactions are described. In order to form a basis for the generalization to

† Lectures given at NATO Advanced Study Institute on Nonlinear Equations in Physics and Mathematics, Istanbul, August, 1977.

A. O. Barut (ed.), Nonlinear Equations in Physics and Mathematics, 375–398.

nonlinear fields, the case of zero interaction is initially discussed from a point of view which dispenses with the superposition of solutions of a linear field equation. This discussion is necessary since nonlinear field equations have no simple superposition principle [5]. However, as it turns out, superposition of solutions of a differential equation is unnecessary even in the linear theory. Instead, the superposition principle of quantum theory is entirely adequate to enable one to construct quantities such as propagators. The generalization to solitary wave fields is straightforward. The solutions of the nonlinear field equations are analogues of linear field creation and annihilation operators. The algorithm developed to construct propagators in the linear theory generalizes exactly to the interacting theory. The first lecture concludes with a discussion of the physical significance of various elements of the nonlinear theory and some of its applications.

In the second lecture a motivation for the study of classical solutions of the field equations is provided. Classical solutions represent a generalized random phase approximation to the quantized fields in the following sense. The expectation values of quantized fields—c number functions—satisfy nonlinear partial differential equations which are formally equivalent to those of the quantum field theory if off diagonal matrix elements of the latter are ignored. In terms of classical solutions, the question of the nonperturbative nature of the solitary waves is discussed with a particular eye to the completeness properties of the solutions of the linear equations. A systematic method for constructing particular solutions of the nonlinear partial differential equations—the method of base equations—is reviewed briefly and several types of new, intrinsically nonlinear waves—collapsons—are described. Finally, some special solutions of a nonlinear Schrödinger equation are discussed.

2. NONPERTURBATIVE, SELF-INTERACTING QUANTUM FIELDS

The field theories discussed in these lectures have field equations which are nonlinear generalizations of the Klein-Gordon equation. Thus, the linear version of the theories will describe particles with no spin, no charge or other internal quantum numbers and mass m ($\hbar = c = 1$). Several types of nonlinear interactions, both polynomial and nonpolynomial, have been investigated [4] but, in this lecture, only the field theories with field equations

$$\partial_\mu \partial^\mu \phi + m^2 \phi + \alpha\phi^{2p+1} + \lambda\phi^{4p+1} = 0$$

$$\left(\partial_\mu \partial^\mu = \partial_t^2 - \vec{\nabla}^2\right) \tag{1}$$

will be considered.

The first law of theoretical physics is "try simple cases first," so consider the case $\alpha = \lambda = 0$—the linear or free-field case. Particular solutions of the Klein-Gordon equation may be written

$$U_{\vec{k}}^{(\pm)}(\check{x}) = A_{\vec{k}}^{(\pm)} e^{\mp i\check{k}\cdot\check{x}} (D\omega V)^{-\frac{1}{2}}, \tag{2}$$

where V is the volume of the system, D is an arbitrary constant, $A_{\vec{k}}^{(\pm)}$ are operators and

$$\check{k}^2 = k_o^2 - \vec{k}^2 = \omega^2 - \vec{k}^2 = m^2 .$$

$$(k_o = +\omega). \tag{3}$$

Since the Klein-Gordon equation is linear, one usually proceeds at this point to the construction, by superposition of $U_{\vec{k}}^{(\pm)}(\check{x})$ for all k, of the general solution. The condition that the fields describe uncharged particles is then imposed by requiring that the general solution be hermitian. However, if one is to generalize by analogy to nonlinear systems, superposition of solutions must be avoided—superposition of solutions of nonlinear equations does not give a new solution [5].

The essential result of hermiticity is that the spinless field should describe particles with no internal degrees of freedom. This can be accomplished by requiring

$$U_{\vec{k}}^{(+)}(\check{x}) = U_{\vec{k}}^{(-)}(\check{x})^{\dagger}, \tag{4}$$

which will be satisfied if

$$A_{\vec{k}}^{(+)} = A_{\vec{k}}^{(-)\dagger} . \tag{5}$$

Furthermore, the commutation relations of $A_{\vec{k}}^{(\pm)}$ are

$$\left[A_{\vec{k}}^{(+)}, A_{\vec{q}}^{(-)}\right] = \delta_{\vec{n}_{\vec{k}},\vec{n}_{\vec{q}}}, \tag{6}$$

where

$$\vec{k} = 2\pi V^{-\frac{1}{3}}\left(n_1\hat{e}_1 + n_2\hat{e}_2 + n_3\hat{e}_3\right). \tag{7}$$

Turning to the nonlinear theory, we wish to find solutions of Equation (1) which are nonlinear analogues of the solutions described in Equations (2)-(7). If we consider a decomposition of the field such that (6)

$$\phi = |\phi| e^{i\chi} \tag{8}$$

we observe at the outset that requiring ϕ to be hermitian imposes a constraint on the phase χ, i.e.,

$$\chi^{\dagger} = -\chi. \tag{9}$$

However, the most general physics is obtained by solving the equations of motion without constraint. Thus, we allow complex (non-hermitian) solutions of Equation (1). One set of solutions is

$$\phi_{\vec{k}}^{(\pm)}(\check{x}) =$$
$$= U_{\vec{k}}^{(\pm)}(\check{x})\left\{\left(1-\alpha U_{\vec{k}}^{(\pm)}(\check{x})^{2p}\Big/4(p+1)m^2\right)^2 - \lambda U_{\vec{k}}^{(\pm)}(\check{x})^{4p}\Big/4(2p+1)m^2\right\}^{-1/(2p)} \tag{10}$$

$$(p \neq 0,\ -1/2,\ -1)$$

where $U_{\vec{k}}^{(\pm)}(\check{x})$ are given by Equations (2)-(7). The expression in brackets is to be understood in terms of its formal series expansion in positive powers of $U_{\vec{k}}^{(\pm)}(\check{x})$.

These fields are clearly suitable for describing a persistent (nonswitchable) interaction. They satisfy the interacting field equations for all times rather than reducing to in or out fields at asymptotically large times. For $\alpha = \lambda = 0$ they reduce to the solutions of the linear equations (although by choice of D other than unity solutions can be constructed which are singular in this limit). They contain either creation or annihilation operators (but not both) and are thus direct nonlinear analogues of $U_{\vec{k}}^{(\pm)}(\check{x})$.

An essential property of these fields is that the phase velocity is constant. Thus, they are analogues of classical solitary waves. Henceforth (with apologies to Scott-Russell), I will refer to them as solitary wave fields, even though, as will be shown subsequently, they are unlocalized. My argument for this departure from the conventional nomenclature is the following. The remarkable and physically significant property of the solitary wave is that, in a system with both nonlinearity and dispersion, it propagates with constant phase velocity. Localization in space is interesting and noteworthy, but one can easily conceive of a wave packet consisting of a superposition of linear waves which is well localized for long times. The distinctive solitary feature is the constant phase velocity. (From the remarks made by Scott-Russell and his choice of the name "wave of translation," it appears to me that he also regarded the constant phase velocity as the significant physical property of the wave.)

A second basis for the analogy with classical solitary waves is that the nonlinear fields are nonperturbative—even though the solutions in Equation (10) are understood in terms of a power series. The solutions in Equation (10) are thus not manifestly nonperturbative, but an equally valid solution is

$$(p \neq 0, -1/2, -1) \tag{11}$$

$$\Psi_{\vec{k}}^{(\pm)}(\check{x}) = r^{-1/(2p)} U_{\vec{k}}^{(\pm)}(\check{x}) \left\{ 1 - \alpha U_{\vec{k}}^{(\pm)}(\check{x})^{2p} \Big/ 2(p+1)m^2 r + U_{\vec{k}}^{(\pm)}(\check{x})^{4p} \right\}^{-1/(2p)}$$

$$r = \left([\alpha/4(p+1)m^2]^2 - \lambda/4(2p+1)m^2 \right)^{\frac{1}{2}} . \tag{12}$$

It is clear that this solution, for $\alpha = \lambda = 0$, depends on the order in which these limits are taken. Thus, one cannot expect to construct such a solution from perturbation theory.

Even if a basis for rejecting the solutions given in Equations (11)-(12) is found, the solution in Equation (10) can still be seen to be nonperturbative. One of the principal assumptions of perturbation theory is that the interacting fields are related to the linear (in, out) fields by a unitary transformation. It only requires a simple calculation to see that $U_{\vec{k}}^{(\pm)}(\check{x})$ and $\phi_{\vec{k}}^{(\pm)}(\check{x})$ are not unitary equivalents. In the next section further evidence for the nonperturbative nature of the solutions will be given.

Since the solitary waves are intrinsically nonlinear, they represent excitations of the system which exist <u>in addition</u> to perturbative solutions. Previous studies in quantum field theory—at least those directed towards calculations of physical interest—have relied on various approximate schemes which are perturbative or semiclassical in nature. Consequently, we can expect new information from the inclusion of solitary waves. Some of the implications will be discussed below.

As might be expected, since the solitary wave fields are not unitary equivalents of the linear fields they are noncanonical in nature. In fact, it is easy to show that the time displacement operator of the solutions in Equation (10) is the free Hamiltonian

$$H = \sum_{\vec{k}} (\omega/2) A_{\vec{k}}^{(-)} A_{\vec{k}}^{(+)}, \tag{13}$$

even though the fields—with their explicit dependence on the coupling constants—are manifestly not free fields. Thus, the canonical equations of motion with H of Equation (13) as time displacement operator are not equivalent to the field equations.

In various classical theories, solitary waves have the property that the amplitude of the wave and the wave velocity are related. In this respect solitary waves of nonlinear generalizations of the Klein-Gordon equation are different (at least, those discussed here are different); the amplitude and wave velocity are independent—in fact, the amplitude contains an arbitrary scaling constant D (but, consistent with the fact that solutions of nonlinear differential equations cannot be multiplied by constants, the constant D does not appear in ϕ in a simple manner).

Finally, the solitary wave fields of Equation (10) have matrix elements which can become singular. For the $\lambda\phi^3$ interaction (in the field equation), the condition for singularity is (for $D = 1$)

$$\rho\lambda/8m^3 \gtrsim 1, \tag{14}$$

where

$$\rho = N/V \tag{15}$$

and N is the number of particles with momentum $\vec{k} \sim 0$ in the system. This singularity of the matrix elements resembles the instability exhibited by some classical solitons as the wave amplitude becomes large [7]. The interpretation given to the condition expressed in Equation (14) is that the series representation of the fields fails for certain critical densities, coupling constants or masses. Furthermore, the representation of the fields given in Equation (10) is invalid for vanishing mass. This result is not surprising. Solitary waves represent a "balance" between nonlinearity and dispersion. Since dispersion is supplied by the $m^2\phi$ term in the field equation, it will no longer counteract the nonlinear growth as m^2 approaches zero [8].

Finally, it is necessary to examine the localization of the solutions. Since $\phi_{\vec{k}}^{(\pm)}(\tilde{x})$ contain oscillating exponentials, one expects that they will describe unlocalized systems and this is indeed the case. The formal series expansion of Equation (10) is

$$\phi_{\vec{k}}^{(\pm)}(\tilde{x}) = \sum_{n=0}^{\infty} C_n^{1/(2p)}(\epsilon) r^n U_{\vec{k}}^{(\pm)}(\tilde{x})^{2pn+1} , \tag{16}$$

where

$$\epsilon = \frac{\alpha}{4(p+1)m^2 r} . \tag{17}$$

$C_n^{1/(2p)}$ is a Laguerre polynomial and r is given by Equation (12). Although this expansion is valid for all p its use will be confined to the case where 2p is integral.

Using the formal series expansion of the fields $\phi_{\vec{k}}^{(\pm)}(\tilde{x})$, it is easy to establish that, in parallel with the linear fields $U_{\vec{k}}^{(\pm)}(\tilde{x})$, the norm of $\phi_{\vec{k}}^{(\pm)}(\tilde{x})$ is independent of position. One has

$$||\phi_{\vec{k}}^{(-)}(\tilde{x})|0\rangle|^2 =$$

$$= \sum_{\substack{n=0 \\ m=0}}^{\infty} \frac{C_n^{1/(2p)} C_m^{1/(2p)}}{(D\omega V)} \left(\frac{r}{(D\omega V)^p}\right)^{n+m} \langle 0| U_{\vec{k}}^{(+)2pn+1} U_{\vec{k}}^{(-)2pm+1} |0\rangle$$

$$= \sum_{n=0}^{\infty} \frac{(C_n^{1/(2p)})^2}{D\omega V} \left[\frac{r}{(D\omega V)^p}\right]^{2n} (2pn+1)! , \tag{18}$$

which is clearly independent of position. In Equation (18) the factor (2pn+1)! arises from use of the commutator, Equation (6).

3. SOLITARY WAVE PROPAGATORS

Now consider the uses of the solutions of the nonlinear field equations. Specifically, one of the principal elements of a quantum field theory is the propagator. In the linear theory the propagator is usually constructed by superimposing the particular solutions $U_{\vec{k}}^{(\pm)}(\tilde{x})$ to obtain the general solution. However, this construction clearly will not generalize to the nonlinear theory since a superposition of solutions is no longer a solution to the nonlinear field equation. Nonetheless, it is possible to form the propagator in the linear theory without using the mathematical principle of superposition of solutions of a linear differential equation.

First, we see that the linear fields $U_{\vec{k}}^{(\pm)}(\check{x})$ can be interpreted as annihilation or creation operators for a system with momentum $\vec{k}$ at position $\check{x}$. Thus, the conditional probability amplitude that a particle with momentum $\vec{k}$, created at position $\check{x}$, is found in the state $|\epsilon\rangle$ is

$$a_{\epsilon\vec{k}}(\check{x}) = \langle\epsilon| U_{\vec{k}}^{(-)}(\check{x})|0\rangle . \tag{19a}$$

Since $|a_{\epsilon\vec{k}}(\check{x})|^2$ is independent of position, this interpretation is entirely consistent with the uncertainty principle—that is, the <u>probability</u> of creating a particle with momentum $\vec{k}$ at position $\check{x}$ is independent of $\check{x}$. Similarly, the conditional amplitude that a particle be annihilated from $|\epsilon\rangle$ at position $\check{y}$ with momentum $\check{q}$ is

$$\tilde{a}_{\epsilon\vec{q}}(\check{y}) = \langle 0| U_{\vec{q}}^{(+)}(\check{y})|\epsilon\rangle . \tag{19}$$

The amplitude for the sequence of events is

$$\wr^{\uparrow}_{\epsilon\vec{k}\vec{q}} = \tilde{a}_{\epsilon\vec{q}}(\check{y})\, a_{\epsilon\vec{k}}(\check{x})\,\theta(y_o - x_o) . \tag{20}$$

The amplitude for the time reversed sequence for negative energy particles is obtained by exchanging $\check{y}$ and $\check{x}$,

$$\wr^{\downarrow}_{\epsilon\vec{k}\vec{q}} = \tilde{a}_{\epsilon q}(\check{x})\, a_{\epsilon k}(\check{y})\,\theta(x_o - y_o) . \tag{21}$$

In the neutral theory using the Stuckelberg-Feynman point of view, the time reversed sequence for negative energy particles (identical with the direct sequence for antiparticles) is entirely indistinguishable in principle from the direct process—hence the total amplitude is the arithmetic average of direct and time reversed

$$\wr_{\epsilon\vec{k}\vec{q}} = \frac{1}{2}\left(\wr^{\uparrow}_{\epsilon\vec{k}\vec{q}} + \wr^{\downarrow}_{\epsilon\vec{k}\vec{q}}\right). \tag{22}$$

Now, according to the rules of quantum theory, one must sum over unobserved quantum numbers and intermediate states. Assuming that the set $|\epsilon\rangle$ is closed, one has

$$\Big\} = \sum_{\epsilon,\vec{k},\vec{q}} \frac{1}{2}\Big(\big\uparrow_{\epsilon\vec{k}\vec{q}} + \big\downarrow_{\epsilon\vec{k}\vec{q}} \Big)$$

$$= \frac{1}{2} \sum_{\vec{k},\vec{q}} \langle 0 | U_{\vec{q}}^{(+)}(\check{y}) U_{\vec{k}}^{(-)}(\check{x}) \theta(y_o - x_o) + U_{\vec{q}}^{(+)}(\check{x}) U_{\vec{k}}^{(-)}(\check{y}) \theta(x_o - y_o) | 0 \rangle$$

$$= \frac{1}{2} \sum_{\vec{k}} (D\omega V)^{-1} \Big(e^{-i\check{k}\cdot(\check{y}-\check{x})} \theta(y_o - x_o) + e^{-i\check{k}\cdot(\check{x}-\check{y})} \theta(x_o - y_o) \Big)$$

$$= i\Delta_F(\check{x}-\check{y}) \tag{23}$$

where the commutator, Equation (6), has been used and D chosen to be unity. Thus, to summarize, we find that the construction of the propagator in the linear theory is entirely independent of the mathematical principle superposition of solutions of a linear differential equation. Instead, this construction employs the superposition principle of quantum theory for probability amplitudes and the rules of quantum theory for combining these amplitudes.

Now, the superposition principle of quantum theory is independent of the form (or existence!) of field equations. For example, the field theory describing K mesons with the internal quantum number strangeness entering dynamically is unknown. Nonetheless, it is well established that K^o and $\overline{K}^o$ form eigenstates of CP, K_S^o and K_L^o, which are linear superpositions of K^o and $\overline{K}^o$. Consequently, the physical principles underlying the construction of the propagator are equally applicable for both linear or nonlinear fields.

The fields $\phi_{\vec{k}}^{(\pm)}(\check{x})$, containing annihilation or creation operators of the linear theory, are interpreted as creation or annihilation operators for solitary waves. That is, by analogy with Equation (19a), the conditional probability amplitude for a solitary wave with quantum numbers $\vec{k}$ created at $\check{x}$ to be found in $|\epsilon\rangle$ is

$$a_{\vec{k}\epsilon}^{sol}(\check{x}) = \langle \epsilon | \phi_{\vec{k}}^{(-)}(\check{x}) | 0 \rangle .$$

Proceeding as in the construction of the propagator for the linear theory, one obtains the solitary wave propagator

$$P_{(\check{x}-\check{y})}^{sol} = (\tfrac{1}{2}) \sum_{\vec{k},\vec{q}} \langle 0 | \phi_{\vec{k}}^{(+)}(\check{y}) \phi_{\vec{q}}^{(-)}(\check{x}) \theta(y_o - x_o) + \phi_{\vec{k}}^{(+)}(\check{x}) \phi_{\vec{q}}^{(-)}(\check{y}) \theta(x_o - y_o) | 0 \rangle =$$

$$= \sum_{n=0}^{\infty} \frac{i(C_n^{1/(2p)})^2}{(2\pi)^4} \left(\frac{r}{V^p}\right)^{2n} \frac{(2pn+1)!\,(2pn+1)^{2pn-2}}{D^{2pn+1}}$$

$$\int \frac{d^4q\, e^{-i\check{q}\cdot(\check{y}-\check{x})}(\check{q}^2-(2pn+1)^2m^2+i\epsilon)^{-1}}{(\vec{q}^{\,2}+(2pn+1)^2m^2)^{pn}}$$

$$= i \sum_{n=0}^{\infty} \left(C_n^{1/(2p)}\right)^2 \left(\frac{r}{V^p}\right)^{2n} \frac{(2pn+1)!\,(2pn+1)^{2pn-2}}{D^{2pn+1}}$$

$$\left(-\vec{\nabla}^2 + (2pn+1)^2m^2\right)^{-pn} \Delta_F\left(\check{x}-\check{y};(2pn+1)^2m^2\right). \tag{24}$$

In the final step of Equation (24), replacing the sum on $\vec{k}$ by an integral requires letting V become infinite. The factors V^{-2pn} have been left explicit in order to demonstrate the invariance of the propagator. Since $V\omega$ is invariant and all other terms are manifestly Lorentz invariant, the total solitary wave propagator is invariant. With this demonstration, the factors V^{-2pn} can be absorbed in re-defined coupling constants.

From inspection of the configuration space solitary wave propagator or its representation in momentum space,

$$P^{sol}_{(\check{k})} = i \sum_{n=0}^{\infty} \frac{(C_n^{1/(2p)}(\epsilon))^2 (rV^{-p})^{2n}(2pn+1)!\,(2pn+1)^{2pn-2}D^{-2pn-1}}{[\vec{k}^2+(2pn+1)^2m^2]^{pn}[\check{k}^2-(2pn+1)^2m^2+i\epsilon]} \tag{25}$$

we see that the solitary wave fields describe a many-particle system with mass spectrum

$$m_n = (2pn+1)m. \tag{26}$$

The leading expression in the propagator is the Feynman propagator of the linear theory while the higher mass terms, due to the presence of the factor $[\vec{k}^2+(2pn+1)^2m^2]^{-pn}$, contribute functions in which the singularity of Δ_F is "smeared" (the factor $[\vec{k}^2+(2pn+1)^2m^2]^{-pn}$ is effectively an integral operator in configuration space). Depending on the choice of p, this smearing effect quickly reduces the strength of the singularity of each successive term so that, for some n, the contributions are no longer singular.

The factor (2pn+1)! causes the series to be divergent. However, it is an asymptotic series. For large n and fixed $\vec{k}$, the nth term diverges while for fixed n and large $\vec{k}$, the nth term vanishes. The Borel summability of a solitary wave propagator for the $\lambda\phi^3$ interaction has been discussed previously [9] (the commutators were chosen differently in this case) but I have not examined the Borel sum of the propagator in Equation (25).

Since the momentum space propagator has poles at masses given by Equation (26) and clearly changes sign as a function of $|\check{k}|$, it must vanish between each pair of the infinite sequence of poles (see Figure 1). While the positions of the poles are independent of the coupling constants and depend on the interaction only through the exponent p, the position of the zeroes will, in general, depend on the coupling constants. Furthermore, the probability that a particular mass state will appear, proportional to the residue of the propagator at that mass, will also depend on the coupling constant. As mentioned previously, the solitary wave field is a kind of coherent wave operator, creating a coherent state similar to those observed in electromagnetic systems. However, in the latter, since the photon mass is zero, there is no mass splitting arising from the multiparticle contributions.

The $|\vec{k}|$ dependence of the solitary wave propagator is also very significant. For large, space-like momenta the higher terms of the series become small relative to the initial term (more precisely, the nth term is small relative to the (n-1)th, etc.). Consequently, at large momentum transfer or short distances, the

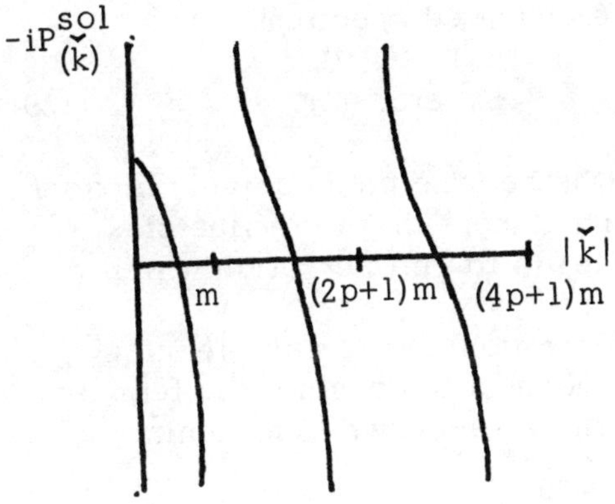

Figure 1. $-iP^{sol}_{(\check{k})}$ vs. $|\check{k}|$.

interaction becomes unimportant—the solitary wave propagator reduces to the linear field propagator. This is a solitary wave version of asymptotic freedom.

The importance of the concept of intrinsic nonlinearity of the solitary waves must be underscored. Since the solitary waves are nonperturbative, they represent new modes of the system. These modes must be counted independently when constructing the total propagator (although the possibility of overlap with linear modes is not excluded). Thus, the question of making a transition from a point $\check{x}$ to a point $\check{y}$ is answered by the total propagator

$$P^{tot}(\check{y}-\check{x}) = P^{pert}(\check{y}-\check{x}) + P^{sol}(\check{y}-\check{x}) + \dots \quad (27)$$

In making this calculation we must account for any overlap among the various contributions. The adjustable constant D in the solitary wave propagator is available for this purpose. One of the outstanding questions concerns the other terms + . . . Equation (27). What other modes must be included to completely determine the interacting field theory? In several applications of the theory I have assumed that there are no additional terms in Equation (27), but I am unable to substantiate this assumption.

The first application of the theory is to the experimental sequence of neutral, spin zero mesons. The mass formula of Equation (26) contains two parameters p and m. These have been evaluated by letting $m_n = m_{\pi^o}$ for n=o and $m_n = m_\eta$ for n = 1 [10]. The resulting spectrum is

$$m_n = 945(\text{MeV}/c^2),\ 1350,\ 1755,\ \text{etc.}, \quad (28)$$

which is to be compared with the observed spectrum

$$m = 958(\eta'),\ 1310(E),\ 1745,\ \text{etc. (?)}. \quad (29)$$

Furthermore, the theoretical widths become progressively broader for increasing n. It is interesting to note that the value of p(=3/2) arising from this theory leads to an interaction which is parity nonconserving.

A second application is to fermion-fermion and fermion-antifermion scattering [11]. The idea is to consider the fermions to interact with mesons through the well-known Hamiltonian

$$H^{int} = ig\bar{\psi}\gamma^5\psi\phi. \quad (30)$$

The interaction is treated perturbatively in g, but in the ensuing matrix elements the total propagator is to be inserted (the solitary wave propagator is treated as a finite sequence of terms). The lowest order cross section for $f-\overline{f}$ scattering consists of a sequence of peaks superimposed on a decreasing background—the peaks arising from the poles of the solitary wave propagator (which occur slightly off the real energy axis). The peaks are sharp due to the fact that the solitary wave propagator vanishes between pairs of poles.

This same idea has been used to calculate nucleon-nucleon potentials [12]. The solitary wave propagator contributes a sequence of terms to the potential which is similar to phenomenological potentials based on one boson exchange. The significant difference arises from the fact the OBE potentials contain a large number of undetermined parameters, while the solitary wave potentials contain only a few. The resulting potential has the general form

$$V^{nn} = V_o e^{-mr}/r + \sum_{n=1}^{\infty} f_n(r) e^{-(2pn+1)rm}/r, \tag{31}$$

where f_n is a polynomial determined by the solitary wave propagator. The comparison with experiment is in progress [13].

4. SOLITARY WAVES AND OTHERS

In this lecture I will be primarily concerned with classical solutions of nonlinear generalizations of the Klein-Gordon equation. The motivation for studying C number functions can be seen by taking the $\lambda\phi^3$ interaction as an example. If one takes the expectation value of the nonlinear differential equation for the quantized field, the interaction contributes a term of the form

$$\langle A|\lambda\phi^3|A\rangle = \sum_{I,I'} \lambda \langle A|\phi|I\rangle\langle I|\phi|I'\rangle\langle I'|\phi|A\rangle, \tag{32}$$

where complete sets of intermediate states $|I\rangle$ and $|I'\rangle$ have been inserted. If we now ignore all off-diagonal elements of ϕ in a kind of generalized random phase approximation, the expectation value of ϕ will satisfy a nonlinear differential equation which is identical, in appearance, with that satisfied by the quantized field.

I have studied several nonlinear generalizations of the Klein-Gordon equation, including the sine Gordon and double sine equations, and have found a variety of exact solutions which are solitary waves, solitons and others. The method of solution will be discussed briefly below, but first I want to discuss, in the context of C number functions, the nonperturbative nature of some of these solutions. In particular, the question of completeness of the solutions of the linear equation will be considered.

Consider a nonlinear generalization of the Klein-Gordon equation—to be specific, take the nonlinearity to be $\lambda\phi^3$ where ϕ is now a classical function. A simple view of perturbation theory is, for

$$\partial_\mu \partial^\mu \phi + m^2\phi + \lambda\phi^3 = 0, \tag{33}$$

$$\phi = \sum_{n=0}^{\infty} \lambda^n \phi^{(n)}. \tag{34}$$

So there is a hierarchy of equations of the form

$$\partial_\mu \partial^\mu \phi^{(0)} + m^2\phi^{(0)} = 0, \tag{35}$$

$$\partial_\mu \partial^\mu \phi^{(1)} + m^2\phi^{(1)} + \lambda\phi^{(0)\,3} = 0, \quad \text{etc.} \tag{36}$$

An essential ingredient in the perturbation theory solution of Equation (33) is that the set of solutions $\{\phi^{(0)}\}$ is complete with respect to a scalar product. Furthermore, this scalar product is independent of time. As is well known, if A^O and B^O are two solutions of the Klein-Gordon equation, Equation (35), the desired scalar product is

$$(A^O, B^O) = \int d^3x \left[\bar{A}^O \partial_O B^O - (\partial_O \bar{A}^O) B^O\right] \tag{37}$$

where bar denotes complex conjugate. The time independence of this scalar product follows from Equation (35), written for A^O and B^O. One has

$$\int\left\{\partial_O\left[\bar{A}^O \partial_O B^O - (\partial_O \bar{A}^O) B^O\right] - \vec{\nabla}\cdot\left[\bar{A}^O \vec{\nabla} B^O - (\vec{\nabla}\bar{A}^O) B^O\right]\right\} = 0; \tag{38}$$

thus, if A^O and B^O vanish or are periodic on the surface, one has

$$\partial_O \int d^3x \left[\bar{A}^O \partial_O B^O - (\partial_O \bar{A}^O) B^O\right] = 0. \tag{39}$$

Now, suppose we remove the superscripts so that A and B are solutions of the nonlinear equations. Following the same steps, the right side of Equation (39) is replaced by

$$\partial_o \int d^3x \left[\bar{A}\partial_o B - (\partial_o \bar{A})B\right] = -\lambda \int d^3x(\bar{A}B^3 - \bar{A}^3 B), \tag{40}$$

when A and B are assumed to satisfy the same boundary conditions as the solutions of the linear equation. This integral cannot be expected to vanish in general. Consequently, there is no conserved (time independent) scalar product and the concept of completeness is lost. Thus, completeness is an idea which properly belongs to the perturbative solutions based on the linearized equations and the resulting hierarchy of higher order (but still linear) differential equations. We may find many solutions of the nonlinear equation which have no expansion in terms of solutions of the linear equation without compromising this concept. Effectively, this means that the basic perturbative scheme developed to analyze interacting field theories can be retained without substantial modifications.

Now, suppose, despite the previous remarks, we attempt to use the completeness properties of the solutions of the linear equation and its scalar product to expand one of the nonlinear solutions for the $\lambda\phi^3$ interaction, say

$$\begin{aligned}\phi &= Be^{-i\check{k}\cdot\check{x}}\left(1 - \lambda B^2 e^{-2i\check{k}\cdot\check{x}}/8m^2\right)^{-1} \\ &= \sum_{n=0}^{\infty} (\lambda/8m^2)^n B^{2n+1} e^{-i(2n+1)\check{k}\cdot\check{x}} ,\end{aligned} \tag{41}$$

with

$$\begin{aligned}\check{k}^2 &= k_o^2 - \vec{k}^2 = \omega_{\vec{k}}^2 - \vec{k}^2 = m^2 , \\ k_o &= +\omega_{\vec{k}} .\end{aligned} \tag{42}$$

Then, choosing the solutions of the linear equation to be

$$U_{\vec{q}} = e^{-i\check{q}\cdot\check{x}} \Big/ (2\pi)^{3/2}\omega_q^{1/2} \tag{43}$$

with

$$\check{q}^2 = q^{o2} - \vec{q}^2 = \omega_q^2 - \vec{q}^2 = m^2$$
$$(q_o = \pm\omega_{\vec{q}}), \tag{44}$$

one has

$$\alpha_{\vec{k}}(\vec{q}) = (U_{\vec{q}}, \phi)$$
$$= i(2\pi)^{-3/2} \sum_{n=0}^{\infty} \frac{(\lambda/8m^2)^n B^{2n+1}}{\omega_{\vec{q}}^{\frac{1}{2}}} \delta[\vec{q}-(2n+1)\vec{k}]$$
$$\cdot\, e^{i[\omega_{\vec{q}}-(2n+1)\omega_{\vec{k}}]x_o} \left\{\pm\omega_{\vec{q}} + (2n+1)\omega_{\vec{k}}\right\}. \tag{45}$$

Now the delta functions in Equation (45) require

$$\vec{q} = (2n+1)\vec{k}, \tag{46}$$

hence

$$q_o^2 - \vec{q}^2 = (2n+1)^2\vec{k}^2 + m^2 - (2n+1)^2\vec{k}^2 = m^2. \tag{47}$$

But, in order for $\alpha_{\vec{k}}(\vec{q})$ to be independent of time, one must have

$$q_o = +\omega_{\vec{q}} = (2n+1)\omega_{\vec{k}}, \tag{48}$$

hence

$$\check{q}^2 = (2n+1)^2 m^2. \tag{49}$$

Consequently, the requirements on $\check{q}$ cannot be met and ϕ has no expansion in terms of the solutions of the Klein-Gordon equation. If we relax the requirement that u be a solution of the Klein-Gordon equation with mass m and form the four-dimensional Fourier transform, we find

$$\phi_{\vec{k}}(\check{x}) = \int d^4q\, e^{i\check{q}\cdot\check{x}} \beta_{\vec{k}}(\check{q}), \tag{50}$$

$$\beta_{\vec{k}} = \sum_{n=0}^{\infty} C_n(\check{q},\vec{k})\,\delta(\check{q} - (2n+1)\check{k}), \tag{51}$$

for which

$$\check{q}^2 = (2n+1)^2\check{k}^2 = (2n+1)^2m^2. \tag{52}$$

Now there is no problem with satisfying the conditions on the four momentum $\check{q}$, but the functions u no longer satisfy a single Klein-Gordon equation. Instead, in the sequence we find solutions of the set of equations

$$\partial_\mu \partial^\mu u + (2n+1)^2 m^2 u = 0. \tag{53}$$

In the quantum theory this property is reflected in the poles of the propagator.

With some understanding of the nature of the solutions of the nonlinear equations in mind, it is worthwhile to devote some space to the method of construction of these solutions. The starting point is a brief paper by E. Pinney [14] concerning solutions of the ordinary nonlinear differential equation

$$\frac{d^2y}{dt^2} + q(t)y + cy^{-3} = 0. \tag{54}$$

Pinney showed that some solutions can be constructed in terms of solutions u and v of the linear differential equation

$$\frac{d^2}{dt^2}\begin{pmatrix}u\\v\end{pmatrix} + q(t)\begin{pmatrix}u\\v\end{pmatrix} = 0. \tag{55}$$

If one has

$$W = u\frac{dv}{dt} - v\frac{du}{dt} \neq 0 \tag{56}$$

and

$$abW^2 = -c \tag{57}$$

where a and b are constants, then a particular solution is

$$y = (av^2 + bv^2)^{\frac{1}{2}}. \tag{58}$$

J. L. Reid has generalized this equation and its solutions extensively [15] and together we generalized further to nonlinear partial differential equations [16]. Recently, I have studied solutions of the nonlinear partial differential equations

$$\partial_\mu \partial^\mu \chi + m^2\chi + \alpha\chi^{2p+1} + \lambda\chi^{4p+1} = 0 \tag{59}$$

and

$$\partial_\mu \partial^\mu \phi + (M^2/g)\left[\sin(g\phi) + (b/2)\sin(2g\phi)\right] = 0 \tag{60}$$

by the Pinney technique (known more generally now as the method of base equations [17]). Without going into detail, it is easy to see how to start the generalization of the solution to Equation (54). That is, let, in solutions and differential equation,

$$-2 \longrightarrow p \tag{61}$$

so Equation (54) becomes

$$\frac{d^2y}{dt^2} + q(t)y + cy^{2p+1} = 0, \tag{62}$$

and with the <u>same</u> u and v as in Equations (56) and (57)

$$y = \left(au^{-p} + bv^{-p}\right)^{-1/p}. \tag{63}$$

There is a little more work to do to make the function in Equation (63) a solution of Equation (62); this is left as an exercise.

The solutions obtained are particular solutions [17] containing one arbitrary parameter. As discussed in the previous section, solutions of the nonlinear partial differential equations are valid for all times and are new, nonperturbative functions. In most cases the solutions have plane symmetry, although spherically symmetric solutions have also been constructed [18].

The solutions used in the previous lecture to discuss quantized fields are examples of solutions of Equation (59). For special values of the coupling constants, these solutions are related to solutions of the double sine Gordon equation, Equation (60). This equation is an example of a class of equations which are currently being studied in connection with the self-induced transparency of nonlinear optical systems [19].

One of the very interesting problems in the study of nonlinear equations such as Korteweg-deVries, sine Gordon and multiple sine Gordon is to develop criteria to predict the existence of multiple soliton solutions. The Korteweg-deVries and sine Gordon equations, in 1+1 dimensions, have an infinite number of conservation laws [19]. This has been suggested as a criterion for the existence of multisoliton solutions. On the other hand, it has been established recently [20], also in 1+1 dimensions, that the double sine Gordon equation does not have an infinite set of conservation laws. I have found multidimensional multisoliton solutions of several nonlinear partial differential equations including Equations (59) and (60) [21]. The most striking feature of these solutions is that the multiple soliton collapses to a single-peaked soliton when the dimension reduces to 1+1. In outline form, these results are obtained in the following way. Consider first the double sine Gordon equation. Particular solutions of this equation have been constructed by letting

$$\phi = (2/g)\sin^{-1}\psi, \tag{64}$$

where ψ satisfies the pair of equations

$$\partial_\mu \partial^\mu \psi + (M^2+b)\psi + D\psi^3 + B\psi^5 = 0 \tag{65}$$

and

$$(\partial_\mu \psi)(\partial^\mu \psi) = (M^2+3b+D)\psi^2(1-\psi^2) - (2b-B)\psi^4(1-\psi^2), \tag{66}$$

with

$$D = -2(M^2+b), \tag{67}$$

$$B = 3b. \tag{68}$$

Now, Equation (65) is a special case of Equation (59). Consequently, if we find solutions of the latter which also satisfy the constraint equation, Equation (66), using Equation (64) we can construct solutions of the double sine Gordon equation. Particular solutions of Equation (59) are

$$\chi = UP^{-1/(2p)}, \tag{69}$$

where

$$P = \left(1 - \alpha U^{2p}/4(p+1)m^2\right)^2 - \lambda U^{4p}/4m^2(2p+1), \tag{70}$$

provided

$$\partial_\mu \partial^\mu U + m^2 U = 0 \tag{71}$$

and

$$(\partial_\mu U)(\partial^\mu U) + m^2 U^2 = 0. \tag{72}$$

Plane wave solutions of Equations (71)-(72) are easily

constructed. One has

$$U = \sum_{i=1}^{N} a_i e^{\alpha_i \check{k}_i \cdot \check{x}} , \tag{73}$$

subject to the conditions

$$\alpha_i \alpha_j \check{k}_i \cdot \check{k}_j + m^2 = 0, \tag{74}$$

$$\alpha_i = \left(-m^2/\check{k}_i^2\right)^{\frac{1}{2}} \qquad (\check{k}_i^2 \neq 0) \tag{75}$$

and the a_i are constants.

The conditions in Equation (74) are very restrictive. They imply that

$$\check{k}_i \neq -\check{k}_j, \tag{76}$$

$$\vec{k}_i \cdot \vec{k}_j \neq 0, \tag{77}$$

and

$$N \leqq 2q - 1 \tag{78}$$

where q is the dimensionality of spacetime. Furthermore, it is easy to show that for $q = 2$, Equation (74) restricts N to be unity. Thus, for the special case of $1 + 1$ dimensions or when all $\vec{k}_i$ are parallel or antiparallel, the functions in Equation (73) collapse to a single function. The resulting functions are single peaked solitons or solitary waves. For this reason I have referred to these waves as collapsons. It is also noteworthy that the solutions of Equations (71)-(72) have their form limited by the dimensionality of spacetime through the condition in Equation (78).

It is evident from the form of U in Equation (73) that the solutions of Equation (69) consist of superpositions of waves. It is straightforward to show that these waves are localized and, in some regions of space are plane waves propagating with constant phase velocity. As the waves pass through each other, they experience a constant phase shift. Thus, they are multidimensional examples of multisolitons.

From the form of the solution χ of Equation (69) these multisolitons vanish as the arguments of the exponentials tend to infinity. In addition, for the special case of Equation (59),

$$\partial_\mu \partial^\mu \eta + m^2 \eta + \lambda \eta^3 = 0, \tag{79}$$

there are kink-like collapsons which reduce to constants asymptotically. These are [22]

$$\eta = (-m^2/\lambda)^{\frac{1}{2}} (U^2-1)(U^2+1)^{-1} \tag{80}$$

where U has the form given in Equation (73), except now

$$\alpha_i = \left(m^2 / 2\check{k}_i^{\,2} \right)^{\frac{1}{2}} \qquad (\check{k}_i^{\,2} \neq 0) \tag{81}$$

and all $\check{k}_i$ are space-like. Equations (76)-(78) also hold while Equation (74) is replaced by

$$\alpha_i \alpha_j \check{k}_i \cdot \check{k}_j - m^2/2 = 0. \tag{82}$$

These kink-like collapsons can also be used to construct multi-soliton solutions of the multidimensional sine Gordon equation [22].

As a final example of a nonlinear theory for which there are exact solutions, I will discuss briefly the nonlinear Schrödinger equation

$$i \frac{\partial \psi}{\partial t} + \sum_{i=1}^{n} \frac{\partial^2 \psi}{\partial x_i^2} + a|\psi|^{2p} \psi + b|\psi|^{4p} \psi = 0. \tag{83}$$

This is the wave equation for a unit mass system in which there are p+1 and 2p+1 body contact potentials. If we take

$$\psi = |\psi| e^{-iEt} = Re^{-iEt}, \tag{84}$$

then Equation (83) becomes

$$\sum_{i=1}^{n} \frac{\partial^2 R}{\partial x_i^2} + ER + aR^{2p+1} + bR^{4p+1} = 0. \tag{85}$$

This is a special case of Equation (59) and some of its solutions can be obtained by the method of base equations [17]. It is particularly interesting in one-space dimension, where a solution which vanishes at $x = \pm\infty$ is

$$R = \left[\frac{-a}{2(p+1)E} + \left\{ \left(\frac{a}{2(1+1)E} \right)^2 - \frac{b}{(2p+1)E} \right\}^{\frac{1}{2}} \cosh[2p(x+x')] \right]^{-1/(2p)} \tag{86}$$

where x' is a constant. For $E<0$, solutions of the Schrödinger equation usually give a discrete energy spectrum. However, in this case, if we examine the conditions on a, b and E (< 0) such that R is real and nonsingular, we find

$$a^2(2p+1)/4(p+1)b \leqq E < 0, \tag{87}$$

thus, there is a band of allowed energies [23].

5. CONCLUDING REMARKS

The main point of these two lectures is that persistent self-interactions of a quantum field theory can be successfully analyzed on the basis of an analogy with classical nonlinear field theories. The central element of the analogy is the existence of operator solutions of the field equations which are generalizations of the creation or annihilation operators of the linear theory. These intrinsically nonlinear fields are analogues of classical solitary waves and are nonperturbative solutions of the field equations. This last characteristic is very important since it allows one to retain the formal structure of perturbation theory. The outstanding question is the delimitation of the theory—what, if any, other types of solutions of the field equations must be included to complete the theory?

A second topic considered makes direct contact with the classical nonlinear theories; namely, particular solutions of the classical analogues of the quantized field equations—constructed by the method of base equations—are, in some cases, solitons or multisolitons. The new types of solitons, which I have called collapsons, are examples of multisolitons in multidimensional spacetime. Since some of these satisfy field equations which, in 1+1 dimensions, do not have an infinite number of conservation laws, an interesting question is raised—namely, is there a connection between conservation laws and multisolitons? Of course, this question has been asked before. Perhaps we are converging on an answer.

ACKNOWLEDGMENTS

At the outset I wish to thank my colleagues in the Institute for their stimulating attention. Special thanks go to Martin Kruskal for pointing out an error in the original lectures.

A portion of this work was completed at Stanford Linear Accelerator Center. I wish to thank Sid Drell and the Theory Group for their hospitality and many people, including Stan Brodsky, Karel Gaemers, Benedikt Humpert, Hannu I. Miettinen, Philip Mannheim, George Pocsik, H. C. Tze, and Marvin Weinstein for interesting comments and suggestions.

I thank CERN, where preparation of these notes was begun, for its hospitality.

Finally, I am grateful to my colleagues B. V. Bronk and H. W. Graben for their interest.

REFERENCES

1. N. J. Zabusky and M. D. Kruskal, P.R.L. 15, 240 (1965). For a review of classical applications, see A. C. Scott, F. Y. F. Chu, and D. W. McGlaughlin, Proc. IEEE 61, 1443 (1973).
2. G. Rosen, Jour. Math. Phys. 6, 1269 (1965); J. L. Gervais and A. Neveu, Phys. Reports 23C, 3 (1976); S. D. Drell, Quark Confinement Schemes in Field Theory, SLAC-PUB-1683, 1975 Erice lectures.
3. S. Coleman, 1975 Erice lectures; R. Jackiw, Rev. Mod. Phys. 49, 681 (1977); A. O. Barut, these proceedings.
4. P. B. Burt and J. L. Reid, Jour. Phys. A5, L88 (1972), Jour. Phys. A6, 1388 (1973); P. B. Burt, Acta Physica Polonica B7, 617 (1976).
5. "Simple superposition" means addition of solutions. More general superposition principles for nonlinear systems have been discussed by S. E. Jones and W. F. Ames, Jour. Math. Anal. Appl. 17, 484 (1967) and W. F. Ames (to be published). See also H. D. Wahlquist and F. B. Estabrook, Phys. Rev. Lett. 31, 1386 (1973).
6. This decomposition is clearly very complicated. For a discussion of such operators in momentum space, see W. Heitler, Quantum Theory of Radiation, Oxford, 1944.
7. R. C. Davidson, Methods in Nonlinear Plasma Theory, Academic Press, New York, 1972.
8. Dispersion enters the Klein-Gordon equation and its nonlinear generalizations in a way which is different from that in, for example, a dielectric. There, when the phase velocities of the individual waves are dependent on frequency and wave vector, one has an infinite set of distinct

wave equations. A(n) (approximate) Physical superposition principle allows one to construct wave packets from solutions of each of these equations. Since the phase velocities are frequency and wave vector dependent, the resulting wave packet is dispersive. On the other hand, for a given wave vector, the frequency of a solution of the Klein-Gordon equation is, due to the presence of the mass term, not a linear function of $\vec{k}$. The Klein-Gordon equation is linear—hence, solutions for different $\vec{k}$ can be superimposed and are dispersive. This dispersion persists in the perturbative solutions of the nonlinear Klein-Gordon equations.

9. P. B. Burt, Jour. Phys. A7, 356 (1974).
10. P. B. Burt, Lett. Nuovo Cim. 13, 26 (1975).
11. P. B. Burt and M. Sebhatu, Lett. Nuovo Cim. 13, 104 (1975).
12. M. Sebhatu, Nuovo Cim. 33A, 568 (1976); Lett. Nuovo Cim. 16, 463 (1976);
    M. Sebhatu and P. B. Burt, "On Two Nucleon Solitary Wave Exchange Potentials" (1977), to be published.
13. M. Sebhatu, to be published.
14. E. Pinney, Proc. Amer. Math. Soc. 1, 681 (1950).
15. J. L. Reid, Proc. Amer. Math. Soc. 27, 61 (1971).
16. J. L. Reid and P. B. Burt, Jour. Math. Anal. Appl. 47, 520 (1974);
    P. B. Burt and J. L. Reid, Jour. Math. Anal. Appl. 53, 43 (1976).
17. P. B. Burt, Exact Solutions of Nonlinear Generalizations of the Klein-Gordon and Schrödinger Equations, Jour. Math. Anal. Appl., in press (1977).
18. P. B. Burt and G. Pocsik, Lett. Nuovo Cimento 17, 329 (1976);
    P. B. Burt, Lett. Nuovo Cimento 18, 547 (1977).
19. R. K. Dodd and R. K. Bullough, Proc. Roy. Soc. (London) A351, 499 (1976).
20. R. K. Dodd and R. K. Bullough, Proc. Roy. Soc. (London) A352, 481 (1977).
21. P. B. Burt, Exact Multiple Soliton Solutions of the Double Sine Gordon Equation, Proc. Roy. Soc. (London), in press (1977).
22. P. B. Burt, Kink-Like Multisolitons in Polynomial Field Theories, to be published (1977).
23. Similar results have been obtained for more general interactions, F. Calogero (private communication).

BOUND STATES OF FERMIONS IN EXTERNAL AND SELF-CONSISTENT FIELDS†

J. Rafelski

CERN
Geneva

PREFACE. The properties of spin-$\frac{1}{2}$ fields in strong external potentials are investigated, motivated by the question 'What happens to the electron orbitals as the charge of the nucleus is increased without bounds?' Interest in this problem and in the related Klein paradox extends back nearly to the beginnings of relativistic quantum mechanics. However, the correct interpretation of the theory for overcritical potentials, where the parts of the complete set of single-particle solutions associated with particles and antiparticles are no longer distinct, was given only recently. The understanding of the spectrum of the Dirac equation is essential in order to obtain an appropriate physical description in the frame of quantum field theory. The strong binding by more than twice the rest mass of the particles in overcritical external potentials leads to spontaneous particle emission, accompanied by creation of a charged lowest energy state, i.e., a charged vacuum. In the limit of ever-increasing background (nuclear) charge, the charge of the vacuum can be described with the help of a relativistic Thomas-Fermi approximation. As a result, an average constant potential is found within the range of the background charge. This in turn permits the understanding and resolution of Klein's paradox. All these theoretical concepts can be tested by the observation of positron production from the decay of the neutral vacuum in heavy-ion collisions, such as uranium on

† Lectures given at NATO Advanced Study Institute on Nonlinear Equations in Physics and Mathematics, Istanbul, August 1977.

A. O. Barut (ed.), Nonlinear Equations in Physics and Mathematics, 399–467.

uranium, in which overcritical potential strength is exceeded. This is an excellent experimental system, since, as the distance between the ions changes, the overcritical potential is switched on and off.

A related problem is that of strongly interacting quantum fields. The investigations described here are motivated by the picture of quark bags, in which light quanta provide all the internal structure. It is assumed that a hadron is a bound state of a few (very) heavy particles (free quarks), which can be characterized by the properties of quasi-fermions (bound quarks) bound in an average gluon potential. From the point of view of the quasi-fermions, this can be treated as an external potential problem—but without an external source; the quasi-fermion amplitudes are the source of the average gluon field. These concepts lead to a c-number theory of interacting fields, amended by suitable boundary conditions.

The investigations of the region of validity of this approach to the description of strongly bound states (in three-space dimensions) are as yet inconclusive. I believe that if the bound state has a significantly smaller mass than the free Fermi field masses the self-consistent method, described above, will be qualitatively correct.

The properties of the solutions of fermion fields interacting with meson fields are interesting in themselves, apart from the point of view presented above. Several general properties of such solutions are considered with the help of field theory analogues of virial relations. A description of the algorithm used to generate examples of the self-consistent solutions in three-space dimensions is also presented. Since it is difficult to associate the properties of strongly bound states with those of weakly interacting quasi-particles, several options for the candidates for the relevant interaction are considered. The present investigations deal only with Abelian fields; non-Abelian analogues of Abelian interactions may have quite new and surprising properties.

1. SOLUTIONS OF THE DIRAC EQUATION

In this section the recent work that deals with the solutions of the Dirac equation in strong external potentials is reviewed.

1.1 The Dirac equation

The Dirac equation for a particle of mass m moving in an (external) potential V takes the form [1]:

$$i\frac{\partial\psi}{\partial t} = (\vec{\alpha}\cdot\vec{p} + \beta m + V)\psi. \tag{1.1}$$

We take the matrices α and β in a special representation

$$\vec{\alpha} = \begin{bmatrix} 0 & \vec{\sigma} \\ \vec{\sigma} & 0 \end{bmatrix}, \quad \beta = \begin{bmatrix} \mathbb{1} & 0 \\ 0 & -\mathbb{1} \end{bmatrix}. \tag{1.2}$$

The potential V in Equation (1.1) transforms under Lorentz transformations as the time component of a four-vector. The solution of the Dirac equation may be written for a spherically symmetric potential $V(|\vec{r}|)$:

$$\psi(\vec{r}) = \begin{bmatrix} g(r)\ \chi_\kappa^\mu(\hat{r}) \\ if(r)\ \chi_{-\kappa}^\mu(\hat{r}) \end{bmatrix}, \tag{1.3}$$

where $\chi_\kappa^\mu(\hat{r})$ is an eigenfunction of the total angular momentum operator. It is a combination of spherical harmonics and Pauli spinors. Substituting Equation (1.3) into the Dirac equation, and using the relations [2]

$$\vec{\sigma}\cdot\vec{\ell}\,\chi_\kappa^\mu = -(\kappa+1)\chi_\kappa^\mu, \tag{1.4}$$

$$\vec{\sigma}\cdot\vec{r}\,\chi_\kappa^\mu = -r\chi_{-\kappa}^\mu, \tag{1.5}$$

we find the coupled radial equations for the functions f and g

$$\left(\frac{d}{dr} - \frac{\kappa-1}{r}\right)f(r) + (\epsilon - V(r) - m)g(r) = 0, \tag{1.6a}$$

$$\left(\frac{d}{dr} + \frac{\kappa+1}{r}\right)g(r) + (V(r) - m - \epsilon)f(r) = 0. \tag{1.6b}$$

1.2 Bound states in a Coulomb potential

The solutions of the Dirac equation for a Coulomb potential V of a point nucleus

$$V_p = -\frac{Z\alpha}{r} \tag{1.7}$$

are well known. We recall here only that when $Z\alpha$ approaches the value one, all s states approach the eigenvalue $\epsilon_{ns} = 0$ and the derivatives $\partial\epsilon_{ns}/\partial Z$ become infinite. Beyond this point no usual s states can be found; the remainder of the usual spectrum is incomplete. Thus, in the case of a point nucleus, additional requirements are necessary to obtain a complete set of solutions of the Dirac equation [3]. The physical choice is to consider a nucleus of finite size. Then, if one so desires, the limit of the point nucleus may be taken. However, since the actual nuclei have finite size, this is not a necessary step. The electrostatic Coulomb potential of a finite-size nucleus consisting of a uniformly charged sphere is:

$$V(r) = \begin{cases} -\frac{3}{2}\,(Z\alpha/R)\,(1-\frac{1}{3}(r^2/R^2))\,, & 0 < r < R \\ -\,Z\alpha/r & R < r < \infty\,. \end{cases} \tag{1.8}$$

Most of the recent calculations [4], [5] have relied on numerical integration techniques both inside and outside the nucleus of Equation (1.6), although in the special case (1.8) one can proceed further analytically. Results for the energy eigenvalues are shown in Figure 1.1. The eigenvalues decrease monotonically as the charge increases. None of the eigenvalues or the wave functions exhibit any unusual behaviour at $Z\alpha = 1$. The points at which the levels join the lower continuum are well isolated. The critical value of $Z = Z^{cr}$ is ~ 170, where the 1s level joins the lower

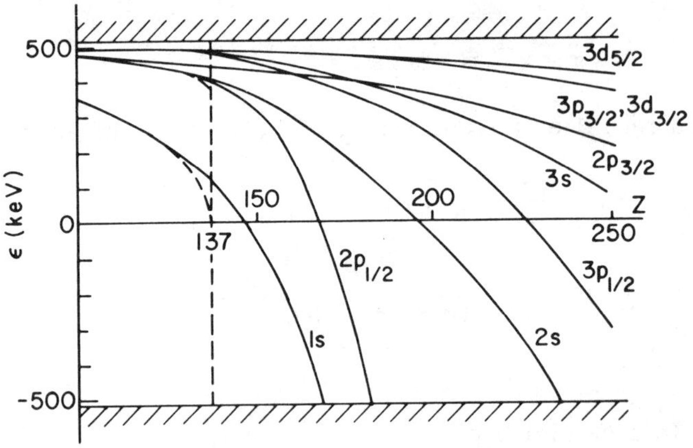

Fig. 1.1. Z-dependence of the eigenvalues for electrons bound to a uniformly charged sphere. The dashed curves refer to results for a point nucleus.

continuum. The $2p_{\frac{1}{2}}$ level joins the lower continuum at about Z=183. The exact location of Z^{cr} is an important question for the experimental verification of the theory. One has to consider in this context the effects of other atomic electrons (screening) as well as the effects of vacuum polarization [5].

1.3 Scalar potentials

The above bound state eigenvalues are based on electrostatic potentials. We consider next an example of a particle of mass m bound by a potential ϕ that transforms as a scalar under Lorentz transformations:

$$i\,\frac{\partial\psi}{\partial t} = \left(\vec{\alpha}\cdot\vec{p} + \beta(m+\phi)\psi\right). \tag{1.9}$$

Here it is possible to consider the potential as given by $\phi=-g/r$. Then one can find an analytic expression [6] for the energy eigenvalues

$$\epsilon_{\pm} = \pm m\left[1 - \frac{g^2}{(N+\sqrt{\kappa^2+g^2})^2}\right]^{\frac{1}{2}}, \tag{1.10}$$

where $N=0, 1, 2, \ldots$. The dependence of several of the lower eigenvalues on the strength of the potential g is shown in Figure 1.2. In this figure and Equation (1.10), the symmetry between positive and negative energies is apparent. From Equation (1.10) it is clear that $\epsilon_{\pm}\to 0$ for all states as $g\to\infty$. After

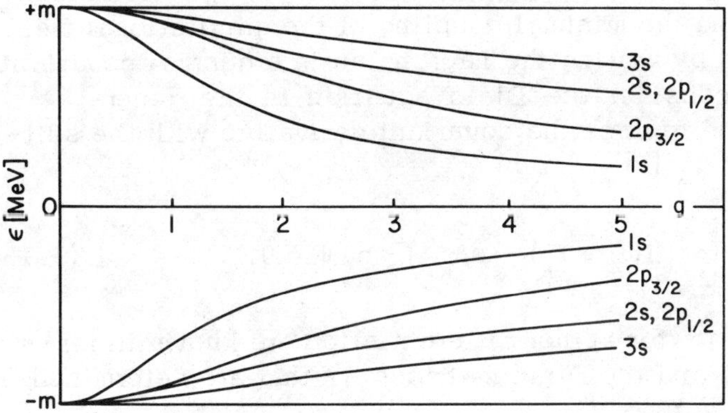

Fig. 1.2. Energy eigenvalues based on a scalar potential with strength g.

second quantization, the negative energy states are associated with antiparticles. Thus, unlike the case of strong electrostatic potentials, the parts of the spectrum associated with particles and antiparticles are always separated for finite g. This is the crucial difference between electrostatic potentials and scalar potentials.

1.4 Gravitational potential

Let us now consider the gravitational field as an external field. It is very much like any other external field. Any time-independent, spherically-symmetric mass distribution has a metric tensor g_{ik} of the particular form [7]

$$g_{ik} = \begin{pmatrix} e^{\nu} & & & \\ & -e^{\lambda} & & \\ & & -r & \\ & & & -r^2\sin^2\theta \end{pmatrix} \tag{1.11}$$

where the two functions e^{ν} and e^{λ} are determined from solutions of Einstein's equations of motion. For the particular case of a Schwarzschild metric in empty space, we have

$$e^{\nu} = e^{-\lambda} = 1 - \frac{2\tilde{M}}{r}. \tag{1.12}$$

Here $\tilde{M} = (\kappa M/c^2)$ where κ is the gravitational constant ($\kappa = 6.67 \times 10^{-8}$ dyne cm^2/g^2) and M is the gravitating mass.

Given g_{ik} (to be understood as the external gravitational field), we can find the minimal coupling of the gravitational field to any other field by writing the Lagrangian in a general covariant form. We now consider the Dirac equation in the general covariant form and replace the covariant derivative with the suitable connections Γ_κ [8]

$$\left[\gamma^{\kappa}\left(i(\partial_\kappa + \Gamma_\kappa) - eA_\kappa\right) - m\right]\psi = 0. \tag{1.13}$$

The connections Γ_κ for spinor differentiation are known in terms of the metric g_{ik} and the Dirac matrices γ_i that are determined with the aid of the g_{ik}:

$$\{\gamma_i, \gamma_k\} = 2g_{ik}. \tag{1.14}$$

The equations of motion become [8]

$$\left(e^{-\lambda/2}\left(\frac{d}{dr}+\frac{1}{r}\right)-\frac{\kappa}{r}\right)f(r)+\left(e^{-\nu/2}\epsilon - m\right)g(r) = 0,$$
$$\left(e^{-\lambda/2}\left(\frac{d}{dr}+\frac{1}{r}\right)+\frac{\kappa}{r}\right)g(r)+\left(-e^{-\nu/2}\epsilon - m\right)f(r) = 0, \qquad (1.15)$$

which should be compared with Equation (1.6). Inclusion of the vector potential is straightforward, involving the replacement of ϵ by $\epsilon - V$.

Equation (1.15) is the Dirac equation in interaction with the external gravitational field. There are no discrete solutions of this equation with the Schwarzschild metric, Equation (1.12). It can be shown that the external field is of supercritical strength. Moreover, the singular behaviour is different from that encountered for point nuclei. In the case of gravitational interaction all bound states meet at $\epsilon = 0$ as the functions e^{ν} and e^{λ} approach the Schwarzschild values, in a suitable parametrization. The parametrization that has been employed can be associated with the metric of an imcompressible fluid model for the constant mass density star, suitably augmented to avoid infinite pressure.

A typical spectrum of the Klein-Gordon equation in an external gravitational potential, which behaves as Equation (1.15), is shown in Figure 1.3 as a function of the mass radius R (which has been chosen unrealistically small only for numerical

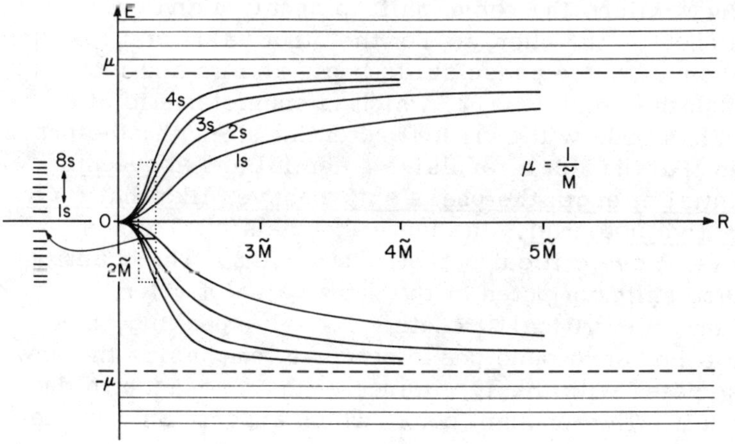

Fig. 1.3. Solution of the Klein-Gordon equation with external gravitational field. The eigenstates collapse to $\epsilon = 0$ as the Schwarzschild metric is approached.

convenience, $\tilde{M} = 1/\mu$, μ is the meson mass). We notice the collapse of the discrete solutions as the Schwarzschild metric solution is approached.

1.5 Continuum states of the square-well potential

Let us turn now to the discussion of the peculiarities of continuum solutions of the Dirac equation for a(n) (electrostatic) square-well potential

$$V = -V_0 \theta(R-r). \tag{1.16}$$

Our main interest will be confined to the lower continuum. The central issue of this discussion is the appearance of a resonance in the lower continuum as the potential strength is increased beyond the critical value. This resonance is the analytic continuation of the bound state into the lower continuum. Requiring that the solutions inside and outside the (attractive) square-well potential are continuous at the surface $r=R$, the phase shift δ can be computed from the relation [9]

$$\cot(pR+\delta) = \frac{(\epsilon+m)p'}{(\epsilon+m+V_0)p} \cot p'R + \frac{V_0}{pR(\epsilon+m+V_0)}. \tag{1.17}$$

Here $p' = \sqrt{(\epsilon+V_0)^2 - m^2}$, $p = \sqrt{\epsilon^2 - m^2}$. Phase shifts determined from this equation are shown for several values of the potential strength in Figure 1.4. For $R = m^{-1}$ and $V_0 = 4.29$ m, a value slightly less than the critical value, the phase shift is negative and exhibits no sharp dependence on the energy. For the other value of V_0, which are greater than V_0^{cr}, the phase shifts begin at zero and increase sharply through the point $\delta = \pi/2$, which is characteristic of a resonance. Thus, below the critical potential strength, the phase shift behaves like that for a repulsive potential and above the critical potential strength the phase shift behaves like that for a strong, attractive potential. In Figure 1.5 the $\sin^2(\delta-\delta_0)$ is shown for several overcritical potential strengths. The quantity δ_0 is the phase shift computed at the same value of ϵ for $V_0 = 4.29$ m, an undercritical strength. Loosely speaking, the subtraction of the background phase serves to emphasize the new features associated with the appearance of the resonance in the lower continuum. The resonant peak, which appears only in the overcritical cases, moves further into the lower continuum and becomes broader as the potential strength is increased.

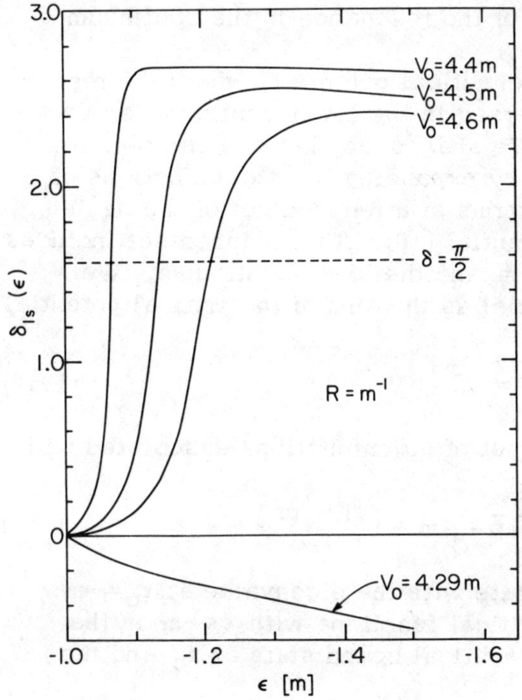

Fig. 1.4. Phase shifts for square-well potentials with strengths near the critical potential.

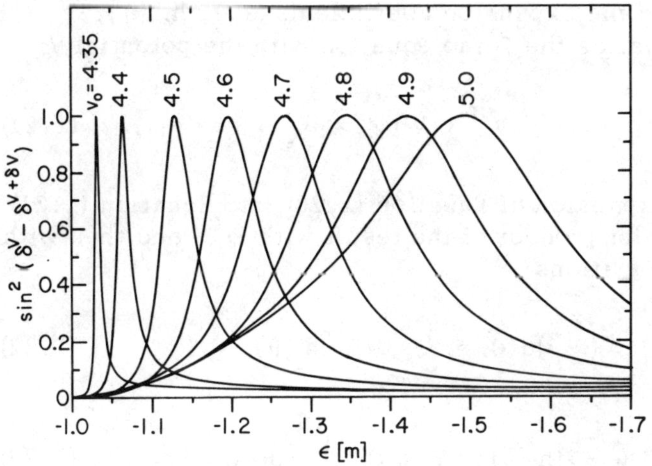

Fig. 1.5. The energy dependence of $\sin^2(\delta - \delta_0)$ for several square wells with strengths near the critical potential.

1.6 Perturbative treatment of the resonance in the continuum

Let us consider an overcritical potential, where the resonance is not so deeply immersed in the lower continuum as to have become too diffuse. We shall show that one can obtain the properties of the resonance by expanding the eigenfunctions of the overcritical problem in terms of a reduced set of the basis functions of the critical potential [10]. This reduced set includes only the bound state at $\epsilon = -m$ and the lower continuum. We write the overcritical potential as the sum of the critical potential and a piece V' as follows:

$$V = V^{cr} + V', \tag{1.18}$$

and introduce the complete set of eigenfunctions associated with V^{cr}:

$$\epsilon\psi_\epsilon^{cr} = \left(\vec{\alpha}\cdot\vec{p} + \beta m + V^{cr}\right)\psi_\epsilon^{cr}, \tag{1.19}$$

which includes the bound state with the eigenvalue at $\epsilon_0 = -m$.

We expand the overcritical functions with $\epsilon < -m$ in the reduced basis involving the critical bound state at ϵ_0 and the lower continuum:

$$\psi_\epsilon = a(\epsilon)\psi_0^{cr} + \int_{-\infty}^{-m} d\epsilon' h_{\epsilon'}(\epsilon)\psi_{\epsilon'}^{cr}, \quad \epsilon < -m. \tag{1.20}$$

To determine the expansion coefficients $a(\epsilon)$, $h_{\epsilon'}(\epsilon)$, we recall that ψ_ϵ satisfies the Dirac equation with the potential V:

$$(\vec{\alpha}\cdot\vec{p} + \beta m + V^{cr} + V')\psi_\epsilon = \epsilon\psi_\epsilon, \quad \epsilon < -m. \tag{1.21}$$

Substituting the expansion of Equation (1.20) into Equation (1.21) and taking the scalar product of the result with ψ_0^{cr} and then with $\psi_{\epsilon'}^{cr}$ leads to the equations

$$(\epsilon - \bar{\epsilon})a(\epsilon) = \int_{-\infty}^{-m} d\epsilon' V_{\epsilon'}^{*} h_{\epsilon'}(\epsilon), \tag{1.22}$$

$$(\epsilon - \bar{\epsilon})h_{\epsilon'}(\epsilon) = V_{\epsilon'} a(\epsilon) + \int_{-\infty}^{-m} d\epsilon'' U_{\epsilon'\epsilon''} h_{\epsilon''}(\epsilon), \tag{1.23}$$

where V_ϵ records the influence of the bound state on the continuum,

$$V_\epsilon = \int d^3x\, \psi_\epsilon^{\dagger cr}(\vec{x})V'(\vec{x})\psi_o^{cr}(\vec{x}), \tag{1.24}$$

and $U_{\epsilon\epsilon'}$ includes the effects of mixing among the continuum states,

$$U_{\epsilon\epsilon'} = \int d^3x\, \psi_\epsilon^{\dagger cr}(\vec{x})V'(\vec{x})\psi_{\epsilon'}^{cr}(\vec{x}). \tag{1.25}$$

The quantity $\bar{\epsilon}$ includes the diagonal shift in the energy of the former bound state, namely

$$\bar{\epsilon} = \epsilon_o + \int d^3x\, \psi_o^{\dagger cr}(\vec{x})V'(\vec{x})\psi_o^{cr}(\vec{x}). \tag{1.26}$$

In solving for the coefficients $a(\epsilon)$ and $h_{\epsilon'}(\epsilon)$ we also require the relation

$$\delta(\epsilon-\epsilon') = a^*(\epsilon)a(\epsilon') + \int_{-\infty}^{-m} d\epsilon''\, h^*_{\epsilon''}(\epsilon)h_{\epsilon''}(\epsilon'), \tag{1.27}$$

which arises from writing the normalization condition for ψ_ϵ in terms of the reduced basis.

The solution of Equations (1.22), (1.23), and (1.27) is straightforward. Ignoring the rearrangement of the continuum, Equation (1.25), the results are

$$a(\epsilon) = \frac{V_\epsilon^*}{(\epsilon - \bar{\epsilon} - F(\epsilon))}, \tag{1.28}$$

$$h_{\epsilon'}(\epsilon) = \frac{V_{\epsilon'}a(\epsilon)}{\epsilon - \epsilon' + i\eta} + \delta(\epsilon - \epsilon'), \tag{1.29}$$

where η is a small positive number and $F(\epsilon)$ is given by

$$F(\epsilon) = \int_{-\infty}^{-m} d\epsilon' \frac{|V_{\epsilon'}|^2}{\epsilon - \epsilon' + i\eta}. \tag{1.30}$$

At the complex value ϵ_1 determined from the nonlinear equation

$$\begin{aligned} \epsilon_1 &\equiv \epsilon_r + i\epsilon_i = \bar{\epsilon} + F(\epsilon_1) \\ &= \bar{\epsilon} + \int_{-\infty}^{-m} d\epsilon' \frac{|V_{\epsilon'}|^2(\epsilon_r - \epsilon')}{(\epsilon_r-\epsilon')^2 + \epsilon_i^2} - i\int_{-\infty}^{-m} d\epsilon' \frac{|V_{\epsilon'}|^2\epsilon_i}{(\epsilon_r-\epsilon') + \epsilon_i^2}, \end{aligned} \tag{1.31}$$

the amplitude a(ε) has a pole. In the narrow resonance approximation we find

$$\epsilon_i \approx -\pi|V_{\epsilon_r}|^2 \qquad (1.32)$$

and

$$\epsilon_r \approx \bar{\epsilon} + P\int_{-\infty}^{-m} d\epsilon' \frac{|V_{\epsilon'}|^2}{\epsilon_r - \epsilon'} \quad . \qquad (1.33)$$

Here the quantity P denotes the Cauchy principal value.

1.7 Continuum solutions, Coulomb potential

As in the case of bound states, we can obtain the solutions numerically by integrating the Dirac equation at a given energy ε. Then from the well-known asymptotic behaviour of the continuum solutions a phase shift can be deduced. However, in the case of the long-range Coulomb potential the phase shift also includes spurious contributions that must be carefully eliminated. Further details will not be given here. All these points have been carefully considered elsewhere [11], where the numerical calculations discussed here were also presented. The appearance of the resonance in the lower continuum for supercritical potentials can be well anticipated from the square-well solutions.

For the system Z = 184 ($_{92}U + _{92}U$), the resonance in the lower continuum is shown in Figure 1.6. It is centred at $\epsilon_r = -926$ keV with $\Gamma = 4.8$ keV (the full width at half maximum).

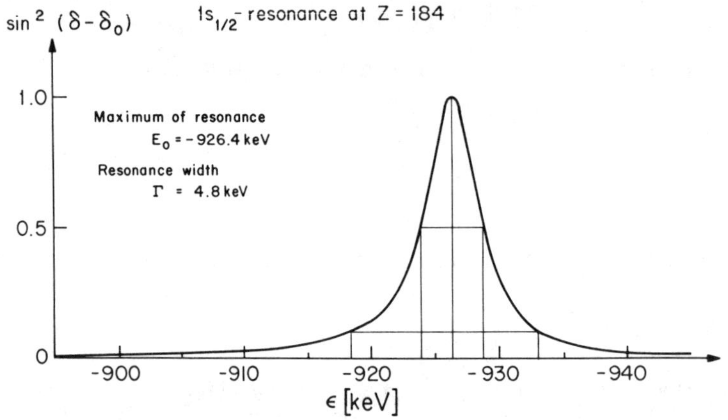

Fig. 1.6. The energy dependence of $\sin^2(\delta - \delta_0)$ in an overcritical electrostatic potential Z = 184.

The results for the positions of the $1s_{\frac{1}{2}}$ and $1p_{\frac{1}{2}}$ resonances as functions of the nuclear charge Z are shown in Figure 1.7. The slopes of the curves near the upper continuum are the same as the slopes of the corresponding curves in Figure 1.1 near the critical value of the potential. The position and width of the resonance can be parametrized as follows:

$$\epsilon_r \cong -m - (Z - Z^{cr})\delta - (Z - Z^{cr})^2 \tau \qquad (1.34)$$

$$\Gamma = 2|\epsilon_i| \cong (Z - Z^{cr})^2 \gamma , \qquad (1.35)$$

provided that $(Z - Z^{cr})/Z < 1$. The second of these expressions is applicable only if $Z > Z^{cr} + 3$. For values of Z nearer Z^{cr}, it is necessary to include a damping factor which takes into consideration that the probability of finding low-energy positrons near the nucleus is small when $Z \approx Z^{cr}$. For the $1s_{\frac{1}{2}}$ state the following values for δ, τ, and γ have been found: $\delta \cong 29$ keV, $\tau \cong 0.33$ keV, $\gamma = 0.04$ keV. The motivation for writing the results in the form of Equations (1.34) and (1.35) was to make contact with the perturbative development of Section 1.6. The form of the expression (1.33) suggests strongly the functional form for ϵ_r, since V' is proportional to Z - Z' and $|V_{\epsilon'}|^2 \sim (Z - Z^{cr})^2$. Similarly, Equation (1.32) suggests Equation (1.35), provided that the matrix-element continuum bound state is Z-independent. This is the case when the above condition $Z > Z^{cr} + 3$ is fulfilled.

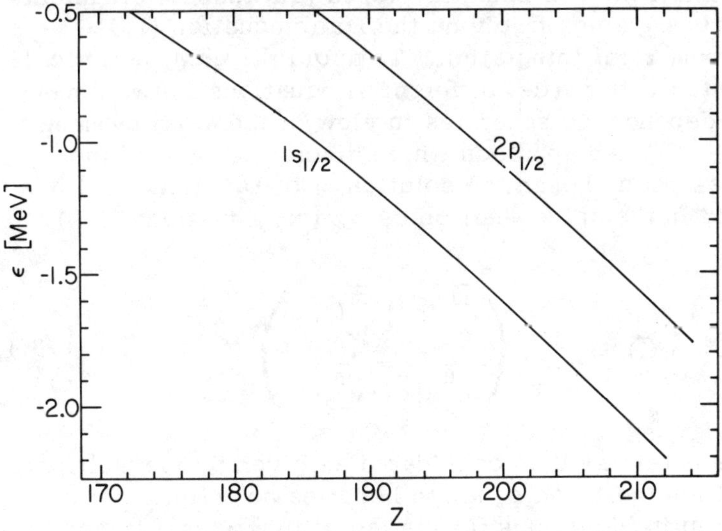

Fig. 1.7. The positions of the $1s_{\frac{1}{2}}$ and $2p_{\frac{1}{2}}$ resonances as functions of the nuclear charge.

1.8 The two-centre Dirac equation

In order to find out about the motion of electrons in the vicinity of the (moving) heavy ions separated by the distance R, we have to consider the combined Coulomb potential of both nuclei. The problems involved in solving the associated two-centre Dirac equation [12], [13] are probably an order of magnitude larger than those discussed above. This is because no operator has been found for the relativistic two-centre problem that plays the role of the square of the angular momentum J^2, and which would then allow the separation of the eigenstate problem into radial-like and angular-like one-dimensional differential (or matrix) equations.

The two-centre Coulomb potential V in cylindrical coordinates ρ and z is given by

$$V(\rho, z) = V_1 + V_2 =$$

$$= -\frac{Z_1\alpha}{[(z+R/2)^2+\rho^2]^{\frac{1}{2}}} - \frac{Z_2\alpha}{[(z-R/2)^2+\rho^2]^{\frac{1}{2}}} . \tag{1.36}$$

Suitable modifications for finite size and electron screening can be added when desired. The nuclei are located at $z=\pm R/2$, $y=0$, $x=0$.

A computer code has been developed [13] that for all intents and purposes is capable of solving the Dirac equation (1.1) exactly by a numerical integration. To avoid the great numerical effort associated with partial differential equations in two dimensions [the φ-dependence separates in view of the axial symmetry of Equation (1.36)], an approach which discretizes one of the dimensions has been chosen. A solution ψ of Equation (1.1) has been expanded in the spinor harmonics χ_κ [cf. Equations (1.4), (1.5)]

$$\psi(\vec{r}) = \sum_{\kappa,\overline{m}} \begin{pmatrix} f_\kappa^{\overline{m}}(|\vec{r}|)\chi_\kappa^{\overline{m}}(\hat{r}) \\ ig_\kappa^{\overline{m}}(|\vec{r}|)\chi_{-\kappa}^{\overline{m}}(\hat{r}) \end{pmatrix} . \tag{1.37}$$

Similarly, the potential V is considered as given by the multipole expansion. Then the Dirac equation becomes an infinite set of coupled differential equations for the radial functions $f_n^{\overline{m}}$ and $g_n^{\overline{m}}$. The coupling between different functions is a consequence of the multipole character of the potential V. A typical eigenvalue spectrum is shown in Figure 1.8 for several collision systems as a function of the heavy ion separation R.

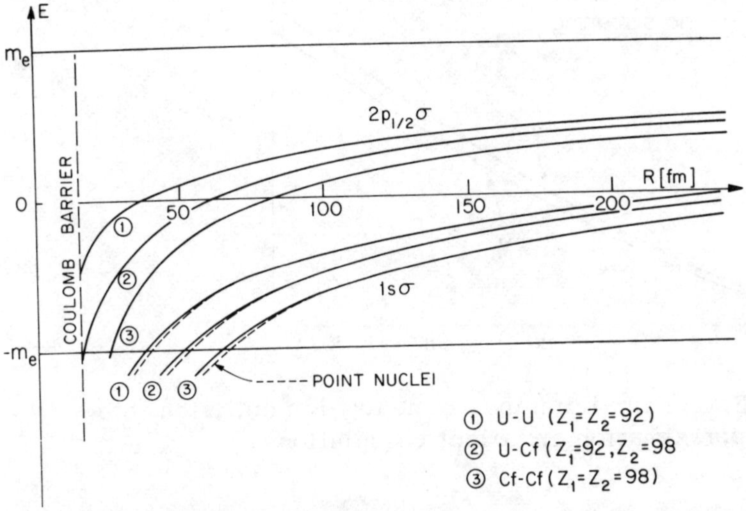

Fig. 1.8. Quasi-molecular correlation diagrams for several supercritical systems; both point- and finite-size nuclei are considered.

There is one particular aspect of the solutions of the two-centre problem which has the greatest impact on all calculations involving spontaneous positron production. The characteristic parameter is the critical distance R^{cr} between the heavy ions at which the binding of the electron exceeds twice its mass. The simplest method to estimate R^{cr} is to solve the Dirac equation with the monopole approximation for the two-centre potential V, that is, with

$$V_o = \begin{cases} -2Z\alpha/r, & r > R/2, \\ -4Z\alpha/R, & r < R/2, \end{cases} \tag{1.38}$$

which corresponds to a model in which the nuclear charge is distributed on a shell with the diameter corresponding to the distance between the nuclei. The critical distance in this approximation is a lower bound on the critical distance calculated with the full potential. The results of such monopole calculations are shown in Figure 1.9—together with the exact solution obtained by the numerical integration of the Dirac equation.

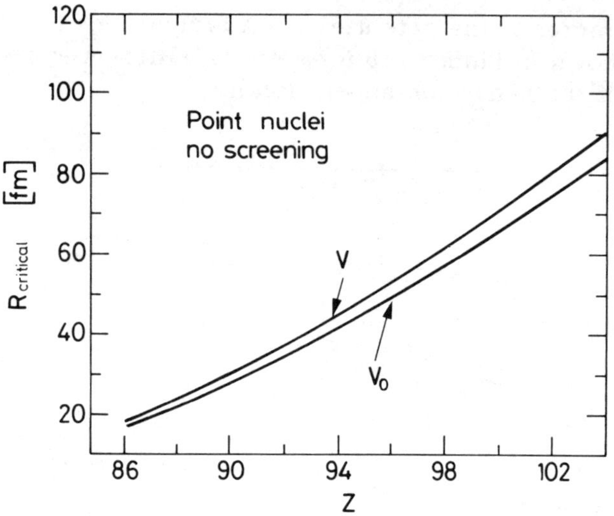

Fig. 1.9. The critical distance in heavy-ion collision; the monopole approximation and exact calculations.

In this section the surprising character of the continuum solutions of the Dirac equation has been described. We have found that a repulsive supercritical potential leads to resonant states (repulsive for 'negative frequency' positron states); this indicates that the relativistic equation of motion contains an attractive force irrespective of the sign of the potential. A related development has been reported for a Dirac particle with an anomalous magnetic moment coupled to the electromagnetic field [14].

2. QUANTUM FIELD THEORY OF SPIN-$\frac{1}{2}$ PARTICLES IN STRONG EXTERNAL FIELDS

The correct treatment of quantized Fermi fields in supercritical external potentials is presented in this section. In particular, it is demonstrated here that:

a) In overcritical <u>external</u> fields, it is impossible to have a stable, neutral vacuum [15].
b) The neutral vacuum decays by emitting positrons into a new ground state—a charged vacuum [15].
c) The charged vacuum is stable [15].
d) In the <u>self-consistent</u> supercritical potential the neutral vacuum is stable [16].

We first consider the main physical distinction between the case of undercritical and overcritical external potentials. The minimal energy required to make a fermion pair is the energy difference between the lowest bound state and the lower continuum. As the strength of the external potential increases, it decreases. At the critical field strength, where the lowest bound level joins the lower continuum, this energy is zero. There the neutral vacuum, one-pair and two-pair states are all degenerate in energy. As the external field strength is increased beyond the critical value, then the state containing two positrons has the lowest energy, and the neutral vacuum will spontaneously decay by the emission of two positrons.

This change of the qualitative nature of the ground state as the potential increases through its critical strength is analogous to a phase transition, as is shown schematically in Figure 2.1. There, the total energy, including that of the positron of zero kinetic energy, is plotted qualitatively as a function of the charge surrounding the nucleus. While in Figure 2.1a the minimum is found for neutral vacuum, in the case of the supercritical potential strength it shifts in Figure 2.1b to a finite value, chosen here to correspond to the supercritical $1s_{\frac{1}{2}}$ shell only.

In the remainder of this section we will make this picture more quantitative as well as prove the points (a) to (d) stated above.

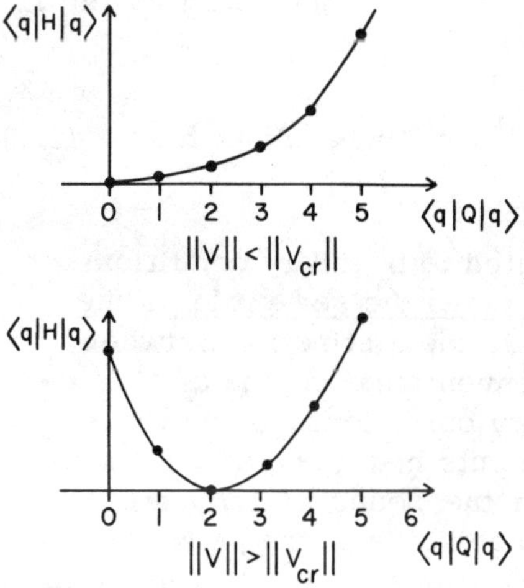

Fig. 2.1. The relation between energy and charge for (a) undercritical and (b) overcritical potentials.

2.1 The Green's function and its analytic structure

The quantized Fermi field describes both particles and antiparticles simultaneously. However, the distinction between the two is not as simple as it might appear.

For the free particles and the weak field case, the distinction is straightforward. All positive energy states may be identified as particle states and all negative energy continuum states may be identified as antiparticle states. When the potential becomes strong enough, so that some of the bound state eigenvalues become negative (as in Figures 1.1 and 1.6), then one has to make a decision as to which category contains these states. The theory can be treated without further modifications by identifying localized negative energy eigenstates with either particles or antiparticles as long as the field strength is subcritical.

In order to treat the critical potential strength and beyond, we must first recall the properties of the Feynman propagator.

Let us consider the case of a particle moving in a time-independent potential V_μ. A Green's function satisfies the equations

$$\left(i\gamma\cdot\frac{\partial}{\partial x} - \gamma\cdot V - m\right) G(x,x') = \delta^4(x-x') \tag{2.1}$$

$$\left(-i\frac{\partial}{\partial x'}\cdot\right)G(x,x')\gamma - G(x,x')(\gamma\cdot V+m) = \delta^4(x-x') . \tag{2.2}$$

Because of the time independence of the potential, we may introduce a partial Fourier transformation

$$G_C(x,x') = \int_C \frac{d\omega}{2\pi}\, e^{-i\omega(t-t')} G(\vec{x},\vec{x}';\omega) . \tag{2.3}$$

The choice of the contour C is related to boundary conditions satisfied by $G(x,x')$ as $t\to\pm\infty$. It plays the same role as the choice of the Fermi energy and makes the distinction between particles and antiparticles. The conventional choice of C, which leads to the Feynman boundary conditions, is shown in Figure 2.2. There, the two branch cuts beginning at $\omega = \pm m$ as well as the poles associated with the bound states, are shown. For undercritical fields, the choice of the Fermi energy is to some extent arbitrary. In weak fields—that is, if all of the eigenstates are near the continuum—then one is free to choose the Fermi energy anywhere in the gap between the lowest bound state and the other continuum. The conventional choice is $\epsilon_F = 0$.

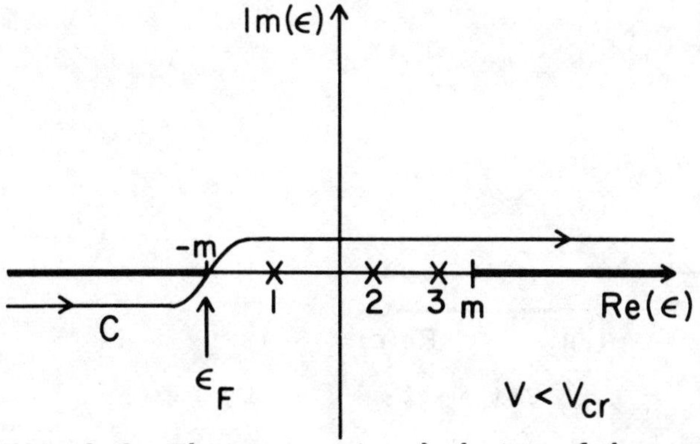

Fig. 2.2. The conventional choice of the contour in the complex ω plane. The contour C crosses the real axis at the Fermi energy ϵ_F.

Increasing the strength of the potential so that the lowest eigenvalue has the value zero poses no problem. One can simply lower the Fermi surface until $\epsilon_F = -m$.

The integrand of Equation (2.3) may be represented as a sum over the entire spectrum of eigensolutions of the Dirac equation, namely,

$$G(\vec{x},\vec{x}';\omega) = \sum_{\epsilon} \frac{\psi_\epsilon(\vec{x})\bar{\psi}_\epsilon(\vec{x}')}{\omega - \epsilon} . \tag{2.4}$$

Substituting this expression into Equation (2.3) and using the contour of Figure 2.1 leads to the representation

$$\begin{aligned} S_F(x,x') &= -i\theta(t-t') \sum_{\epsilon > \epsilon_F} \psi_\epsilon(\vec{x})\bar{\psi}_\epsilon(\vec{x}')e^{-i\epsilon(t-t')} \\ &\quad + i\theta(t'-t) \sum_{\epsilon < \epsilon_F} \psi_\epsilon(\vec{x})\bar{\psi}_\epsilon(\vec{x}')e^{-i\epsilon(t-t')} \end{aligned} \tag{2.5}$$

for the Feynman propagator S_F.

In the discussion of the solutions of the Dirac equation we have learned that a bound state may become a resonance in supercritical potentials. The behaviour of $G(x,x';\omega)$ in the complex ω plane reflects this property. As the potential strength is increased from an undercritical value to an overcritical value, the pole associated with the lowest bound state in Figure 2.2 moves off the real axis and into the upper half of the complex plane as

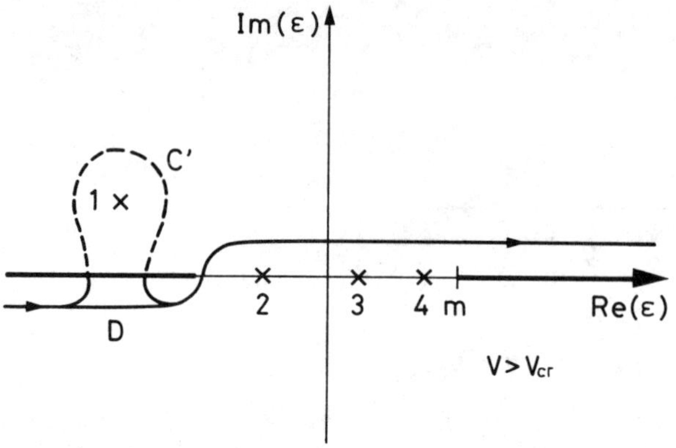

Fig. 2.3. The contours in the overcritical case.

shown in Figure 2.3. It is important to appreciate that this singularity is on the second sheet [17]. The contour C can now either be deformed into the contour C', so as to continue to embrace the pole, or the contour D is chosen, where the Fermi energy remains at -m. The path C' corresponds to the choice of the neutral vacuum as the reference state, while contour D is associated with a charged vacuum state (to be defined further below).

Substituting the first term of Equation (1.20) into Equation (2.4) for G, we find that

$$G_D(\vec{x},\vec{x}';\omega) \sim \int_{-\infty}^{\infty} d\epsilon \, \frac{|a(\epsilon)|^2}{\omega-\epsilon-i\eta} \, \psi_o^{cr}(\vec{x})\bar{\psi}_o^{cr}(\vec{x}') + \ldots, \tag{2.6}$$

where only the interesting part of G has been kept and η is negative when $\omega<-m$ and positive when $\omega>m$, as required by the choice of contour D. The modifications of Equation (2.6), due to the continuum part of Equation (1.20), are small. From Equation (1.28) it is apparent that $a(\epsilon)$ carries the singularity associated with the resonance, that is the pole shown in the upper-half plane of Figure 2.3. This pole, however, occurs on the second sheet and the only contribution to the integral of Equation (2.6) arises from the pole at $\epsilon=\omega-i\eta$ (provided that $\omega<-m$). Thus the result of the integration is

$$G_D(\vec{x},\vec{x}';\omega) \sim i \, \frac{\Gamma\theta(-m-\omega)}{(\omega-\epsilon_r)^2+\Gamma^2/4} \, \psi_o^{cr}(\vec{x})\bar{\psi}_o^{cr}(\vec{x}') + \ldots, \tag{2.7}$$

where we have treated the resonance approximately as discussed in Section 1.6. A very different result is obtained if we choose the contour C'. Then the pole at $\epsilon = \epsilon_{res} + \Gamma/2$ makes a contribution of the form

$$G_{C'}(\vec{x},\vec{x}';\omega) \sim \frac{\psi_o^{cr}(\vec{x})\bar{\psi}_o^{cr}(\vec{x}')}{\omega-\epsilon_r-i\Gamma/2} + \cdots, \tag{2.8}$$

which is characteristic of a complex eigenvalue, a reflection of the lack of stability of the state of reference defined by the choice C' in the case of external fields. To summarize, every time a bound state descends into the negative energy continuum, we must redefine the Green's function so as to include only the remaining poles on the real axis. This is done by maintaining the fixed shape D of the contour. As described further below, this implies a change in the charge of the vacuum each time a pole crosses the fixed integration path D.

2.2 The reduced Hamiltonian and the instability of the neutral vacua in overcritical external fields

With the known analytical solutions for eigenstates of the Dirac equation even for the overcritical potentials, we can write the overcritical Hamiltonian in the reduced basis. We begin with $\hat{\psi}$ in the complete set of basis functions of Equation (1.19), which is separated into three categories for convenience,

$$\hat{\psi}(\vec{x}) = \sum_{p\neq 1s} \hat{b}_p^{cr}\psi_p^{cr}(\vec{x}) + \hat{b}_o^{cr}\psi_o^{cr}(\vec{x}) + \int_{-\infty}^{-m} d\epsilon\, \hat{d}_\epsilon^{cr\dagger}\psi_\epsilon^{cr}(\vec{x}). \tag{2.9}$$

Substituting this relation into the normal-ordered Hamiltonian yields

$$\hat{H} = \sum_{p\neq 1s} \epsilon_p \hat{b}_p^{cr\dagger}\hat{b}_p^{cr} + \epsilon_o \hat{b}_o^{cr\dagger}\hat{b}_o^{cr} - \int_{-\infty}^{-m} d\epsilon\, \epsilon\, \hat{d}_\epsilon^{cr\dagger}\hat{d}_\epsilon^{cr} + \int d^3x : \hat{\psi}^\dagger(\vec{x})V'(\vec{x})\hat{\psi}(\vec{x}) : \tag{2.10}$$

where the last term arises because the basis functions are not eigenfunctions of the Dirac operator. If the resonance is not too deeply embedded in the lower continuum, then the essential features are contained in a reduced Hamiltonian,

$$\hat{H}_r = \hat{H}_o + \hat{H}_I, \tag{2.11}$$

where

$$\hat{H}_o = \bar{\epsilon}\,\hat{b}_o^{cr\dagger}\hat{b}_o^{cr} - \int_{-\infty}^{-m} d\epsilon\,\epsilon\,\hat{d}_\epsilon^{cr\dagger}\hat{d}_\epsilon^{cr}, \tag{2.12}$$

is the diagonal piece and

$$\hat{H}_I = \int_{-\infty}^{-m} d\epsilon\left(\hat{b}_o^{cr\dagger}\hat{d}_\epsilon^{cr\dagger}V_\epsilon^* + \text{h.c.}\right), \tag{2.13}$$

is the nondiagonal part. The first term of Equation (2.12) includes the diagonal piece of the last term of Equation (2.10), and the effects of the matrix elements $U_{\epsilon\epsilon'}$ have again been neglected. It is clear that H_I represents the effects of creating or destroying an electron and a positron.

The distinction between the neutral state in the presence of overcritical fields and the ordinary neutral state in undercritical potentials is subtle. In the undercritical region, one may obtain the neutral vacuum state from the state containing a single 1s electron with the following projection:

$$|\tilde{0}\rangle = \int d^3x\,\psi_{1s}^\dagger(\vec{x})\hat{\psi}(\vec{x})\,|1s\rangle, \tag{2.14}$$

where $\psi_{1s}(\vec{x})$ is the wave function for the 1s state. Both the 1s state and the vacuum are stable. In the overcritical case, the neutral state may be obtained from the state containing the electron charge as follows:

$$|0\rangle = \int_{-\infty}^{-m} d\epsilon\,a(\epsilon)\int d^3x\,\psi_\epsilon^\dagger(\vec{x})\hat{\psi}(\vec{x})\,|\tilde{q}\rangle, \tag{2.15}$$

where $\psi_\epsilon(\vec{x})$ is a(n) (overcritical) lower continuum eigenfunction and $a(\epsilon)$ is the amplitude (1.28) associated with the resonance. Only the charged vacuum state $|\tilde{q}\rangle$ in Equation (2.15) is stable (disregarding the spin degeneracy). The state $|0\rangle$ is not an eigenstate of the Hamiltonian, since a superposition of solutions of the Dirac equation is involved in its description.

Let us consider a gedankenexperiment in which the neutral state (2.15) is prepared at time $t=0$. Subsequently two positrons will be produced spontaneously and the charge of the region surrounding the nucleus will equal that of two electrons in order to conserve charge. (One might be tempted to view this process as pair production, but we feel that characterizing it as the spontaneous decay of the neutral vacuum is more appropriate in view of the above discussion.) To describe this process, we make the following ansatz for its time dependence:

$$|t\rangle = y(t)|0\rangle + \int_{-\infty}^{-m} d\epsilon\, w_\epsilon(t)\hat{d}_\epsilon^{cr\dagger}\hat{b}_0^{cr\dagger}|0\rangle \qquad (2.16)$$

where $|y(t)|^2$ represents the probability of finding the neutral vacuum and $|w_\epsilon(t)|^2$ gives the probability of finding a positron with energy ϵ. The inclusion of spin degeneracy is straightforward and will not be discussed here. The functions y and w_ϵ satisfy the initial conditions that we prescribe:

$$y(0) = 1, \qquad w_\epsilon(0) = 0. \qquad (2.17)$$

The time evolution of $|t\rangle$ is governed by the Schrödinger equation

$$(\hat{H}_0 + \hat{H}_I)|t\rangle = i\frac{\partial}{\partial t}|t\rangle. \qquad (2.18)$$

Substituting the relations of Equations (2.12), (2.13), (2.16) into this expression and taking the scalar products with $\langle 0|$ and $\langle 0|b_0^{cr}d_\epsilon^{cr}$ yields the coupled differential equations,

$$i\frac{\partial}{\partial t}y(t) = -\int_{-\infty}^{-m} d\epsilon\, V_\epsilon w_\epsilon\,, \qquad (2.19)$$

$$i\frac{\partial}{\partial t}w(t) = (\bar{\epsilon}-\epsilon)w_\epsilon(t) - V_\epsilon^* y(t). \qquad (2.20)$$

The Fourier representations of y and w_ϵ take the form

$$y(t) = e^{-i\bar{\epsilon}t}\int_{-\infty}^{\infty} d\epsilon' e^{i\epsilon' t} a(\epsilon')a^*(\epsilon'), \qquad (2.21)$$

$$w(t) = e^{-i\bar{\epsilon}t}\int_{-\infty}^{\infty} d\epsilon' e^{i\epsilon' t} a(\epsilon')h^*(\epsilon'), \qquad (2.22)$$

in view of the conditions (2.17) and the relation of Equations (2.19) and (2.20) to Equations (1.22) and (1.23). The most straightforward method of evaluating these integrals is to use the methods of complex analysis. The solutions are:

$$y(t) = e^{-i\bar{\epsilon}t}\left[\theta(t)e^{i\epsilon_1^+ t} + \theta(-t)e^{i\bar{\epsilon}_1 t}\right], \tag{2.23}$$

$$w(t) = \frac{V_\epsilon^*}{\epsilon_1^+ - \epsilon}\,\theta(t)\left[e^{i(\epsilon_1^+-\epsilon)t} - e^{i(\epsilon-\bar{\epsilon})t}\right]$$

$$+ \frac{V_\epsilon^*}{\epsilon_1^- - \epsilon}\,\theta(t)\left[e^{i(\epsilon_1^--\epsilon)t} - e^{i(\epsilon-\bar{\epsilon})t}\right], \quad \epsilon_1^\pm = \epsilon_r \pm i\epsilon_i \tag{2.24}$$

where we have written the equations in a form which shows that we have obtained solutions for which the origin of time is a centre of symmetry. It is apparent that the amplitude y decays according to the law

$$y(t) = e^{i(\epsilon_r - \bar{\epsilon})}\, e^{-\Gamma|t|/2}, \tag{2.25}$$

and thus the probability of finding the neutral vacuum decays exponentially with time as $|y(t)|^2 = \exp-(\Gamma|t|)$. The spectrum of positrons is given by

$$|w_\epsilon(\pm\infty)|^2 = \frac{|V_\epsilon|^2}{(\epsilon-\epsilon_r)^2 + \Gamma^2/4} = |a(\epsilon)|^2, \tag{2.26}$$

which has usual Breit-Wigner form since the variation of $|V_\epsilon|^2$ near the resonance is unimportant. This establishes our claims that the neutral vacuum is unstable and spontaneously decays emitting positrons.

Let us now consider the charged state $|q\rangle \approx \hat{b}_0^{cr\dagger}|0\rangle$. Its time dependence is also determined by the Hamiltonian of Equations (2.12) and (2.13). However, in this case, the nondiagonal part of the Hamiltonian will not have any effect—all the relevant matrix elements being zero. Thus the time dependence of the charged state is

$$|q(t)\rangle = e^{-i\bar{\epsilon}t}|q\rangle. \tag{2.27}$$

Therefore no dramatic consequences ensue upon lowering a fully occupied state into the lower continuum. The total energy of the charged state is $\langle q|H_r|q\rangle = \bar{\epsilon}$, which is lower than the energy of of the neutral vacuum $\langle 0|H_r|0\rangle = 0$, and in fact exceeds it by more than the rest mass of the positron. This verifies our claim that the charged state is stable.

With this we have concluded our proof that, in order to have a stable reference state, we must choose the charged state and henceforth refer to it as the charged vacuum. This means that the Fermi surface must always be chosen just above the lower continuum.

2.3 Self-consistent potentials

From the preceding part of this section, we have learned about the unexpected properties of the supercritically bound states in external potentials. We now will consider the potential as not being prescribed externally. As the source of the self-consistent potential we choose the Dirac wave functions that are computed with such a potential. What we have in mind would conventionally be called Hartree approximation. It is important to realize that classical solutions of interacting field theories may have a meaning as approximation to quantum field theories even in régimes where the Hartree approximation should fail, for example in the limit of strong coupling and few bound particles (see also Section 4).

When fermions move in a self-consistent (shell) potential, then the total energy of the bound state is given by the sum of the Dirac eigenenergies and of the energy associated with the shell potential, sometimes also called "correlation energy." The condition for the transition to a state equivalent to the charged vacuum is now

$$\epsilon_o + E_{meson} < 0 \tag{2.28}$$

and not

$$\epsilon_o < 0 \tag{2.29}$$

as was the case when external fields were considered. It must be expected that the critical potential strength can be exceeded by the self-consistent potential and a resonance be embedded in the negative energy continuum of the Dirac equation. In view of Equation (2.28) we recognize that even if the self-consistent field is supercritical, the neutral vacuum may remain stable. This last remark can only be appreciated after a thorough discussion of the case of an external potential.

For the sake of argument, let us assume now that we have found a theory such that Equation (2.28) is satisfied. We then would find that a charged ground state has lower energy than the neutral state—thus the latter state is the 'false vacuum' [18] which under certain circumstances could undergo a transition into the 'true vacuum.' The properties of such a macroscopic transition have been studied by Coleman in a model theory [18]. Fortunately, we will find in Section 4.6 that it is hard to satisfy condition (2.28). It will be shown there in a model theory that the opposite of Equation (2.28) is the case; that is, we find $\epsilon_0 + E_{meson} > 0$, although $\epsilon_0 < -m$, beyond a certain coupling strength. This means that despite the resonance of the Dirac equation we do not have a resonance of the complete Hamiltonian. However, as we shall see, the understanding of the resonant Dirac equation solution is a necessary prerequisite in order to obtain the correct description of the strongly interacting fields.

3. SUPERCHARGED VACUUM AND KLEIN'S PARADOX

We have seen that in the vicinity of a sufficiently large assembly of charged particles (held together in one place, for example, by the strong interaction), the electron-positron field becomes overcritical. Here the response of the overcritical electron field to an ever-increasing strength of the external potential [19] is described. In particular, we focus on two gedankenexperiments. In the first one, we consider a spherically symmetric external potential created by a large (nuclear) charge $Z > Z^{cr}$. Then we will turn to the discussion of Klein's paradox and its resolution in the theory of supercritical fields. In both cases we will employ the Thomas-Fermi approximation.

3.1 Relativistic Thomas-Fermi approximation

In the relativistic Thomas-Fermi approximation the density of electrons is related to the Fermi momentum $k_F(x)$ by

$$\rho = \frac{e}{3\pi^2} k_F^3 . \tag{3.1}$$

The effect of the spin degeneracy is included in Equation (3.1). The relativistic relation between the Fermi energy E_F and Fermi momentum is

$$k_F^2 = \left[(E_F - V)^2 - m^2\right] \theta(E_F - V - m), \quad (3.2)$$

where V is the electrostatic potential. The step function ensures that k_F^2 is a positive quantity. E_F is the Fermi energy of the system considered.

From Equations (3.1) and (3.2) we now obtain for the charge density of the ground state characterized by a choice of the Fermi energy E_F:

$$\rho = \langle 0 | \hat{\rho}(x) | 0 \rangle$$

$$= (e/3\pi^2)\left[(E_F - V)^2 - m^2\right]^{3/2} \theta(E_F - V - m). \quad (3.3)$$

Introducing the total charge density ρ_T, which is composed of the external "nuclear" part ρ_N and the electronic part,

$$\rho_T = \rho_N + \rho, \quad (3.4)$$

and using Coulomb's law

$$\Delta V(\vec{r}) = -e\rho_T(\vec{r}), \quad (3.5)$$

we find a self-consistent nonlinear differential equation for the average potential V, that depends on the choice of the ground state $|0\rangle$ through the value of the parameter E_F.

$$\Delta V(\vec{r}) = -e\rho_N(\vec{r}) - \left(\frac{e^2}{3\pi^2}\right)\left[(E_F - V)^2 - m^2\right]^{3/2} \theta(E_F - V - m). \quad (3.6)$$

As long as the nuclear system ρ_N is isolated from external sources of electrons, the proper choice of E_F is $E_F = -m$. If this condition is relaxed and an inexhaustible supply of electrons is available, we must account only for the kinetic energy of these electrons. Thus for neutral atomic systems one must take $E_F = m$, which gives the usual Thomas-Fermi model, in the limit $|-2mV| > |V^2|$. Equation (3.6) is very useful and may be applied to describe many interesting problems in atomic physics.

3.3 The charged vacuum in the Thomas-Fermi approximation

The point at which the 1s wave function joins the continuum solutions of negative frequency has been determined [5] to be about Z = 172, under "realistic" assumptions and extrapolations of the known properties of nuclear and electromagnetic interactions. Similarly, the next critical point, for which the $2p_{\frac{1}{2}}$ state is expected to join the continuum is about Z = 183. At that point the charge of the vacuum becomes four. Soon, as we increase the nuclear charge, higher angular momentum states will also join the lower continuum, and the charge of the vacuum will rise even faster. Thereafter, the self-interaction of the vacuum charge will become an important aspect. In particular, we can already foresee that the vacuum charge may screen a substantial part of the nuclear charge and prevent more states from joining the lower continuum, or, from another point of view, the repulsive interaction of an electron with the surrounding vacuum charge may become comparable with the attractive force of the nuclear charge.

It has been proposed [19] to treat that complex situation by use of the relativistic Thomas-Fermi approximation. The charge density of the vacuum is equal to the charge density carried by all of the states that have joined the lower continuum. In the Thomas-Fermi model, this sum over all of these states is represented by an integral over all the momentum states within the Fermi sphere of radius k_F as defined by Equation (3.2) with the Fermi energy fixed at $E_F = -m$. This means that only the states accessible to spontaneous decay are filled. Inserting $E_F = -m$ into Equation (3.6) yields

$$\Delta V(\vec{r}) = -e\rho_N(\vec{r}) - \left(\frac{e^2}{3\pi^2}\right)(2mV+V^2)^{3/2}\,\theta(-V-2m). \qquad (3.7)$$

We now proceed to discuss the solution of Equation (3.7). Since the charge density of the vacuum must be confined to the vicinity of the external charge, we require a solution such that

$$V(r) \underset{r\to\infty}{=\longrightarrow} -\gamma\alpha/r. \qquad (3.8)$$

(α is the fine structure constant.) For every choice of Z, γ is determined by the boundary condition on the electrostatic potential at the origin

$$\left.\frac{\partial V}{\partial r}\right|_{r=0} = 0. \qquad (3.9)$$

Equations (3.7) and (3.9) are therefore eigenvalue equations for γ, the unscreened part of the nuclear charge, and $Z-\gamma$ gives the charge of the vacuum.

Neglecting at first the inhomogeneity of the solution, we find that $V(0) = V_o$ is determined from the condition,

$$\rho_T = \rho + \rho_N = 0, \tag{3.10}$$

in the limit of large Z, i.e., when the distribution of nuclear charge is large as compared with 1/m, then

$$V_o = \left\{ m - \left[m^2 + (3\pi^2 e\rho_N/e^2)^{\frac{2}{3}} \right]^{\frac{1}{2}} \right\}. \tag{3.11}$$

Integration of Equation (3.7) is straightforward [19]. An equal number of protons and neutrons and normal nuclear density have been assumed for the nuclear-charge distribution. The results for γ are plotted in Figure 3.1. From the figure, one can see that γ increases monotonically with Z and that γ/Z decreases as Z increases. In fact, from the requirement that $\underline{V_o}$ remains constant with growing Z, at the surface of the nuclear charge distribution we find:

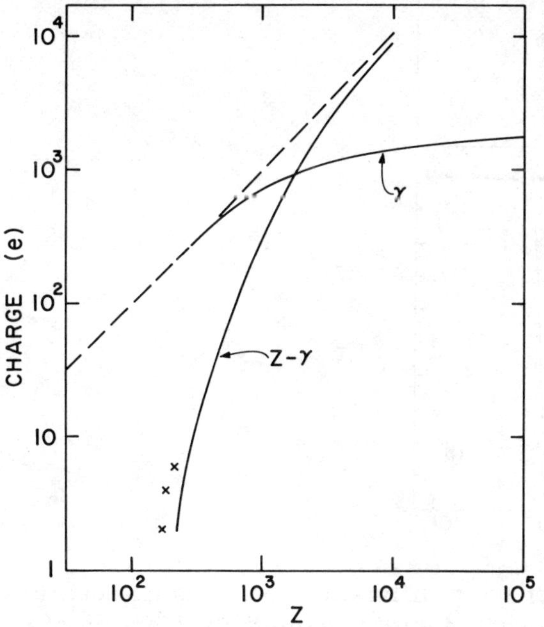

Fig. 3.1. The unscreened charge γ and the total charge of the vacuum $(Z-\gamma)$ as a function of Z. The crosses denote points from single-particle calculations. The dashed line denotes the nuclear charge Z.

$$V_o \sim \frac{\gamma(Z)}{R(Z)}$$

and, since $R(Z) \sim Z^{\frac{1}{3}}$, we find

$$\gamma(Z) \sim Z^{\frac{1}{3}} . \qquad (3.12)$$

The single-particle results are denoted by crosses in Figure 3.1 and agree reasonably well with an extrapolation of the Thomas-Fermi results into the realm of small values of $Z-\gamma$. We note here that the condition for the validity of the Thomas-Fermi approximation is $Z-\gamma \gg 1$. In Figure 3.2 we consider the approach to infinite nuclear matter. As shown there, both the

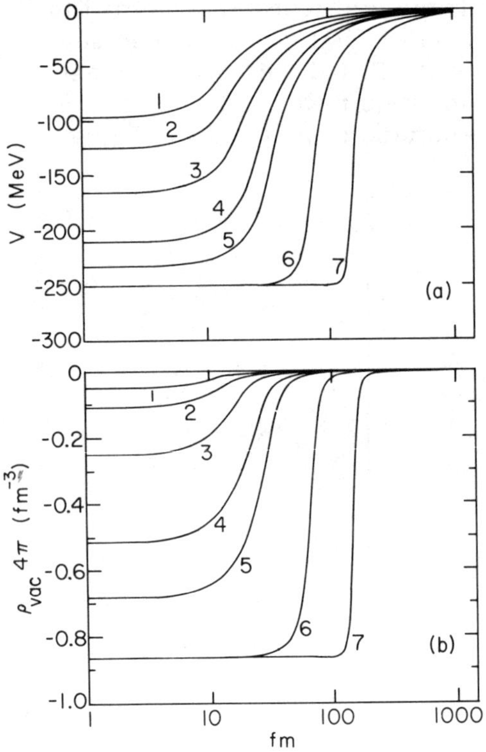

Fig. 3.2. The solutions of the relativistic Thomas-Fermi equation (3.7) for selected values of the nuclear charge as a function of r. Curve 1—Z = 600; curve 2—Z = 1000; curve 3—Z = 2000; curve 4—Z = 5000; curve 5—Z = 10,000; curve 6—Z = 10^5; curve 7—Z = 10^6. a) The self-consistent potential; b) the corresponding charge distribution of the vacuum.

vacuum charge density and the potential approach a limit as Z increases from 10^4 to 10^5. The radial total charge density, calculated from the right-hand side of Equation (3.7), is shown in Figure 3.3. The results are scaled with γ so that each curve is normalized to unity. We see that the charge densities become more and more nearly those of a surface dipole with clearly defined regions of positive and negative charge.

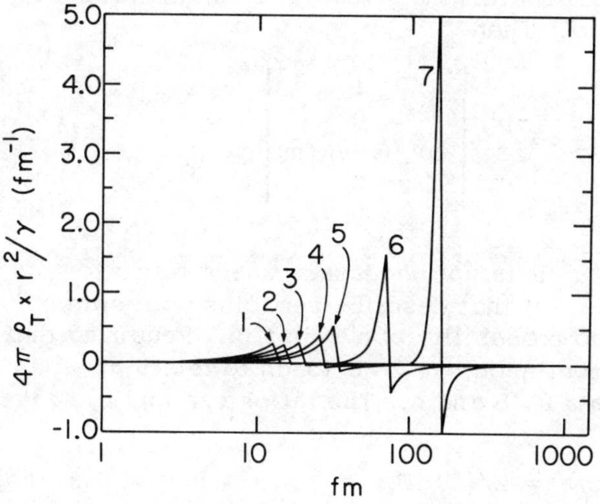

Fig. 3.3. The total charge densities, scaled with γ. Same selected values of nuclear charge as in Fig. 3.2.

In accordance with our expectations, the charge generated by successive levels joining the lower continuum is sufficient to screen most of the bare nuclear charge as it increases without bound. This property is evident in the fact that for $Z > 10^5$ the self-consistent potential does not change within the nuclear matter distribution.

3.3 Klein's paradox

Let us consider an electron of momentum p and energy $\epsilon = \sqrt{p^2 + m^2}$ with spin up that is incident from the left on an electrostatic square-well barrier $V_0 > 0$. The discontinuous form of the potential requires that region I outside the potential well and region II inside the potential well be treated separately. In region I the solution of the Dirac equation is

$$\psi_I(z) = ae^{ipz}\begin{bmatrix}1\\0\\p/(\epsilon+m)\\0\end{bmatrix} + be^{-ipz}\begin{bmatrix}1\\0\\-p/(\epsilon+m)\\0\end{bmatrix}, \tag{3.13}$$

where the second part of the wave function describes the reflected wave. The form of the wave function in region II depends upon the magnitude of the potential strength. We first consider values of V_o that are not too large. Then

$$\psi_{II}(z) = ce^{-p'z}\begin{bmatrix}1\\0\\ip'/(\epsilon-V_o+m)\\0\end{bmatrix} \tag{3.14}$$

where $p' = \sqrt{m^2-(\epsilon-V_o)^2}$. It is not necessary to add terms to Equations (3.13) and (3.14) that describe particles with spin down, since there is no probability of a spin flip. Requiring that the wave function be continuous at $z=0$ leads to relations between the coefficients a, b and c. The ratios c/a and b/a take the form

$$c/a = 2/(1+i\Gamma) \tag{3.15a}$$

$$b/a = (1-i\Gamma)/(1+i\Gamma), \tag{3.15b}$$

where

$$\Gamma = [(\epsilon+m)(m-\epsilon+V_o)]^{\frac{1}{2}}\cdot[(\epsilon-m)(m+\epsilon-V_o)]^{-\frac{1}{2}} \tag{3.15c}$$

is real. The incident current may be calculated from $j_i = \psi_i^\dagger\alpha_3\psi_i$, where the subscript i denotes only the first part of the wave function of Equation (3.13). This current is equal to $2p|a|^2/(\epsilon+m)$. The ratio of the reflected current to the incident is $|b|^2/|a|^2$, which is equal to one because of the form of Equation (3.15b). The transmitted current is equal to zero. Thus all of the incident current is reflected, and the situation is analogous to nonrelativistic quantum mechanics.

Let us now consider what happens as V_o is increased beyond $\epsilon+m$. The wave function must be written as:

$$\psi_{II}(z) = de^{ip''z}\begin{bmatrix} 1 \\ 0 \\ p''/(\epsilon+m-V_o) \\ 0 \end{bmatrix}, \qquad \epsilon + m < V_o \tag{3.16}$$

where $p'' = \sqrt{(V_o - \epsilon)^2 - m^2}$. The continuity condition now leads to

$$d/a = 2/(1 - \Gamma') \tag{3.17a}$$

$$b/a = (1 + \Gamma')/(1 - \Gamma'), \tag{3.17b}$$

$$\Gamma' = [(\epsilon+m)(V_o - \epsilon + m)]^{\frac{1}{2}}\,[(\epsilon - m)(V_o - \epsilon - m)]^{-\frac{1}{2}}. \tag{3.17c}$$

The transmitted current is equal to $2p'|d|^2/(\epsilon + m - V_o)$, <u>which is negative</u>, and the magnitude of the reflected current is larger than that of the incident current. The transmission coefficient, which is the ratio of the transmitted current to the incident current, is given by

$$T = -4\Gamma'/(1 - \Gamma')^2. \tag{3.18}$$

In Figure 3.4 the negative of the transmission coefficient is shown as a function of energy for $V_o = 3m$ and for $V_o = 10m$. We note that the transmitted current may be much larger than the incident current. This paradox was first noted by Klein [20]. Only in the context of a single-particle interpretation does this appear paradoxical. When one appreciates that electrons and positrons are inextricably connected in the Dirac theory, it is natural to identify the negative current in region II with the appearance of positrons. The increase of the reflected current over the incident current is necessary to conserve charge. The reflected current plus the transmitted current is always equal to the incident current.

The results we have obtained above would imply that the scattering of electrons of a repulsive, strong potential barrier induces pair production. As we have seen in the previous section, <u>one has to consider the spontaneous pair production beforehand</u>. Furthermore, as shown in Section 3.2, it is quite possible that a saturation in the pair production will occur.

In order to make a connection between our investigation of Klein's paradox and the supercritical potentials, we redefine the reference point of the energy: We consider the potential to be strongly attractive in region II ($V = -V_o < 0$). Then we expect that

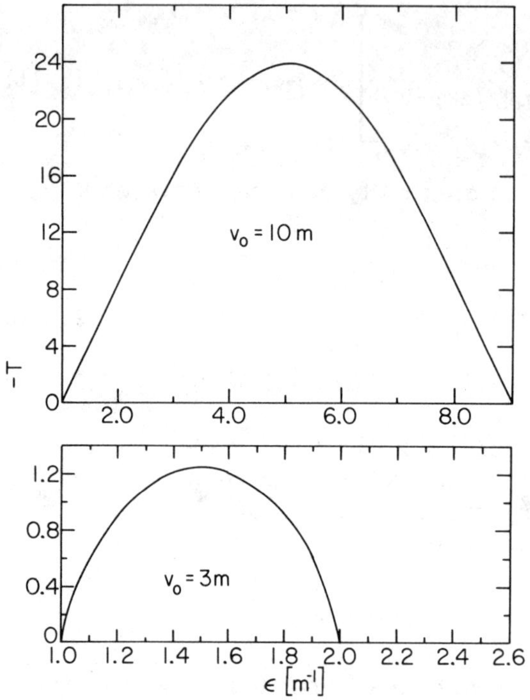

Fig. 3.4. Transmission coefficients as functions of energy.

within the range of this region (which may be infinite) all supercritical states are spontaneously filled with 'electrons,' while the positrons are emitted to infinity. Now Klein's gedankenexperiment consists of scattering positrons of the (filled) attractive potential well. Since no final states are available for electrons within the well, we find that the transmission coefficient vanishes. Thus no matter can be transmitted into the region of the potential. This result must be, strictly speaking, supplemented by a proper quantum field theoretic formulation, which we will, however, not go into now. Here the nonlinear system will be considered with a special emphasis on those aspects connected with the self-consistent treatment of the vacuum charge.

When the external charge is localized to the left of $z=0$,

$$\rho_N(\vec{x}) = \theta(-z)\rho_N^{0}(z), \tag{3.19}$$

then it is apparent that a solution for V must induce a vacuum charge confined also to the left. The requisite value of the Fermi energy is

$$\epsilon_F = -m. \tag{3.20}$$

Therefore all supercritical states satisfying

$$V + m < \epsilon < -m, \tag{3.21}$$

within the range of the potential well are occupied by electrons. The associated electron charge density, Equation (3.3), becomes in this case:

$$e\rho = -\theta(-z)\,\frac{e^2}{3\pi^2}\,(2mV + V^2)^{3/2}\,\theta(-V-2m). \tag{3.22}$$

The potential well can now be obtained solving Equation (3.5). Since we want a potential step, all of the external charge must be screened; that is,

$$\int\left(\rho_N(\vec{x}) + \rho(\vec{x})\right)d^3x = 0. \tag{3.23}$$

This can be achieved only with an oscillating electron charge, which in turn implies that the self-consistent potential must have an oscillating part. Thus a potential step with the deepness described as in Equation (3.11), $V \approx -V_0$, can be only an approximate solution of the self-consistent Thomas-Fermi equation.

Before closing this section a comment can be made concerning the validity of our assertions for macroscopic charge distributions. In the microscopic charge distributions, like those associated with nuclei, an overcritical value of the total charge must be first exceeded in order for spontaneous positron production to begin. In infinite charge distributions any arbitrarily weak ρ_N is permissible, since over infinite volume sufficient strength for overcriticality will be attained. However, the barrier separating the particle and antiparticle states may become much too wide to allow the spontaneous decay of the neutral vacuum. The most common situation of this nature occurs with the Van de Graaff generator. There the potential step of several MeV is spread over macroscopic distances. As a consequence, the decay time of the neutral vacuum is so large that 'sparking' of the vacuum should never be observed in the vicinity of such generators.

4. STRONG FIELDS IN QUANTUM FIELD THEORY

In this section the case of a completely self-consistent potential as introduced in Section 2.3 is considered. Although the study of the self-consistent solutions of interacting classical field theory is an interesting topic in itself, a deeper motivation is provided by the possibility that, in the strong coupling limit ($g^2 \gg 1$), the classical solution contains relevant information about the interacting quantum system. The conventional perturbative treatment in quantum field theory proceeds via an expansion in the vicinity of noninteracting fields. It is argued below that the self-consistent solutions of c-number equations provide the natural basis for an expansion around a large coupling constant (as far as bound states are considered), provided that the c-number solutions satisfy

$$M << Nm_q . \tag{4.1}$$

Here m_q is the mass of the single-particle states in the absence of an interaction, N is the number of bound particles and M is the bound state mass obtained from the self-consistent solution of c-number fields. Equation (4.1) thus presupposes that the interaction produces strongly bound states of N particles.

At this point we recognize the connection with the developments of previous sections, where the properties of strong binding in external potentials are investigated. This knowledge is now required in order to understand the properties of the self-consistent strongly bound solutions. In particular, we have learned that when the Dirac binding energy exceeds twice the rest mass, qualitatively new effects can occur.

4.1 Self-consistent 'quark bags'

As indicated in the above discussion, we take the self-consistent solution of the coupled Dirac-meson fields to represent a bound state of strongly interacting fermions. This physical picture corresponds to the idea of a quark trapped in a self-consistent potential well [21]-[23]. If taken seriously, this means that the solutions of the interacting fields contain information about experimentally measurable quantities. A preferred model then can emerge that has the 'best' phenomenological properties. I have found that such a 'pet' theory is that of interacting fermion-scalar gluon fields

$$\mathcal{L} = \bar{\Psi}(\gamma\cdot p-m)\Psi + g\Phi\bar{\Psi}\Psi + \partial_\mu\Phi\partial^\mu\Phi - (U(\Phi) - U(0)). \qquad (4.2)$$

In the first part of this subsection we will make a connection between the classical theory and quantum field theory, using as an example the Lagrangian (4.2), where we sometimes also write for the gluon potential U the simplest term

$$U(\Phi^2) \longrightarrow \tfrac{1}{2}\mu^2\Phi^2. \qquad (4.3)$$

In order to treat the strongly interacting ($g^2 \gg 1$) limit of the quantum field theory, induced by the Lagrangian Equation (4.2), let us now consider quasi-particles. These are entities carrying, typically, the same internal quantum numbers, for example spin, as the particles, but subject to a quite different interaction from the free modes. The advantage that is gained consists basically in the modification of the interaction; in particular, quasi-particles interact weakly and we may use the well-proved perturbative methods to treat the residual interaction. A quasi-fermion state with quantum numbers $|p\rangle$ can be obtained most readily from the quantum operator $\hat{\Psi}(\vec{x},0)$ by projecting with the associated amplitude $\psi_p(\vec{x})$ (a circumflex over a quantity indicates its q-number character)

$$|p\rangle = \int \hat{\Psi}^\dagger(\vec{x},0)\psi_p(\vec{x})|0\rangle d^3x. \qquad (4.4)$$

Here $|0\rangle$ denotes the ground state. The equations of motion for the amplitude ψ_p will be determined a posteriori. It is advantageous to introduce the quasi-particle and antiparticle operators in the usual manner:

$$\beta_p^\dagger = \int \hat{\Psi}^\dagger(\vec{x},0)\psi_p(\vec{x})d^3x, \qquad (4.5)$$

$$\gamma_n^\dagger = \int \psi_n^\dagger(\vec{x})\hat{\Psi}(\vec{x},0)d^3x. \qquad (4.6)$$

The completeness of the basis $\{\psi_p, \psi_n\}$ is reflected in

$$\delta^3(\vec{x}-\vec{x}')\delta_{\alpha\beta} = \sum_n \psi_n^\alpha(\vec{x})\psi_n^{\beta\dagger}(\vec{x}') + \sum_p \psi_p^\alpha(\vec{x})\psi_p^{\beta\dagger}(\vec{x}'). \qquad (4.7)$$

Here (α,β) are the internal structure indices, for example spinor indices, while the sums also stand for possible integrals over continuous parts of the spectra. We now see that the internal structure of the quasi-particles is that of the particles and is a consequence of the (anti-)commutation relations of the fields $\hat{\Psi}$. Any quasi-particle state may be constructed from a product of the quasi-fermion operators acting on the ground state. Thus, for example, the three quasi-fermion baryon state B is

$$|B\rangle = \sum_{ijk} b^{p}_{ijk}\hat{\beta}^{\dagger}_{i}\hat{\beta}^{\dagger}_{j}\hat{\beta}^{\dagger}_{k} e^{-i\int\hat{\pi}_{\Phi}(\vec{x},0)\varphi_{c}(\vec{x})d^3x} |0\rangle \tag{4.8}$$

with suitable coefficients b_{ijk} that generate the symmetry properties of the state. The remaining factor in Equation (4.8) will be explained in a moment.

The 'bag' field for the quarks is generated by the quasi-fermionic source

$$S = \langle B|\hat{\bar{\Psi}}\hat{\Psi}|B\rangle . \tag{4.9}$$

The presence of a strong source, the strength being determined to a large extent by g^2, generates an average gluon "shell" potential φ_c

$$\varphi_c = \langle B|\hat{\Phi}|B\rangle . \tag{4.10}$$

To achieve this property, we have introduced the exponential in Equation (4.8), which makes $|B\rangle$ a coherent state for the meson field, where

$$\pi_{\Phi} = \frac{\delta\mathcal{L}}{\delta\Phi} \tag{4.11}$$

and

$$[\hat{\Phi}(\vec{x},0),\hat{\pi}_{\Phi}(\vec{x}',0)] = i\delta^3(\vec{x}-\vec{x}') . \tag{4.12}$$

Equipped with the trial state $|B\rangle$ (and similarly a meson state $|M\rangle$), we can minimize the invariant mass that is associated with them, as a function of the unknown functions φ_c and ψ_p. The Lorentz invariant momentum P_μ of our trial state $|B\rangle$ is given by

$$P_\mu = \int \langle B|\hat{T}_\mu^{\ \nu}(x)|B\rangle d^4 0_\nu \quad , \tag{4.13}$$

where $T_\mu^{\ \nu}$ is the energy momentum tensor following from the Lagrangian $\mathcal{L}$ under consideration. The hypersurface in the four-dimensional space is denoted by $d^4 0_\nu$. Note that we have chosen, in the above discussion,

$$d^4 0_\nu = n_\nu d^3 x, \tag{4.14}$$

with $n_\nu = (1,0,0,0)$ associated with the laboratory frame of reference. We have written the fully covariant form of P_μ in Equation (4.13) to show the Lorentz covariance of the bag approach. Furthermore, within the ansatz for $|B\rangle$ we have chosen, we search for solutions with vanishing momentum $\vec{P}$, i.e.,

$$\int \langle B|\hat{T}_o^{\,i}|B\rangle d^3x = 0. \tag{4.15}$$

If our solution satisfies condition (4.15), it is called a baryon in zero average momentum frame. By construction of our trial state we obtain the property

$$\langle B|\hat{\pi}_\Phi(x)|B\rangle = 0, \tag{4.16}$$

which guarantees the vanishing of the average gluon momentum. Similarly, it is easy to show that the sum of quasi-fermion momenta vanishes. This is in agreement with a picture in which quarks follow closely the motion of the gluon bag and vice versa.

In the zero average momentum frame we therefore have

$$P_o = M = \int d^3x \langle B|\hat{T}_o^{\,o}|B\rangle = \int d^3x\, \mathcal{M}_B, \tag{4.17}$$

where for $\mathcal{L}$, defined by Equation (4.2) we find, using the exact symmetry of all quarks in the bag:

$$\mathcal{M}_B = N\psi_i^\dagger(\vec{\alpha}\cdot\vec{p} + \beta(m_q - g\varphi_c)\psi_i$$
$$+ \tfrac{1}{2}|\vec{\nabla}\varphi_c|^2 + U(\varphi_c) - U_o \ , \tag{4.18}$$

where $N = 3$ is the number of (valence) quarks.

A very similar result is also obtained for the mesons:

$$\mathcal{M}_M \equiv \langle M|\mathcal{H}|M\rangle = \psi_i^\dagger[\vec{\alpha}\cdot\vec{p} + \beta(m_q - g\varphi_c)]\psi_i$$
$$- (i \rightleftarrows j) + \tfrac{1}{2}|\vec{\nabla}\varphi_c|^2 + U(\varphi_c) - U_o \,. \tag{4.19}$$

The usual (-) sign for antiquasi-fermion amplitude ψ_j in (4.19) is necessary to change the sign of the negative frequency mode associated with the solution of its Dirac equation. For a scalar potential φ_c the spectrum of the Dirac equation is symmetric between particles and antiparticles; that is,

$$\epsilon_p = -\epsilon_{ap} \tag{4.20}$$

and the wave functions transform like

$$\psi_p = \gamma_5 \psi_{ap}^* . \tag{4.21}$$

Therefore, the invariant mass for all hadrons can be written as a function of the number of quarks in the bag only:

$$M_N = \int d^3x \, \mathcal{M}_N = \int d^3x \Big[N\{\psi_o^\dagger (\vec{\alpha}\cdot\vec{p} + \beta(m_q - g\varphi_c))\psi_o\} + \tfrac{1}{2}(\vec{\nabla}\varphi_c)^2 + (U(\varphi_c) - U_o)\Big] = M_{N,f} + M_g . \tag{4.22}$$

The index 'o' on the spinor wave function indicates that we must choose the lowest energy particle solution (which in the case of scalar interaction is the lowest positive energy eigenvalue solution of the Dirac equation) in order to obtain the smallest value of M_N. The equations of motion satisfied by ψ_o and φ_c follow now from the variational condition:

$$\delta\left(M_N - \epsilon_o N \int d^3x \, \psi_o^\dagger \psi_o\right) = 0, \tag{4.23}$$

where we have introduced the usual Lagrange multiplier in order to normalize the solutions:

$$\int \psi_o^\dagger \psi_o d^3x = 1. \tag{4.24}$$

We find from Equation (4.23) the self-consistent equations of motion

$$[\vec{\alpha}\cdot\vec{p} + \beta(m_q - g\varphi_c)]\psi_o = \epsilon_o\psi_o \quad , \tag{4.25}$$

$$-\Delta g\varphi_c + \frac{\partial U}{\partial \varphi_c} = g^2 N \bar{\psi}_o \psi_o . \tag{4.26}$$

N is equal to three for baryons and two for mesons. When solving the above equations, we consider all dimensional quantities in units of m_q. If we obtain the solutions as a function of g^2, then it is possible to consider the case $N = 1$ only, and the relevant parameter is

$$g'^2 = g^2 N. \tag{4.27}$$

We note that M_N can be written for U given in Equation (4.3) as

$$M_N = N\left(\epsilon_o[g'^2] + \frac{1}{g'^2}\int\left(\tfrac{1}{2}(\vec{\nabla} g\varphi_c)^2 + \tfrac{1}{2}\mu^2(g\varphi_c)^2\right)d^3x\right). \tag{4.28}$$

Thus M_N/N is only a function of g'^2. We thus have seen that self-consistent solutions of classical interacting fields are closely related to the quark-bag solutions of N quarks. In the quark-bag approach the main approximation has been the following: Only the diagonal part of the Hamiltonian matrix of the interacting fields in a trial state has been considered. The off-diagonal parts remain to be diagonalized and the renormalization program carried out.

An argument in favour of the bag approximation, applicable only in the limit of strong coupling, involves the large separation in energy between the mass of the trial state, M_N, and the free fermion mass, m. I believe that the second-order contributions to M from the off-diagonal matrix elements will be of the order of:

$$\delta M \lesssim \left[\int \psi_i^\dagger g\varphi_c \psi_i d^3x\right]^2 \Big/ \left[N m_q - M_N\right]$$

$$\approx \left(\frac{M_N}{N m_q}\right) M_N.$$

If this is the case, then the ratio

$$\epsilon = M/m_q$$

is the small parameter that determines the validity of the bag approximation.

Above we have illustrated in some detail the properties of the bag with a scalar type of interaction. However, in spirit and with minor modifications, our discussion applies to other interactions as well, in particular also to the vector interaction discussed in Section 4.6. A similar, but a more straightforward, derivation of the self-consistent equations is also given there.

Finally, I wish to make two points more explicit:

i) The expectation value (4.13) should include the zero-point energy of the Fermi field:

$$E_o = -\frac{1}{2}\left(\sum_p \epsilon_p[\varphi_c] - \sum_n \epsilon_n[\varphi_c]\right),$$

which is a (divergent) implicit functional of φ_c. It is possible to include in the function $U(\varphi_c)$ the necessary counterterms—the renormalized zero-point energy then is still a functional of φ_c and should, in principle, be included in Equation (4.19). However, such a contribution can be absorbed in the function $U(\varphi_c)$. Since we have, in principle, no idea about the form of U, we need not consider this point in further detail for the moment.

ii) It is instructive to notice that the choice of φ_c that is self-consistent in the sense of Equations (4.25) and (4.26) makes the remaining interaction purely off-diagonal. The full Hamiltonian of q-number fields is (up to the necessary symmetrization)

$$\hat{H} = \hat{H}_D + \hat{H}_\Phi + \hat{H}_I$$

with

$$\hat{H}_D = \int d^3x\, \hat{\Psi}^\dagger(\vec{\alpha}\cdot\vec{p} + \beta m)\hat{\Psi}$$

$$\hat{H}_\Phi = \int d^3x\, \tfrac{1}{2}(\hat{\pi}^2 + (\vec{\nabla}\hat{\Phi})^2 + (U(\hat{\Phi}) - U_o)^2)$$

$$\hat{H}_I = \int d^3x(-g\hat{\Phi})\hat{\bar{\Psi}}\hat{\Psi}.$$

Now let us add and subtract a term in $\hat{H}$

$$\hat{H} = (\hat{H}_D + \hat{H}_{ex}) + \hat{H}_\Phi + (\hat{H}_I - \hat{H}_{ex}),$$

where

$$\hat{H}_{ex} = \int d^3x(-g\varphi_c)\hat{\bar{\Psi}}\hat{\Psi}\ .$$

The approach presented above corresponds to such a choice of a trial state that the <u>diagonal</u> matrix elements of the new interaction

$\hat{H}_I' = \hat{H}_I - \hat{H}_{ex}$ vanish, while the Hamiltonian

$$\hat{H}_o = (\hat{H}_D + \hat{H}_{ex}) + \hat{H}_\Phi$$

is diagonal.

4.2 Numerical methods

Our aim in this section is to illustrate the numerical methods involved in minimization of the mass (or Hamiltonian) M_N, Equation (4.28), as a function of the fields ψ_o and φ_c, with the constraint that ψ_o be the lowest frequency eigenmode of the quasi-fermion field.

Great care should be exercised in all numerical work, since the classical Dirac spectrum $\{\epsilon_i\}$ is not bounded below. Therefore, M_N can be made arbitrary negative by a choice of ψ_{trial} that includes large negative energy components. In practical calculations only eigenmodes of Equation (4.25) for a given φ_c should be allowed. Let us go into this remark in more detail. Since the set of eigensolutions $\{\psi_k\}$ generated by φ_c is complete, a trial function ψ_t could be expanded as

$$\psi_t = \sum_k a_k^t \psi_k. \tag{4.29}$$

Then we would find that the quantity we wish to minimize, Equation (4.23), is

$$(\mathrm{II} - \epsilon_t N) - N \sum_k |a_k^t|^2 (\epsilon_k - \epsilon_t). \tag{4.30}$$

With the unbound-below spectrum of the Dirac equation we thus find that a minimization with respect to parameters of the trial function could not be successful, since $\epsilon_k - \epsilon_t$ can be smaller than zero.

From this consideration we recognize the need to constrain the number of allowed degrees of freedom of the Dirac field as arose naturally in the bag approximation. We must exclude from our considerations the possibility of a transition that a classical Dirac particle can undergo into a (classically) empty state of arbitrary large negative frequency. In a particular application this means that the Dirac equation must always be solved exactly for some prescribed potential, and an eigenstate ψ_o obtained. We can thus consider the Hamiltonian to be given by

$$H^S = \epsilon_o[\varphi_c] + \int d^3x \left[\tfrac{1}{2} (\nabla\varphi_c) + U(\varphi_c) - U_o \right] \tag{4.31}$$

where ϵ_o is given implicitly by the solution of Equation (4.25) with the eigenvalue condition

$$\lim_{|\vec{x}|\to\infty} \psi_o(\vec{x}) = 0. \tag{4.32}$$

Any trial value of φ_c in Equation (4.31) will give an upper limit for H^s.

But how do we actually obtain a self-consistent solution for φ_c with ψ_o constrained as described above to the lowest particle solution? It is very inviting to proceed with an iterative algorithm in which, for a given spherical $\varphi_{c,n-1}$ and fixed g and μ/m, a solution ψ_o of Equation (5.25) is obtained as described for spherical potentials in Section 1. From Equation (4.26) we then determine a new function $\varphi_{c,n}$. In practical calculations it turns out that the radius of convergence of such an iteration decreases rapidly as a function of g^2, making it virtually impossible to obtain a solution for $g^2 \gtrsim 2$, a region in which we are mainly interested.

I have developed [24] an algorithm that has so far shown an unrestricted radius of convergence when applied to equations of the type (4.23) to (4.26). Let us introduce for the purpose of clarity two additional functions

$$\phi = g\varphi_s \tag{4.33}$$

$$f = \varphi_s/g. \tag{4.34}$$

Equations (4.25) and (4.26) can now be written in the generalized self-explanatory fashion

$$D[\phi]\psi = 0, \tag{4.35}$$

$$Kf = \rho_s[\phi], \tag{4.36}$$

$$\phi = g^2 f. \tag{4.37}$$

Here D and K are linear operators, while the source ρ_s is $\bar{\psi}_o\psi_o$. Next we note that, using an arbitrary linear functional L, we can write Equation (4.37) in the form

$$g^2 = L\phi/Lf. \tag{4.38}$$

We will return to the discussion of the actual form of L further below.

Our algorithm is: Given a function ϕ_{n-1} the scalar density is found solving Equation (4.35). Then we obtain from Equation (4.36)

$$f_n = K^{-1}\rho_s[\phi_{n-1}] \tag{4.39}$$

and Equation (4.38) gives us in turn

$$g_n^2 = L\phi_{n-1}/Lf_n. \tag{4.40}$$

The algorithm loop is closed by the use of Equation (4.37)

$$\phi_n = g_n^2 f_n. \tag{4.41}$$

Note that we have iterated the coupling constant g^2. For a given starting value ϕ_o a convergent point

$$g^2 = \lim_{n\to\infty} g_n^2 \tag{4.42}$$

may be found that is dependent on the actual form of the initial function ϕ_o. I have found empirically that there is a one-to-one correspondence between

$$\phi_c = \lim_{n\to\infty} \phi_n \tag{4.43}$$

and g^2, Equation (4.42), in the case of Equations (4.25), (4.26) and other similar theories discussed further below.

Now, as to the choice of the operator L, the difference between ϕ_{n-1} and $g_n^2 f_n = \phi_n$ needs to be small, in the sense that we are searching for the minimum of the integral

$$I_n = \int w(r)[\phi_{n-1} - g_n^2 f_n]^2 dr \tag{4.44}$$

as a function of g_n^2. Here w(r) is a weighting function that may be dependent on ϕ_{n-1}. This prescription gives us

$$L\phi = \int w(r)\phi dr. \tag{4.45}$$

The choice $w(r) = 1$ is usually sufficient to have a convergent algorithm. However, $w(r) = \rho_s[\phi_{n-1}]$ gives a quicker convergence in most cases. I would like to conjecture that the algorithm described above should be convergent in most similar cases of physical interest, provided that a suitable choice for the operator L has been made.

Finally, a few words concerning stability of the (radial) solutions φ_c. To prove the stability of the solutions it would be necessary to show that H^s, Equation (4.31), is increased in second order by arbitrary variations of φ_c. The necessary constrained variational calculation requires a major computational effort and has not been carried out in that form. However, I have tried a large number of $\delta\varphi_s$ chosen at random. The change in M was positive in every case.

4.3 The virial approach to classical field equations

Before actually discussing the solutions of particular models, let us see how much we can learn about the general properties of self-consistent solutions from the equations of motion [25]. Basic to our considerations will be a (virial) relation describing the kinetic energy of the fermion field ψ in some prescribed potential V(x), where V is a matrix in the spinor space and can, in principle, consist of all possible couplings (S, V, P, A, T). Let H_D be the Dirac operator

$$H_D = \vec{\alpha}\cdot\vec{p} + \beta m + V. \tag{4.46}$$

Then we have

$$[\vec{x}\cdot\vec{p}, H_D] = i\vec{\alpha}\cdot\vec{p} - i\vec{x}(\vec{\nabla}V). \tag{4.47}$$

Taking the expectation value of Equation (4.46) between localized eigenfunctions ψ_i of H_D, we find the <u>virial</u> equation for the Dirac field:

$$\int d^3x\left[\psi_i^\dagger\vec{\alpha}\cdot\vec{p}\,\psi_i\right] = \int d^3x\left[\psi_i^\dagger\vec{x}\cdot(\vec{\nabla}V)\psi_i\right]. \tag{4.48}$$

This is a very useful relation which is valid only when ψ_i is an eigenfunction of H_D. We record that all of the ψ_i considered above are localized discrete eigenstates in configuration space. We now apply Equation (4.48) to the Lagrangian field theory describing scalar, meson field in interaction with the Dirac field: The Lagrangian has been given in Equation (4.2) and the equations of motion by Equations (4.25) and (4.26) with $N=1$ and $\dot{\varphi}_c=0$. Then Equation (4.48) for the kinetic energy of the Dirac field becomes

$$\int d^3x \left[\psi_i^\dagger \vec{\alpha}\cdot\vec{p}\,\psi_i\right] = \int d^3x \left[(-\vec{x}\cdot\vec{\nabla}\varphi_c)g\bar{\psi}_i\psi_i\right]. \tag{4.49}$$

We find after some manipulations that

$$\int d^nx \left[\psi_i^\dagger \vec{\alpha}\cdot\vec{p}\,\psi_i\right] = \int d^nx \left[\frac{n-2}{2}|\vec{\nabla}\varphi_c|^2 + n(U-U_o)\right] + ST, \tag{4.50}$$

with a vanishing surface term for localized solutions, with

$$n \equiv \vec{\nabla}\cdot\vec{x} \tag{4.51}$$

as the dimensionality of the space. Normally $n=3$, but to dramatize the uniqueness of the three-dimensional, physical space, let us consider the number of dimensions as a parameter.

The Hamiltonian associated with the Lagrangian (4.2) is, with a time-independent φ_c:

$$H^s = \int d^nx \Big[\psi_o^\dagger(\vec{\alpha}\cdot\vec{p} + \beta m)\psi_o - g\varphi_c\bar{\psi}_o\psi_o + \tfrac{1}{2}|\vec{\nabla}\varphi_c|^2 + U(\varphi_c) - U_o\Big]. \tag{4.52}$$

We may use the above result, Equation (4.50), together with the equations of motion (4.25) and (4.26) to obtain the following expression for the Hamiltonian:

$$E^s_{eff} = \int d^nx \left\{\frac{n-3}{2}|\nabla\varphi_c|^2 + (n+1)[U(\varphi_c)-U_o] + \left(\frac{m}{g} - \varphi_c\right)\frac{\partial U}{\partial\varphi_c}\right\}. \tag{4.53}$$

As is well known, the above equation (4.53) cannot be used as a basis of a variational principle. For known φ_c, which minimizes Equation (4.52) or better (4.31), it gives the proper value of H^s. Therefore we may view E^s_{eff} as an expression defining the total energy.

A striking feature of Equation (4.53) is the fact that for $n<3$ (n is the number of space dimensions) the effective kinetic energy of the scalar field becomes negative definite. It just vanishes for $n=3$. Thus the energy content of E^s_{eff} in a normal number of space dimensions is only implicitly dependent on the derivatives of the field φ_c. The negative definite kinetic energy for $n < 3$ signals a possible instability of the Hamiltonian; the energy of the solution could be reduced by a small noncontinuous variation of φ_c.

Another notable feature is that any φ_c^4 proportional term in U cancels out in three space dimensions (in n-dimensional space φ_c^{n+1} cancels). This is an important feature, since φ_c^4 is commonly held responsible for the stability of a theory with spontaneously broken symmetry [22], [26]. In the case of a one-dimensional world ($n = 1$), where the φ_c^2 contribution vanishes, this is obviously a necessary term to stabilize the theory. It would therefore seem that a φ^6 plays the role of the φ^4 term in three space dimensions, as compared with one-dimensional models.

A similar derivation can be carried out in the case of a Lorentz vector field A_μ of mass μ_V in interaction with the Dirac field. For the Lagrangian

$$\mathcal{L}_{A'} = -\tfrac{1}{2}\partial_\mu A^\nu \partial^\mu A_\nu + \tfrac{1}{2}\mu_V^2 A_\mu A^\mu - g_V A_\mu \bar{\Psi}\gamma^\mu \Psi + \Psi(\gamma\cdot p - m)\Psi, \tag{4.54}$$

we find

$$E_{\text{eff}}^{A'} = -\int d^n x \left[\frac{n-3}{2}(\nabla A_\mu)^2 + \frac{n-1}{2}\mu_V^2 A^2 - m\bar{\psi}_o\psi_o\right], \tag{4.55}$$

which is negative definite, considering the longitudinal component A_o only. We note that the longitudinal part of the vector-type interaction is repulsive in the particle-particle channel and attractive in the particle-antiparticle channel.

The situation changes when the sign in the part of the Lagrangian corresponding to the free vector field is changed:

$$\mathcal{L}_A = \tfrac{1}{2}\partial_\mu A^\nu \partial^\mu A_\nu - \tfrac{1}{2}\mu_V^2 A^2 - g_V A_\mu \bar{\Psi}\gamma^\mu \Psi + \bar{\Psi}(\gamma\cdot p - m)\Psi. \tag{4.56}$$

We then find

$$E_{\text{eff}}^{A} = \int d^n x \left[\frac{n-3}{2}(\nabla A_\mu)^2 + \frac{(n-1)}{2}\mu_V^2 A^2 + m\bar{\psi}_k\psi_k\right], \tag{4.57}$$

which is positive definite, constrained to the longitudinal part of A_μ, provided that the scalar integral $\int d^n x \bar{\psi}_o\psi_o > 0$. The above-described change in sign accomplishes at the same time a change in the 'polarity' of the vector interaction—the longitudinal part is now attractive in the particle-particle channel, while the particle-antiparticle channel becomes repulsive. The perturbative quantum field theory, if based on Equation (4.56), would suffer from the well-known difficulties associated with the possible need for negative metric particles (ghosts) to guarantee a spectrum bounded below. Therefore, such modifications are usually not considered seriously. Such an example considered in the frame of classical field theory can serve as an educational example in order to gain experience with 'attractive' vector-type fields encountered in non-Abelian meson theories (quantum chromodynamics). Returning for a moment to Equation (4.55), it should be mentioned that it is in principle possible to find a solution in which the space-vector part of A_μ dominates—thus allowing a stable solution, even with the conventional choice for the sign of the vector field action.

We now consider the case of the self-interacting Dirac field, where H assumes the form, for scalar-type self-interaction [27]

$$H^p = \int d^n x \left[\psi_o^\dagger (\vec{\alpha} \cdot \vec{p} + \beta m) \psi_o - \frac{G}{2} (\bar{\psi}_o \psi_o)^2 \right]. \tag{4.58}$$

In view of the equations of motion,

$$\left[\vec{\alpha} \cdot \vec{p} + \beta (m - G \bar{\psi}_o \psi_o) \right] \psi_o = \epsilon_o \psi_o, \tag{4.59}$$

the virial relation reads as

$$T_o = \int d^n \psi_o^\dagger \vec{\alpha} \cdot \vec{p} \psi_o = - \frac{G}{2} \int d^n x \left[(\vec{x} \cdot \vec{\nabla}) (\bar{\psi}_o \psi_o)^2 \right]. \tag{4.60}$$

Upon partial integration of the right-hand side, we obtain, up to a vanishing surface term,

$$T_o = -n V_o \quad , \tag{4.61}$$

where

$$V_o = - \int d^n x \frac{G}{2} (\bar{\psi}_o \psi_o)^2. \tag{4.62}$$

The energy for the self-interacting Dirac field can be written now as

$$E^p_{eff} = (n-1)(-V_o) + m_q S_o . \tag{4.63}$$

S is the scalar integral

$$S_o = \int d^n x(\bar{\psi}_o \psi_o) . \tag{4.64}$$

This result, Equation (4.63), corrects a superficial impression that H^p, Equation (4.58), is unbound, since it has the structure $x^2 - x^4$. This means that the kinetic energy of the Dirac field more than offsets the attractive self-interaction V_o. Since $-V_o$ is always positive, we find that for all dimensions the positivity of the solution depends on the sign of the scalar integral. Further, we note that the eigenfrequency can be written as

$$\epsilon_o = (n-2)(-V_o) + mS_o . \tag{4.65}$$

4.4 Properties of bound states of scalar self-consistent bags

Properties of bound states obtained within several models will be described here. In particular, we are interested in the geometric properties of the quasi-fermion amplitudes ψ_o and mass densities $\mathcal{M}$. Furthermore, we consider also g^2 and μ/m systematics of the observable matrix elements of ψ_o, such as the charge radius, the magnetic moment of the proton and axial coupling constant of the semileptonic charged current.

It is my feeling that although the average properties (involving matrix elements) of our variational solutions, belonging to some specific models, may be closer to reality than should be expected, a more detailed description is required for comparison of the theoretical and experimental form factors. In particular, we should recall that the amplitudes ψ_o are, strictly speaking, transition matrix elements between N and $N \pm 1$ fermion states, and not probability amplitudes, as in the case of weakly interacting fields. Also, we have considered here only 'valence' quarks—the sea quarks do also contribute significantly, in particular to the inelastic form factors.

In Figure 4.1 we show some of the typical solutions associated with the Lagrangian (4.2) of the interacting fermion-scalar gluon fields [23] for $U(\varphi^2) = \frac{1}{2}\mu^2\varphi^2$ and $N = 3$. The self-consistent quasi-particle (bound quark) mass

$$m^* = m_q - g\varphi_c \tag{4.66}$$

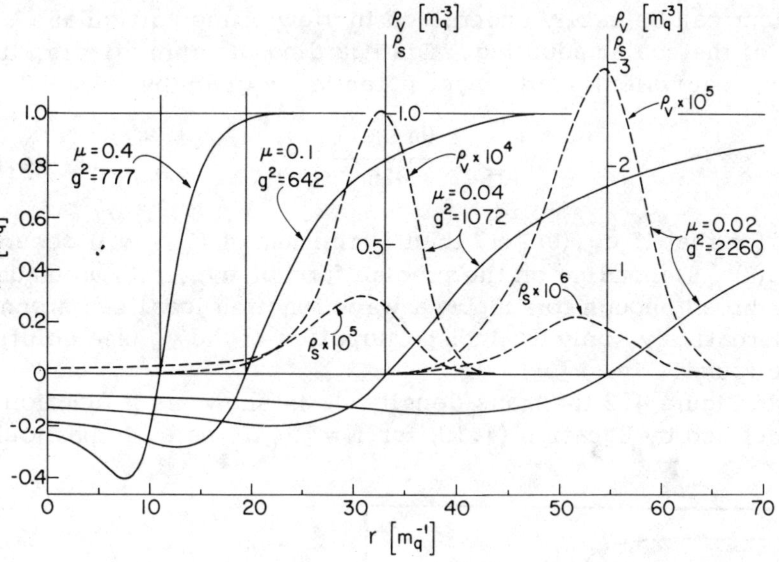

Fig. 4.1. The effective quasi-particle mass $m^* = m_q - g\varphi_c$ (in units of m_q) as a function of r (in units of m_q^{-1}) for $\mu/m_q = 0.02$, 0.1, 0.4 and $g^2 = 2260$, 1072, 642, 777, correspondingly. The dashed curves are the vector and scalar densities for $\mu = 0.4 m_q$, $g^2 = 642$ and $\mu = 0.02 m_q$, $g^2 = 2260$. The vector density is normalized to one.

is plotted for several values of the gluon mass μ/m_q and the coupling constant g^2 as a function of $r = |\vec{x}|$. For $\mu = 0.04 m_q$ and $\mu = 0.02 m_q$ we also show the vector

$$\rho_V = \psi_o^\dagger \psi_o \tag{4.67}$$

and scalar

$$\rho_S = \bar{\psi}_o \psi_o \tag{4.68}$$

densities. We note that, in Figure 4.1, ρ_S is enhanced by a factor of 10 in relation to ρ_V. For μ of the order of the bare quark mass m_q, $m^*(r)$ has a pronounced minimum. For coupling constants smalller than those considered here, i.e., for $g^2 \lesssim 10$, we have also found solutions that were evenly distributed over the volume. The transition from the volume to surface solution as a function of g^2 and μ/m is smooth. Qualitatively this behaviour can be easily understood in view of the particular nature of the scalar coupling. Disregarding the spin effects, the spectrum equivalent Schrödinger potential is given by

$$V_{eff} \sim \frac{(m - g\varphi_C)^2}{2m} . \tag{4.69}$$

We notice that if $g\varphi_C(0) > m$, then a minimum of V_{eff} will occur at $m = g\varphi_C(R)$. Depending on the precise form of $g\varphi_C$, this minimum may be broad enough to support a wave function localized around R. Alternatively, only a minor perturbation of the volume solution may be found.

In Figure 4.2 the mass density $\mathcal{M}$ is shown as a function of r, as defined by Equation (4.19) for N = 1. We note, in particular,

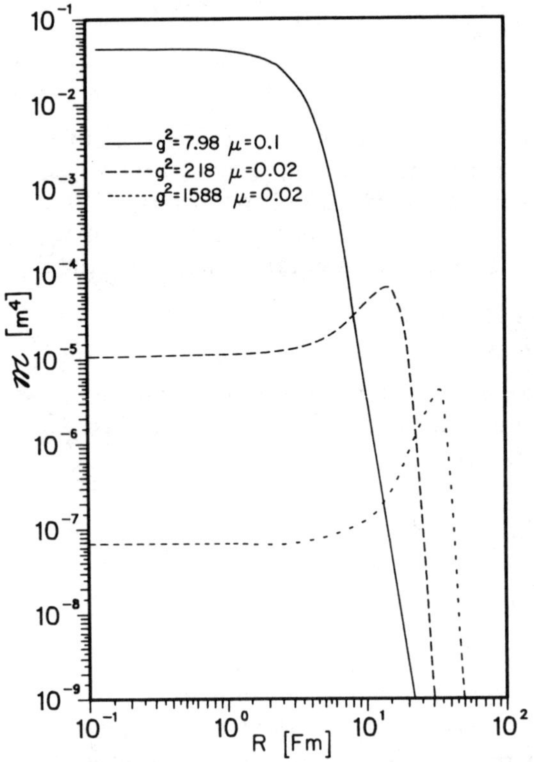

Fig. 4.2. The mass density of the soliton solution as a function of r; several values of the characteristic parameters g^2 and μ in a double-logarithmic plot.

the existence of 'surface' solutions for which $\mathcal{M}$ peaks at a certain distance R for large values of the coupling constant, while there are also volume solutions for smaller coupling constants.

In Figure 4.3 the masses [23] of the bound states M_N are given as functions of g^2 for the gluon masses of 0.4, 0.1, 0.02 (m_q). We see that smaller gluon masses give considerably smaller bag masses. As g^2 rises we find a point g_c^2 such that

$$2M_2 < \mu, \qquad g^2 > g_c^2(\mu/m_q). \tag{4.70}$$

For $g^2 > g_c^2$ it can be argued that no free gluons can exist owing to their strong coupling to the hadronic states. As g^2 is increased with a fixed gluon mass μ, M_N decreases monotonically.

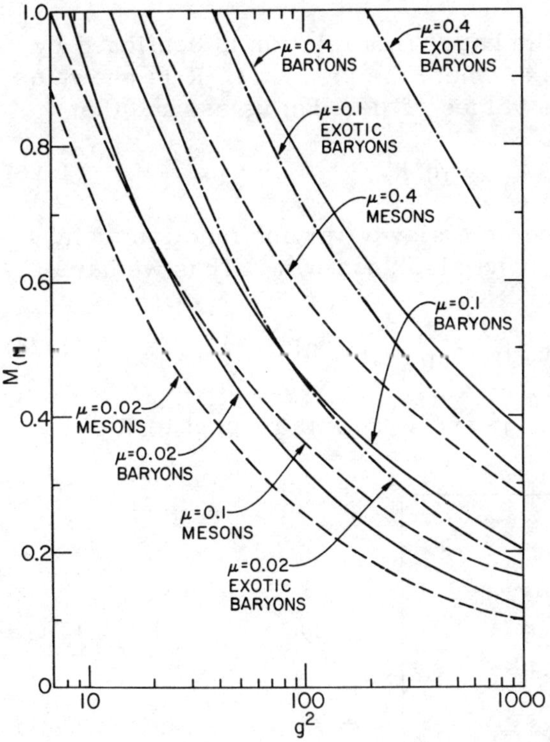

Fig. 4.3. The total mass M_N in units of the quark mass of several quark states; baryons (N=3), mesons (N=2) and exotic baryons (N=6) as a function of the coupling constant g^2 for several values of the gluon mass μ/m_q=0.02, 0.1, 0.4.

However, weakly bound exotic states are found as shown in Figure 4.3. These states are conglomerates of the allowed hadronic states. This lack of saturation is particular for the scalar interaction. When other (repulsive) interactions are included, such states may easily become unbound.

There is an easy way of understanding the functional form of the curves shown in Figure 4.3. The mass consists of two terms, cf. Equation (4.22),

$$M_N = M_f + M_g \quad , \tag{4.71}$$

where M_g is the energy contained in the gluon field φ_c. For large g^2 the eigen-frequency of the Dirac equation solution is

$$\epsilon_o = R^{-1}, \tag{4.72}$$

where R is the radius of the bag. This relation is confirmed by results shown in Figure 4.4, where the product $\epsilon_o \cdot R$ is shown as a function of g^2, for large values of g^2. Let us assume that

$$M_g = \gamma g^2 R^\beta, \tag{4.73}$$

where β may be considered as a slowly varying function of R. The constants γ and β depend also upon μ/m. Thus we have

$$M_N = \frac{N}{R} + g^2 \gamma R^\beta = \frac{N}{R}\left(1 + \frac{g^2}{N}\gamma R^{\beta+1}\right). \tag{4.74}$$

Upon minimization at fixed g^2 and μ/m with respect to R

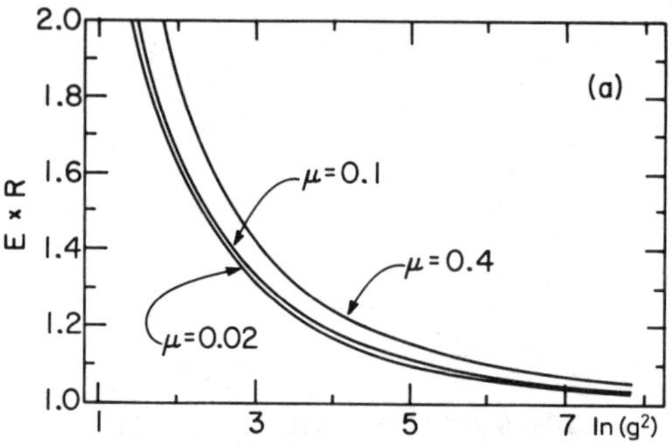

Fig. 4.4. The product of the lowest positive Dirac eigenvalue with r.m.s. radius of the amplitude ψ_o as a function of $\ln g^2$.

$$\frac{\partial M_N}{\partial R} = 0 = -\frac{N}{R^2} + g^2\gamma\beta R^{\beta-1} . \qquad (4.75)$$

We obtain

$$\frac{g^2}{N}\gamma R^{\beta+1} = \beta^{-1} , \qquad (4.76)$$

and for Equation (4.74)

$$M = M_f(1+\beta^{-1}) . \qquad (4.77)$$

Defining

$$R_a = M_f/M_g , \qquad (4.78)$$

we find

$$R_a = \beta . \qquad (4.79)$$

Thus the ratio of the fermionic to the gluonic part of the hadron mass has a geometrical meaning, since the value of β, Equation (4.74), determines the nature of the solution. For $\beta = 3$ we have a volume bag with the glue distributed over the whole hadron. For $\beta = 2$ we have a surface solution, and $\beta = 1$ indicates a string-like solution. As shown in Figure 4.5,

$$P = M_N R/N \cong 1 + 1/R_a \qquad (4.80)$$

tends to the value 1.5 for large g^2, consistent with $R_a \lesssim 2$. We

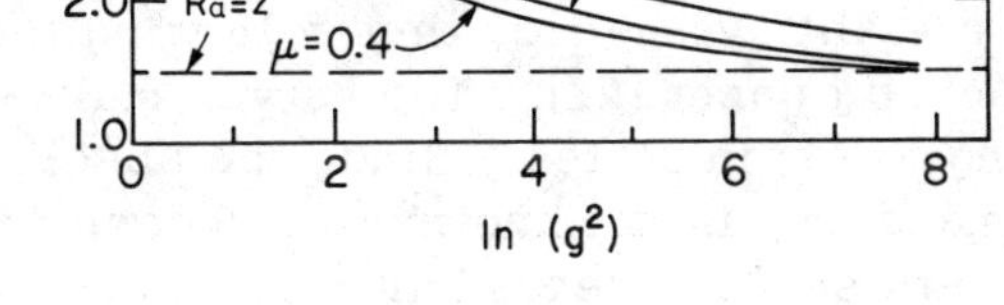

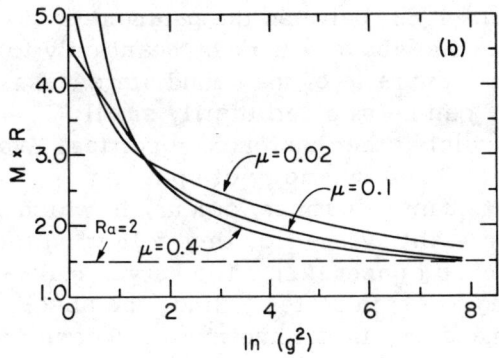

Fig. 4.5. The product of the mass of the quark bag with the r.m.s. radius as a function of $\ln g^2$.

recall that for the SLAC bag [22] $R_a = 2$ (surface bag) and for the MIT bag [27] $R_a = 3$ (volume bag).

We can use relation (4.76) to eliminate R completely from Equation (4.74). Then, with R_a as the second variable apart from g^2 we find

$$M_N \sim N^{R_a/(R_a+1)} (g^2)^{-(R_a+1)^{-1}}, \tag{4.81}$$

where the dependence on N follows from Equation (4.28) which can also be written as

$$M_N(g^2) = NM_1(Ng^2). \tag{4.82}$$

In passing we note that the lack of saturation of the scalar interaction is apparent in Equation (4.81). The energy per quark M_N/N is proportional to $N^{(R_a+1)^{-1}}$ and is therefore decreasing with an increasing number of quarks in the bag (of course, only colour singlets are allowed).

We consider the limit $g^2 \to \infty$, $m_q \to \infty$ in Equation (4.81), while μ remains finite. If there is a relation between the values of m_q and g^2 such as

$$g^2 \sim \left(\frac{m_q}{\mu}\right)^{R_a+1}, \tag{4.83}$$

then M_N stays constant in the limit $(m_q, g^2) \to \infty$ and we have permanent confinement in the sense that, though the free quark mass m_q is infinite, the mass of the bound state remains finite. It is apparent that the relation (4.83) between the parameters of the theory is rather arbitrary. The above remark is meant only to alert the reader to the fact that the ratio of the bound state mass M to the free quark mass m_q can be made arbitrarily small.

We now proceed to calculate other hadronic properties. We take wave functions with exact SU(6) symmetry of spin and flavour, and an extra internal quantum number, colour, in which the baryonic wave functions are antisymmetric, thus following the concepts set forth in the SLAC bag paper [22]. The baryon multiplet $\underline{56}$ has 56 states counting the spin states. Since the absolute value of the free quark mass m_q is not known, we choose to consider only quantities that are scale-independent. The simplest ones are the products of the r.m.s. radius of the baryons and mesons with their masses N·P, Equation (4.80), which are shown in Figures 4.6a and 4.6b. The experimental number to

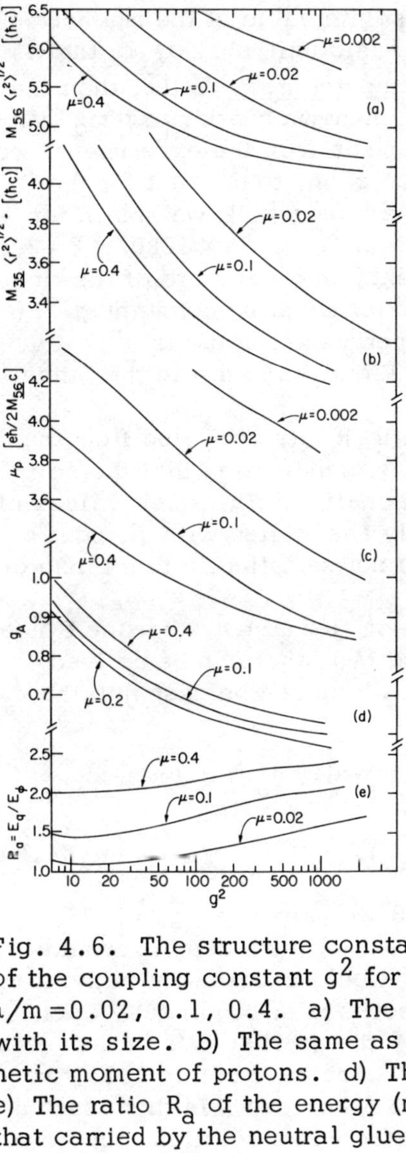

Fig. 4.6. The structure constants of SU(6) hadrons as a function of the coupling constant g^2 for several values of the gluon mass, $\mu/m = 0.02$, 0.1, 0.4. a) The product of the mass of the baryon with its size. b) The same as in (a) for mesons. c) The magnetic moment of protons. d) The axial-coupling constant g_A. e) The ratio R_a of the energy (momentum) carried by the quarks to that carried by the neutral glue.

compare with for baryons is, most likely, the product of the proton charge radius, 0.8 fm, with the average mass of the 56 multiplet, $M_{56} = 1280$ MeV, which is $5.2\hbar c$. We see that this lies well within values spanned by our calculations. We note that 5.2 implies for $N = 3$ and $\epsilon_0 \sim 1/R$ that $R_a \sim 1.7$. Thus it would seem that hadrons are surface-like, or $\epsilon_0 \not\sim 1/R$.

We can also calculate the absolute value of the magnetic moment of the proton, using as the basic unit $e\hbar/2M_{\underline{56}}c$; that is, the Bohr magneton in units of the computed mass (rather than the experimental value). Therefore the unknown quark mass m_q cancels out from our result. For comparison with the experiment, we must scale up the experimental value of μ_p to account for the larger multiplet mass than that of the proton [23]; we obtain for comparison with Figure 4.6c a value $\mu_p^{exp} = 2.79 \times 1280/938 = 3.8$.

We see that the self-consistent bag results are of the right magnitude. We consider also the axial coupling constant g_A of the neutron decay process. The experimental value is $g_A^{exp} = 1.25$. The results shown in Figure 4.6d are in the range $0.6 < g_A < 1$.

Finally, in Figure 4.6e the ratio R_a as computed from the solutions is shown. We find $1 < R_a \lesssim 2$, indicating that the self-consistent quark bag is mostly surface-like. The small values of R_a for small g^2 cannot be as clearly associated with β, Equation (4.74), since relation (4.72) is not satisfied in this range of g^2 (see Figure 4.5).

As the final point on the subject, let us note that there is a simple relation between g_A, μ_p, and M_N, that can be derived using the Dirac equation (4.25) and which can be cast into the form

$$\begin{aligned}\mu_p &= (0.5 + 0.3g_A)\cdot N\cdot(1 + R_a^{-1}) \\ &= 2.625\,(1 + R_a^{-1}).\end{aligned} \tag{4.84}$$

In the second line of Equation (4.84) we have used $g_A = 1.25$, $N = 3$. Thus if we were able to find a $U(\varphi)$ such that μ_p would also have the desired value 3.8, R_a would necessarily turn out to be 2.2. This means that in a 'perfect' scalar bag about 70 percent of the mass would be carried by the quarks. We further note that the bag could be volume-like. Since g_A must be 1.25, we know that a solution with $\epsilon_0 \langle r^2 \rangle^{\frac{1}{2}} > 1$ is needed. In this case the value $R_a = 2.2$ does not imply a surface-like solution. In view of the above remarks, it is quite possible that a theory based on scalar interaction with very good phenomenological properties could be found.

4.5 Properties of solutions of a self-coupled Dirac field

We have recognized that localized, stable solutions of classical field equations are the basic ingredients in the description of extended objects in quantum field theory. We now turn to the discussion of the self-coupled Dirac field $\overline{\psi}$ in three space dimensions governed by the action

$$L = \int d^4x \left[\overline{\psi}\gamma \cdot p\psi - m\overline{\psi}\psi + \frac{G}{2}(\overline{\psi}\psi)^2 \right], \tag{4.85}$$

where m is the mass of the free (G = 0) fermion field. G, like the weak coupling constant, has the dimension of inverse mass squared.

The quantum field theory associated with Equation (4.85) is known to lead to a non-renormalizable perturbation expansion in the vicinity of G = 0. I believe that the difficulties associated with the perturbation expansion do not provide a compelling argument for rejection of the Lagrangian (4.85). In particular, it is quite possible that the properties of the theory change with increasing strength of the coupling G in a nonanalytic fashion. I find [28] that for $G > G_{min} = 4.47/(4\pi m^2)$ localized solutions of the classical field equation arise.

Another way of seeing the self-interaction invokes the infinite gluon mass limit of a scalar meson field discussed in the last section. We have considered the properties of the Hamiltonian (4.58) associated with the self-coupled field already in Section 4.3. We therefore turn now to the discussion of the properties of the solutions obtained by the methods introduced in Section 4.2, with the modification that the equation defining a field ϕ in terms of $\overline{\psi}_o\psi_o$ is very simple now and reads

$$\phi = G\overline{\psi}_o\psi_o . \tag{4.86}$$

From this point on, the iteration procedure can proceed as described previously. Use of numerical continuation of the solutions as a function of the coupling constant increased the convergence speed of the iteration. The solutions ψ_o that will be discussed below are normalized to unity. We note a similar scale invariance as already noted with meson fields, Equation (4.27); given an arbitrary norm

$$Q = \int d^3x\, \psi_o^+\psi_o , \tag{4.87}$$

we can introduce

$$\psi_o' = Q^{\frac{1}{2}}\psi_o. \tag{4.88}$$

Then ψ_o' is a solution of the problem with the parameter

$$G' = G/Q, \tag{4.89}$$

that leads to the total energy

$$H'(G/Q) = H(G)*Q. \tag{4.90}$$

The dimensionless parameter of the model is

$$\beta = m^2 G/4\pi. \tag{4.91}$$

The lowest mass solution associated with lowest positive frequency solution has been computed. The energy (H/m) and the frequency (ϵ_o/m) are shown in Figure 4.7a. We recognize that

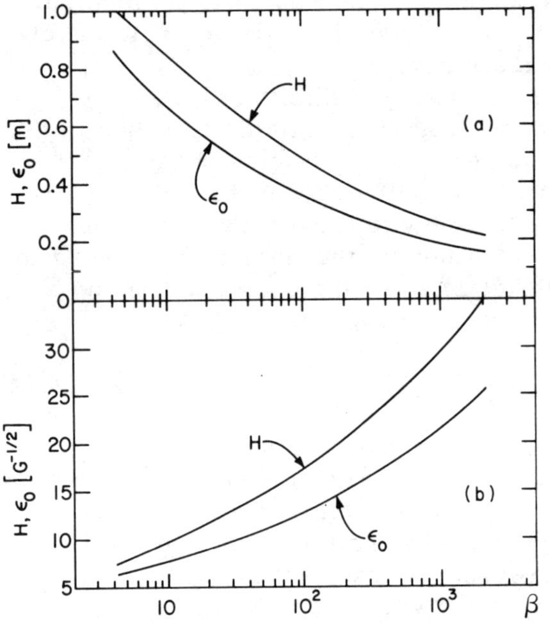

Fig. 4.7. The energy H and the eigenfrequency ϵ_o as a function of the dimensionless coupling constant β: a) in units of m; b) in units of $G^{-\frac{1}{2}}$.

both functions fall monotonically with increasing β. We also find a minimum value of G for which the mass of the bound state is smaller than m. It is $G_{min} = 4.47/(4\pi m^2)$, as mentioned previously. Interestingly enough, the qualitative behaviour of $H(Gm^2)$ changes when H is plotted in units of $G^{-\frac{1}{2}}$, as shown in Figure 4.7b. We find

$$(H/G^{-\frac{1}{2}}) = (4\pi\beta)^{\frac{1}{2}}(H/m), \tag{4.92}$$

which monotonically rises. The lines in Figure 4.7a may be understood also as obtained for fixed m with changing G, while in Figure 4.7b, G is fixed and m varies.

In Figure 4.8 the forms of the solutions for some representative values of $\beta = 22$, 475, 2125 are shown. The effective mass

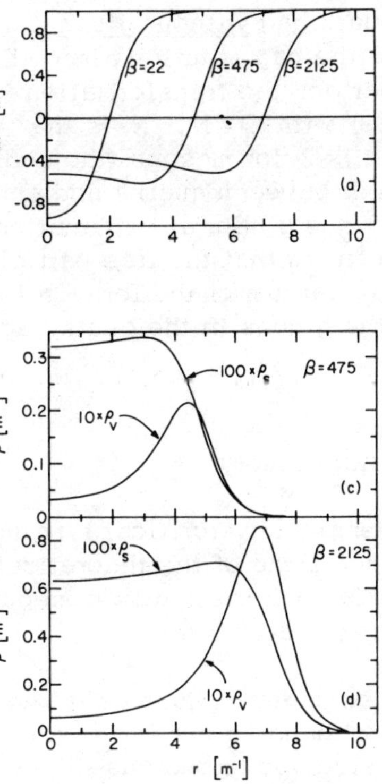

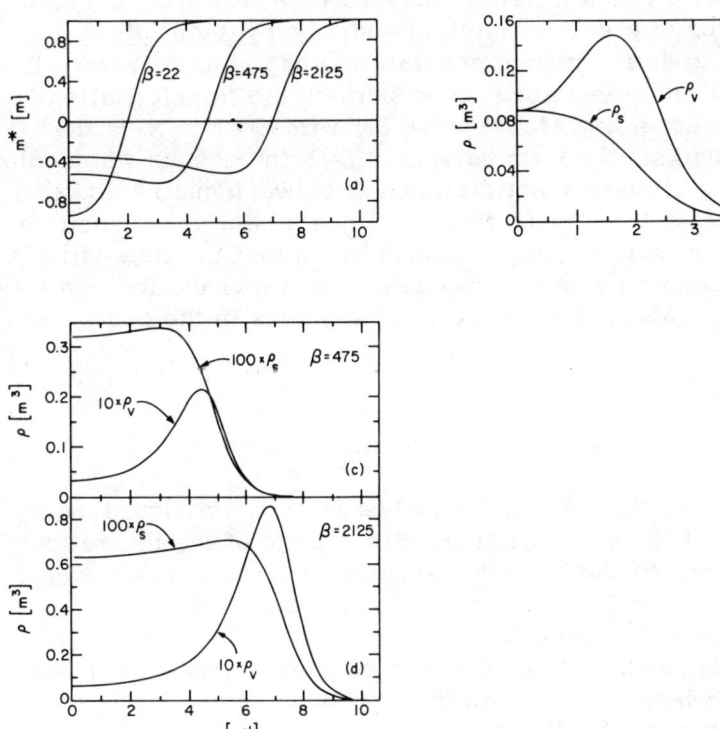

Fig. 4.8. The soliton solutions of self-coupled Dirac fields for $\beta = 22$, 475, 2125. a) The effective mass m^*; b), c), d) the scalar and vector densities.

m^* as a function of r is shown in Figure 4.8a. With increasing coupling strength the size of the solution increases significantly, while the value of m^* at origin decreases. Notably, it remains always negative. In Figures 4.8b, c, and d, the vector ρ_v and scalar ρ_s densities belonging to the same solutions are shown. As β increases, the vector density becomes more and more localized around the point where $m^* = 0$. The scalar density remains distributed over the whole volume of the solution. In that respect, the self-interacting field solutions differ from the solutions involving a gluon field. There we have seen that the scalar density, like the vector density, is localized at the surface of the solutions. The volume character of ρ_s in the present case is necessary to allow the relatively constant effective mass m^* over the space occupied by the solution.

In view of the results presented in Figures 4.7 and 4.8, the self-interacting fermion field discussed here may also be taken as a prototype of a self-consistent quark bag without gluons. All we need to do in order to generate the bag solutions from the soliton solutions given above is to perform the transformation described in Equations (4.87) to (4.89) with $Q = N$. N is the number of quarks: $N = 3$ for baryons, $N = 2$ for mesons (the scalar self-coupling also does not distinguish between quarks and antiquarks). A self-interacting field provides a natural explanation of this phenomenon of quark confinement in that the free particle of mass m cannot be considered as a quantum of the fermion field for $G > G_{min}$. Also, the absence of the gluons in the quark bag is satisfactory.

4.6 Supercritical fields in quantum field theory

In this section we shall consider the supercritical binding in the frame of the bag approach. Since some of the theoretical details concerning the bag approximation differ slightly from our previous considerations, we will briefly rederive the self-consistent c-number equations.

A simple and possibly relevant example of what I have in mind arises when considering strongly bound states in the (classical) field theory of (attractive) massive vector mesons interacting with fermion fields. The additional effects we expect have already been discussed qualitatively in Section 2.3.

We consider a fermion field coupled to a vector meson field with the Hamiltonian [16] (with $A \cdot A \equiv A^2 = A_0^2 - \vec{A} \cdot \vec{A}$)

$$H = \frac{1}{2}\left\{\int d^3x \frac{1}{2}\left[\hat{\Psi}^\dagger, (\vec{\alpha}\cdot\vec{p} + \beta m - g_V(\vec{\alpha}\cdot\hat{\vec{A}} - A^0))\,\hat{\Psi}\right]\right.$$

$$\left. - \frac{1}{2}\int d^3x\left[\hat{\pi}_\mu^{\,2} + (\nabla\hat{A}_\mu)^2 + W(\hat{A}\cdot\hat{A}) - W(0)\right] + \text{h.c.}\right\}, \quad (4.93)$$

where $W(A^2)$ is an arbitrary potential energy of the vector field A_μ, which leads to the equations of motion

$$\left(\vec{\alpha}\cdot\vec{p} + \beta m - g_V(\vec{\alpha}\cdot\hat{\vec{A}} - \hat{A}_o)\right)\hat{\Psi} = i\frac{\partial\hat{\Psi}}{\partial t} \quad (4.94)$$

$$\dot{\hat{\pi}}_\mu - \Delta\hat{A}_\mu + \hat{A}_\mu\frac{\partial W}{\partial A^2} = \frac{g_V}{2}\left[\bar{\hat{\Psi}}, \gamma_\mu\hat{\Psi}\right]. \quad (4.95)$$

We also have the auxiliary condition

$$\langle ps|\partial_\mu\hat{A}^\mu\,|ps\rangle = 0 \quad (4.96)$$

for any physical state $|ps\rangle$.

For strong interactions it is most convenient to choose as the basis for the expansion of the Fermi field $\hat{\Psi}$ the complete set of solutions of the Dirac equation with an average field A_c (p stands for both discrete and continuous indices):

$$\left(\vec{\alpha}\cdot(\vec{p} - g_V\vec{A}_c) + \beta m + g_V A_c^{\,o}\right)\psi_p = \epsilon_p\psi_p. \quad (4.97)$$

The field operator is expanded in the quasi-particle Fock space as

$$\hat{\Psi}(\vec{x},0) = \int_m^\infty \hat{b}_\epsilon\psi_\epsilon(\vec{x})d\epsilon + \sum_{m>\epsilon_p>-m}\hat{b}_p\psi_p(\vec{x}) + \int_{-\infty}^{-m}\hat{d}_\epsilon^{\,\dagger}\psi_\epsilon(\vec{x})d\epsilon, \quad (4.98)$$

whenever the field A_c^μ is undercritical. A different expansion is appropriate, as discussed in Section 2.3, when a resonance of Equation (4.97) is embedded in the negative energy continuum <u>and</u> the total energy remains positive. Then we have

$$\hat{\Psi}(\vec{x},0) = \int_m^\infty \hat{b}_\epsilon\psi_\epsilon d\epsilon + \sum_{m>\epsilon_p>-m}\hat{b}_p\psi_p + \hat{b}_o\int_{-\infty}^{-m}a(\epsilon)\psi_\epsilon d\epsilon$$

$$+ \int_{-\infty}^{-m}\hat{d}_{\epsilon'}^{\,\dagger}\left[\int_{-\infty}^{-m}h_{\epsilon'}(\epsilon)\psi_\epsilon d\epsilon\right]d\epsilon', \quad (4.99)$$

where $a(\epsilon)$ and $h_{\epsilon'}(\epsilon)$ are suitable functions that describe the resonance in a negative energy continuum (see Section 2). If more than one resonance is embedded in the continuum, a generalization of Equation (4.99) is in order.

It is important to stress again the difference between the self-consistent field problem and the case of an external field. The neutral vacuum, characterized by $b_p|0\rangle = d_n|0\rangle = 0$, may be a stable, zero-energy state even should supercritical binding described by (4.99) occur, since the total energy of the excited state $b_0^\dagger|0\rangle$ may be positive upon consideration of the meson energy included in Equation (4.93), in contradistinction to the situation with external potentials.

Our choice of the mean field A_c^μ was directed by our desire to eliminate from the ground state most contributions from the virtual quasi-particle excitations. Therefore $A_c^\mu = \langle ps|A^\mu|ps\rangle$ and we obtain

$$-\Delta A_c^\mu + A_c^\mu \left(\frac{\partial W}{\partial A^2}\right)_{A_c^2} = g_v \langle ps|\tfrac{1}{2}\left[\hat{\bar{\Psi}}, \gamma^\mu \hat{\Psi}\right]|ps\rangle. \qquad (4.100)$$

We consider the state $b_0^\dagger|0\rangle = |p\rangle$ as a trial state. Then Equation (4.100) becomes (neglecting virtual vacuum polarization charge density):

$$-\Delta A_c^\mu + A_c^\mu \left(\frac{\partial W}{\partial A^2}\right)_{A_c^2} = g_v \bar{\psi}_o \gamma^\mu \psi_o. \qquad (4.101)$$

Here ψ_o is the coefficient of b_o in Equation (4.99) or the lowest energy particle wave function in Equation (4.98).

At this point, the interaction must be modified [16] to allow bound fermion states, bound only by 'electric' forces. That is to say, the sign on the right-hand side of Equation (4.101) has to be changed, as discussed in Section 4.3, Equation (4.56):

$$-\Delta A_c^\mu + A_c^\mu \left(\frac{\partial W}{\partial A^2}\right)_{A_c^2} = -g_v \bar{\psi}_o \gamma^\mu \psi_o. \qquad (4.102)$$

This change is motivated by the consideration of possible internal symmetry of the fields of interest—the sign may be understood in analogy with the different possible signs of the z-component of the isospin. This change also means that the energy of the classical bound state becomes

$$E_v = E_{Dirac} + E_{meson} = \int d^3x\, \psi_o^\dagger (\vec{\alpha}\cdot\vec{p} + \beta m - g_v(\vec{\alpha}\cdot\vec{A} - A^o))\psi_o$$

$$+ \frac{1}{2}\int d^3x\left[\left(\nabla A_c^\mu\right)^2 + W(A_c\cdot A_c) - W(0)\right] = \int d^3x\, \mathcal{H}, \qquad (4.103)$$

where we find

$$E_{Dirac} = \epsilon_o. \qquad (4.104)$$

In view of the arbitrary (but motivated) changes involved in writing (4.103), what follows should be understood only as illustrating a type of behaviour in a coupled field theory distinct from that found in the external field problem.

Let us devote the rest of our discussion to the special case

$$W = \mu_v^2 A^2 \qquad (4.105)$$

where μ_v is the vector meson mass. We recall from Equation (4.57) that the total energy is given by

$$E_{eff}^A = \int d^3x\left(\mu_v^2 A^2 + m\bar{\psi}_o\psi_o\right). \qquad (4.106)$$

The energy E_{eff}^A is positive definite if the scalar integral $\int d^3x\,\bar{\psi}_o\psi_o$ is positive. No lower bound for the scalar integral is known when vector fields are present, but in all examples studied numerically it has been found to be positive definite and very small for large g_v^2.

Equations (4.97) and (4.102) have been solved numerically with $W = \mu_v^2 A^2$ for the lowest energy particle state such that the Dirac wave functions (either eigenstates or resonances) are normalized to one. In Figure 4.9 we show the energy of the bound fermion, E_{Dirac} [Equation (4.104)], and the total energy E_v [Equation (4.103)], as a function of the coupling constant g_v for vector meson masses $\mu_v = 0.2$ and $0.4m$. The energy is measured as usual in units of the bare fermion mass m. In this example, we see that the total energy is always positive for coupling constants $g_v^2/4\pi \gtrsim 4$, while the Dirac eigenvalues E_{Dirac} take large negative values. The scalar integral is a negligible contribution to E_v [Equation (4.103)] for strong coupling. Exact agreement of numerical results with those expected from the virial theorem [Equation (4.106)] (as long as $E_{Dirac} > -m$) has been found. This

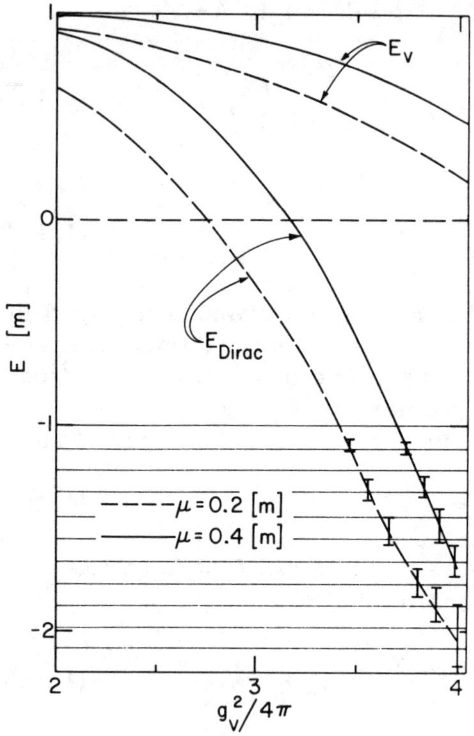

Fig. 4.9. The Dirac eigenvalue and total energies of interacting Dirac-vector meson fields.

is also a good check of the computer codes involved in solving the nonlinear equations (4.97) and (4.102).

The main conclusion that can be drawn from this exercise is the recognition that in this simple model the total energy of the strongly bound state lies above that of the neutral vacuum, i.e., $E_V > 0$ even when the self-consistent potential is supercritical, i.e., $\epsilon_0 < -m$. Therefore the neutral vacuum is the stable ground state of the theory. Further, we note that essential in obtaining a proper solution of Equations (4.97) and (4.102) is a proper treatment of resonant states in the antiparticle continuum when they occur. As described in Sections 1 and 2, the proper description of the system involves then the amplitude:

$$\psi_0 \cong \int_{-\infty}^{-m} a(E)\psi_E dE. \tag{4.106}$$

Without the proper understanding of the supercritical spectrum, a solution would seem discontinuous at the critical points, since we would take the lowest discrete state, for example a 2s state in place of the embedded 1s-state resonance.

Another way of seeing the difference between the presently considered case and the external field problem studied at the beginning of this paper involves the Green's function formulation considered in Section 2.1. There we learned that the pole associated with the critical state cannot drag the integration path off the real axis, as the external potential attains overcritical strength. Now, we have shown that the opposite is true for interacting fields studied in this subsection. The integration path C' of Figure 2.3 should be taken to obtain the Green's function in the case of the interacting fields, instead of the previously employed path D. A quite different result would follow if we had found that E_V in Equation (4.103) is actually a negative quantity at certain values of g_V^2. Then the path D would have to be taken again and we would expect that the vacuum state is really the charged vacuum. In contrast to the external field case we would then be faced with a microscopic phase transition. At every point in space the vacuum would be charged. Thus another procedure would be needed to determine the new ground state, since it is impossible to describe such a global change of the world by the single-mode approximation.

ACKNOWLEDGEMENT

I would like to thank J. S. Bell for reading the manuscript and his suggestions that helped clarify some points of this paper.

REFERENCES

1. P. A. M. Dirac, Proc. Roy. Soc. (London) A117, 610 (1928).
2. M. E. Rose, Relativistic Electron Theory, Wiley, New York, 1961.
3. K. N. Case, Phys. Rev. 80, 797 (1950);
W. M. Frank, D. J. Land, and R. M. Spector, Rev. Mod. Phys. 43, 36 (1971).
4. F. G. Werner and J. A. Wheeler, Phys. Rev. 109, 126 (1958);
D. Rein, Z. Phys. 221, 423 (1969);

W. Pieper and W. Greiner, Z. Phys. 218, 327 (1969).
5. G. Soff, B. Müller and J. Rafelski, Z. Naturforsch 29a, 1267 (1974).
6. G. Soff, B. Müller, J. Rafelski, and W. Greiner, Z. Naturforsch. 28a, 1389 (1973).
7. R. Adler, M. Bazin, and M. Schiffer, General Relativity McGraw-Hill, New York, 1965.
8. M. Soffel, B. Müller and W. Greiner, J. Phys. A10 (GB), 551 (1977).
9. J. Rafelski, L. P. Fulcher, and A. Klein, Fermions and Bosons Interacting with Arbitrarily Strong External Fields, to be published in Phys. Reports.
10. B. Müller, H. Peitz, J. Rafelski, and W. Greiner, Phys. Rev. Letters 28, 1235 (1972);
B. Müller, J. Rafelski, and W. Greiner, Z. Phys. 257, 62 (1972).
11. B. Müller, J. Rafelski, and W. Greiner, Nuovo Cimento 18A, 551 (1973).
12. B. Müller, J. Rafelski, and W. Greiner, Phys. Letters 47B, 5 (1973);
B. Müller and W. Greiner, Z. Naturforsch. 31A, 1 (1976).
13. J. Rafelski and B. Müller, Phys. Rev. Letters 36, 517 (1976);
J. Rafelski and B. Müller, Phys. Letters 65B, 205 (1976).
14. A. O. Barut, this conference.
15. J. Rafelski, B. Müller, and W. Greiner, Nuclear Phys. B38, 585 (1974);
B. Müller, J. Rafelski, and W. Greiner, Z. Phys. 257, 183 (1972);
L. Fulcher and A. Klein, Phys. Rev. D8, 2455 (1973);
L. Fulcher and A. Klein, Ann. Phys. (USA) 84, 335 (1974).
16. J. Rafelski and B. Müller, Phys. Rev. D14, 3532 (1976).
17. M. L. Goldberger and K. M. Watson, Collision Theory, Wiley, New York, 1964.
18. S. Coleman, this conference.
19. B. Müller and J. Rafelski, Phys. Rev. Letters 34, 349 (1975) and to be published;
A. Klein and J. Rafelski, Bose Condensation in Supercritical External Fields: Charged Condensates, Z. Phys. A284, 71 (1978).
20. O. Klein, Z. Phys. 53, 157 (1929).
21. P. Vinciarelli, Nuovo Cimento Letters 4, 905 (1972) and Nuclear Phys. B89, 463 (1975).

22. W. A. Bardeen, M. S. Chanowitz, S. D. Drell, M. Weinstein, and T.-M. Yan, Phys. Rev. D11, 1094 (1975).
23. J. Rafelski, Phys. Rev. D14, 2358 (1976).
24. J. Rafelski, Nuovo Cimento Letters 17, 575 (1976).
25. J. Rafelski, Virial Theorem and Stability of Localized Solutions of Relativistic Classical Interacting Fields, Phys. Rev. D16, 1890 (1977).
26. D. E. L. Pottinger and R. J. Rivers, Nuclear Phys. 117, 189 (1976);
R. Friedberg and T. D. Lee, Phys. Rev. D15, 1694 (1977).
27. A. Chodos, R. L. Jaffe, K. Johnson, C. B. Thorn, and V. F. Weisskopf, Phys. Rev. D9, 3471 (1974).
28. J. Rafelski, Phys. Letters 66B, 262 (1977).

SUBJECT INDEX